The World of Small States

Volume 9

Small states differ considerably in their geography, history, political structures, legal systems and wealth. Nevertheless, because of their size, small states face a set of common challenges including vulnerability to external economic impacts such as changing trade regimes and limited ability to diversify economic activity; limited public and private sector capacity, including the legal and judicial infrastructure. A number of small states have experienced colonization and must accommodate the legacy of one or more forms of colonial law alongside the customary law of the indigenous people. Many small states are islands. These are particularly susceptible to environmental impacts such as natural disasters and climate change. Small states can also be flexible, adaptable, sites of social development and innovation, and have an influence in the world disproportionate to their size.

The importance of research into small states is increasingly recognised by the global legal community . Small states are microcosms which allow us to study and gain insight into the challenges of big states. Their small size makes research easier and the testing of solutions more easily. Small states, however, also have unique problems for which unique solutions must be designed. For example, in a small state with a correspondingly-sized legal profession, ethical guidelines in regard to the appointment of judges have to take into account to the small size of the profession.

The aim of this exciting and unique series is to be the essential compendium for every legal researcher interested in small states but also for practitioners and policy makers working in small state.

More information about this series at http://www.springer.com/series/15142

Stefano Moncada • Lino Briguglio •
Hilary Bambrick • Ilan Kelman • Catherine Iorns •
Leonard Nurse
Editors

Small Island Developing States

Vulnerability and Resilience Under Climate Change

Springer

Editors
Stefano Moncada
Islands and Small States Institute,
Institute for European Studies
University of Malta
Msida, Malta

Lino Briguglio
Islands and Small States Institute
University of Malta
Msida, Malta

Hilary Bambrick
Faculty of Health
Queensland University of Technology
Brisbane, Australia

Ilan Kelman
Institute for Risk and Disaster Reduction and
Institute for Global Health
University College London
London, UK

University of Agder
Kristiansand, Norway

Catherine Iorns
Faculty of Law
Victoria University of Wellington
Wellington, New Zealand

Leonard Nurse
CERMES, Faculty of Science and Technology
University of the West Indies
Cave Hill, Barbados

ISSN 2627-5996 ISSN 2627-6003 (electronic)
The World of Small States
ISBN 978-3-030-82773-1 ISBN 978-3-030-82774-8 (eBook)
https://doi.org/10.1007/978-3-030-82774-8

This Springer imprint is published by the registered company Springer Nature Switzerland AG.
The registered company address is: Gewerbestrasse 11, 6330 Cham, Switzerland

Acknowledgments

We would like to thank all the authors who responded positively to our invitation to prepare manuscripts for this effort. Equally, we wish to express sincere appreciation to all reviewers who contributed to the anonymous peer review process. In addition, we would also like to acknowledge the excellent work undertaken by Dominik Kalweit, William Grech, and the personnel of the Islands and Small States Institute (ISSI) of the University of Malta for their contribution with the management and copy-editing of the whole book process.

Contents

Introduction to the Book

Small Island Developing States: Vulnerability and Resilience Under Climate Change

Stefano Moncada, Hilary Bambrick, Lino Briguglio, Catherine Iorns, Ilan Kelman, and Leonard Nurse

1 The Main Thrust and Rationale of the Book

Scattered around the tropics and sub-tropics are several dozen states and sub-national jurisdictions considered to be Small Island Developing States (SIDS).

The literature on such states and territories frequently says that they exhibit specific socio-economic characteristics: they are said to have small domestic markets and challenging population dynamics, and they are said to exhibit high reliance on external trade and dependence on a narrow range of exports, making them highly exposed to external shocks (Briguglio, 2016). Climate change impacts are also exacerbating SIDS' vulnerability. Many of these states depend on economic activities located in the coastal area, and rely highly on ocean resources, rendering them

S. Moncada (✉)
Islands and Small States Institute, Institute for European Studies, University of Malta, Msida, Malta
e-mail: stefano.moncada@um.edu.mt

H. Bambrick
Faculty of Health, Queensland University of Technology, Brisbane, QLD, Australia

L. Briguglio
Islands and Small States Institute, University of Malta, Msida, Malta

C. Iorns
Faculty of Law, Victoria University of Wellington, Wellington, New Zealand

I. Kelman
Institute for Risk and Disaster Reduction and Institute for Global Health, University College London, London, UK

University of Agder, Kristiansand, Norway

L. Nurse
CERMES, Faculty of Science and Technology, University of the West Indies, Cave Hill, Barbados

S. Moncada et al. (eds.), *Small Island Developing States*, The World of Small States 9, https://doi.org/10.1007/978-3-030-82774-8_1

potentially vulnerable to sea-level rise and other climate change impacts. Additionally, climate change impacts and natural hazards are often interlinked with several other possible SIDS characteristics such as low income, poor infrastructure and weak institutions, among others, which can deepen exposure to climate change impacts and act as obstacles to successful resilience building.

Despite these vulnerabilities, many SIDS have a long history of successful response to environmental changes, resulting in a repertoire that includes cultural practices, traditional knowledge and skills (Nunn & Mimura, 1997), and that enable them to build their resilience—especially in the face of climate change (Nunn, 2007). However, these different knowledge forms are progressively exposed to modernisation and reliance upon Western-led development assistance (Mercer et al., 2007)—arguably weakening the effectiveness of those responses, and in some instances threaten to eliminate them altogether.

While national and international programmes often indicate the need to strengthen resilience, the term has different interpretations and definitions across disciplines and has perhaps been manipulated for diverse agendas and used in academic and policy contexts to manufacture constructs that are often understood, and experienced, differently by island communities (Kelman, 2020). Indeed, though research in SIDS has advanced, there remains little research on how climate and environmental changes are affecting SIDS (Moncada et al., 2018)—including a more accurate understanding of vulnerability and resilience in these contexts (Robinson, 2020).

This book therefore sets out to explore vulnerable and resilient communities in SIDS, how these are and are not impacted by climate change, and how to evaluate mitigation and adaptation activities. It also identifies factors capable of enhancing or inhibiting the long-term ability of people in SIDS to deal with climate change, and critiques discourse, vocabulary, and constructions used around SIDS and their dealing with vulnerability and resilience in the context of climate change.

2 Structure of the Book

The book is structured to capture three distinct approaches to the topic of vulnerability and resilience in SIDS: concepts and dimensions, sectors, and places. The first part, exemplified in the first two chapters relates to concepts, methodologies and, to a lesser degree, measurements of vulnerability and resilience. The second part, which comprises eight chapters, discusses vulnerability and resilience from a sectoral and a dimensional basis, touching upon areas such as information technology, infrastructure, agriculture, fisheries, diplomacy, gender and indigenous and traditional knowledge. The third part, comprising six chapters, is centred around places and case studies, ranging from single countries to comparative analysis among several SIDS, across Caribbean, Pacific and Indian Oceans.

Part I

In her chapter entitled "*States of 'Knowing': Uncertainty, Ambiguity and Risk in SIDS Climate Change Impacts*" **Foley** critically reviews climate change impacts in SIDS thereby shedding light on current status of knowledge (and lack thereof) on climate change impacts in SIDS. She presents a critical analysis of both quantitative (e.g. climate models, meteorological datasets) and qualitative (e.g. historical photography, narrative accounts) approaches used to study climate change impacts, focusing on their applicability in island contexts, where smallness, boundedness and isolation may present challenges not encountered at broader scales. Inherent limitations of various data and methods, alongside the nonlinearity and unpredictability of the climate system, give rise to further uncertainties and ambiguities. In light of this, Foley questions whether climate data can reasonably inform decision-making relating to climate change impacts on SIDS.

In their chapter, **Brown, Tompkins, Hudson, Schreckenberg, and Corbett** note the absence of both political ecology in island studies and islands in political ecology and make the case for adopting a political ecology approach in matters relating to adaptation to environmental and climate change in SIDS. In their work "*Climate and Development Research in Small Island Developing States: The Benefits of a Political Ecology Approach*", they present a summary of the main universal themes prevalent in islands research, notably: sovereignty, migration, disaster risk reduction and natural resource management trade-offs. The chapter argues that a political ecology approach to assessing sustainable development can support a re-conceptualisation of the challenges faced by SIDS, and arguably reshape perceptions of adaptive capacity and opportunities for future adaptation.

Part II

Fink, Klöck, Korovulavula, and Nunn kick off the sectoral and dimensional analysis in their chapter entitled "*Community Participation, Situated Knowledge and Climate Change (Mal)adaptation in Rural Island Communities: Evidence from Artificial Shoreline-Protection Structures in Fiji*". Using the example of seawalls in rural Fiji communities, the authors emphasise the need to combine indigenous and scientific knowledges, stressing that real empowerment requires appropriately skilled persons with both a scientific understanding of climate change, a sense of locality and a vested interest in the long-term security of its inhabitants. The authors highlight that successful climate change adaptation requires informed investigation of the local context, the drivers of change, and local inhabitants' awareness of the consequences of different response measures. Their contribution stresses the importance of creating such situated knowledge through community participation, scientific information on climate change as well as the advantages and disadvantages of various coping strategies, which need to be effectively communicated to community decision-makers and integrated with existing local cultural knowledge.

The chapter by **Dookie** and **Osgood,** entitled "*Widening the Scope of Disaster Preparedness in Caribbean SIDS: Building Resilience Through Improving Climate Information*" offers a review of the detrimental impacts of Tropical Storm Erika in Dominica in August 2015 and details lessons learnt in disaster risk management. The authors argue that despite the frequency and severity of disasters in the insular Caribbean, there are relatively few insights into the necessary relevance and scope of disaster preparedness. The importance of having disaster preparedness plans is emphasised here with a view to creating deeper awareness of impending disasters and response options and timely communication of appropriate climate information. This understanding could facilitate building comprehensive disaster risk resilience and holistic adaptation strategies by highlighting local vulnerabilities and policy gaps. Their findings highlight the need to have a specific focus on storm disaster preparedness, channeling more attention to disaster preparedness and resilience strategies, assisting disaster policy directives and the accompanying resource allocation planning within the Caribbean in other small island states.

Dumas-Johansen and **Thulstrup** focus on agroforestry in their study entitled "*Sustainable Land Use Systems in Natural Resource Policies: The Role of Agroforestry in the Rio Conventions for Small Island Developing States*". The authors provide a review of SIDS national reports to the three Rio conventions,[1] assessing their particular climate vulnerability and the degree to which national governments view agroforestry as a viable strategy for absorbing the impacts of climate change. They find that while many SIDS share similar vulnerabilities, their climates, ecosystems and current agroforestry practices differ substantially. SIDS' priorities range from large scale agroforestry development to the design of national agroforestry programmes. This chapter finds that the benefits of agroforestry are not yet adequately recognised and that there is a need to promote stronger linkages between agroforestry and climate change adaptation and mitigation policies. An analysis of the benefits of integrated agroforestry systems show how this can support SIDS governments in understanding the potential such systems have in addressing social, economic and environmental challenges faced by the agriculture sector. The authors also argue that agroforestry provides a good opportunity for the development of an integrated approach to implementing the Rio conventions at the national level.

The connection between climate change and agriculture is also discussed by **Sultan,** whose chapter entitled "*Impacts of Climate Change on Agriculture and Food Security in a Small Island State*" discusses the issue of food security, focusing on the SIDS of Mauritius. Sultan estimates the economic impacts of climate change on the agriculture for the Island, applying the Ricardian approach on a small territorial scale, based on the microclimate system of the island. Sultan notes that agriculture activities occupy 43.3% of total land surface in the mainland of Mauritius and employs 7.2% of the labour force, and that being self-sufficient in a few food crops, a decline in agricultural output caused by climate change poses a threat, not only to the livelihoods of farmers, but also to food security. Using cross-sectional farm data from a sample of 392 farmers, the research finds that the agriculture sector responds negatively to changes in mean summer temperature and precipitation. The economic impacts of a rise in mean temperature by 1 °C is US$26.6 per acre per year. The findings of this research emphasise the vulnerability of a small island to

[1] For further information about the Rio Conventions access here https://www.cbd.int/rio/.

climate change in the farming sector and advocates the need of investments in appropriate adaptation measures for farmers in Mauritius.

SIDS, through AOSIS, have often taken centre stage in climate change negotiations, as indicated by **Rasheed** in his chapter "*Early Development of Small Island Developing States (SIDS) Climate Governance: Disproportionate Impact on UN Negotiations*". The author discusses how SIDS have influenced the United Nations (UN) climate negotiations from the very beginning. Rasheed's main contribution lies in the understanding of the disproportionate impact of SIDS' climate negotiating role to better understand present and future trends in SIDS climate politics. Rasheed argues that, despite their weak material capabilities to shape international affairs, SIDS have made a notable and disproportionate impact on the UN climate negotiations emphasising their special case, and by using constructivist approach to foreign policy analysis, he finds that ideas about common but differentiated responsibilities promoted in international climate negotiations have shaped SIDS's climate agenda during pre-and post-UNFCCC negotiations and have influenced the design of the UN climate governance system for SIDS.

Tandrayen Ragoobur's chapter "*Social and Economic Vulnerability to Climate Change: A Gender Dimension for Indian Ocean SIDS*" addresses gender specific socio-economic vulnerability to climate change in four Indian Ocean island economies of Comoros, Madagascar, Maldives and Mauritius. The study provides a systematic analysis of gender differences in environmental changes from vulnerability to adaptability and set out gender-inclusive climate change policies. In analysing the disproportionate bearing of environmental changes on women and their differential coping strategies relative to men, various indicators sourced from the main international available databases for the four islands covering 1960 to 2017, using principal component analysis. The principal component analysis extracts the component factors relevant for men and women separately and a Social and Economic Vulnerability Index is developed applying a gender perspective. The results reveal that the human resource dimension (unemployment and education) and demographics have significant bearings on both women's and men's vulnerability to climate change. However, additional key components are observed for women, namely the proportion of women in decision making and health status relating to fertility rates. With differences in component factors, Tandrayen Ragoobur finds that gender mainstreaming in climate change policies and strategies is imperative for the Indian Ocean island economies.

Campbell and Lee's work is entitled "*Vulnerability of Jamaica South Coast Fishing Communities to Coastal Erosion and Flooding*" and examines how several fishing communities across the Caribbean are experiencing rapid changes in the coastal ecosystem services that exacerbate ecological, social and food security challenges. The authors combine coastal and community-based vulnerability approaches to assess the climate change vulnerability of three major Jamaican fishing communities. The authors find that these communities are at the frontline of serious threats from climate change and badly affected by rapid changes in coastal and marine ecosystems. Furthermore, the impacts of coastal erosion on fishing businesses, as well as the vulnerability of fishing infrastructure and assets to coastal flooding are examined in the three sites. The paper concludes by highlighting four key strategies for reducing livelihood vulnerabilities to coastal erosion and flooding: *i.* creating economic alternatives, *ii.* improving ecosystem resilience, *iii.* strengthening co-management approaches, and iv. developing gender-sensitive indicators. The authors contend that their research demonstrates the value of integrative social and coastal vulnerability assessments in assessing the nuanced impacts on fishery businesses, and propose possible solutions in this regard.

The last chapter of this part entitled "*What Drives Climate Change Vulnerability of the Fisheries Sector: A Comparison Between Three Different SIDS Groups*" also tackles the fisheries sector. Here **Monnereau, Mahon, McConney, Nurse, Turner and Valles** investigate the drivers of climate change vulnerability among three different SIDS regional groups (Caribbean, Pacific and Atlantic, Indian Ocean, Mediterranean and South China Sea (AIMS)). They notice that the ability to identify where adaptation is most needed is constrained by the outputs of vulnerability assessments to date, which typically produce single aggregate rankings, providing limited information on the underlying factors of regional vulnerability. SIDS have often been banded together to address common sustainability challenges. However, despite their similarities they are not homogenous and differences between the three SIDS regional groups are to be expected. Using a diverse range of relevant indicators and a conceptual framework, Monnereau et al. compare the vulnerability of the national fisheries sectors in Caribbean, Pacific, and AIMS SIDS at three levels of decreasing data aggregation, namely overall vulnerability; across the three components making up overall vulnerability, i.e. exposure, sensitivity, and adaptive capacity; and across groups of correlated indicators making up each of the aforementioned components. They find that the vulnerability of the fisheries sector in Pacific SIDS is associated with high sensitivity and low adaptive capacity whereas, in contrast, the vulnerability of Caribbean SIDS is driven by high exposure. These results highlight that what drives overall vulnerability differs fundamentally among SIDS and that the use of a single aggregate score to assess overall vulnerability masks much of the information on specific underlying factors that is needed to guide adequate adaptation strategies to climate change in SIDS.

Part III

The third part, including six chapters, is centred around places and case studies, ranging from single countries to comparative analysis among several SIDS, across Caribbean, Pacific and Indian Oceans.

Mycoo explores why the vulnerabilities to climate change of Caribbean SIDS are intensified by spatial planning inadequacies. The author presents a conceptual framework for planning reform that offers possibilities for enhancing resilience to climate change. In her chapter entitled "*Vulnerabilities to Climate Change and Enhancing Resilience in Caribbean Small Island Developing States: A Spatial Planning Framework*", Mycoo examines in details the spatial planning experience of Trinidad and Tobago and St. Lucia, finding that the inherent vulnerabilities of SIDS are exacerbated by weak implementation of spatial development policies and limited capacity to enforce against poor development practices.

The chapter by **Yamamoto** and **Esteban** addresses human mobility in the context of disasters, and how this has been steering the international community to develop initiatives, often based on consensus, with regards to the protection of the affected people. The authors in their chapter entitled "*Human Mobility and Disaster Risk Reduction in Pacific and Caribbean Small Islands*" examine how SIDS governments in the Pacific and the Caribbean have been implementing measures to deal with human displacement in their norms and policies related to disaster, by specifically looking at regional agendas of Pacific and Caribbean SIDS, focusing on disaster displacement policies. While the authors do not believe that islands will disappear in the future, looking at human mobility and building of skills as a form of adaptation strategies can be considered "no-regret" strategies, which can help improve the long-term economies and resilience of populations at risk. For this to happen, they argue, it would undoubtedly be of benefit that surrounding countries establish policies that permit those displaced by episodic disasters to remain on their soil.

Telesford's chapter "*Climate Change, the SDGs and the Case of Grenada's Blue Growth Plan: Towards an Island Socio-Ecological Framework for Development*" focuses on the Caribbean SIDS of Grenada, looking at the proposed Blue Growth plan, which encourages investments in tourism infrastructure along the coast. This raises concerns relating to climate change vulnerability of SIDS. Telesford highlights the importance of a thorough critically assessment of the projects in the proposed blue growth plans, especially given the existing vulnerability of SIDS. In his work he proposes that the Sustainable Development Goals (SDGs) can provide a framework for assessing the implementation of blue growth projects. Using Grenada as a case study, the blue growth investment plan is critiqued from the perspective of building resilience to climate change pressures and sustainable development. The chapter concludes with a discussion on how the framework can be reformulated to enhance policy planning that leads towards sustainable development and resilience building of SIDS.

Dhar, in his chapter entitled "*Identifying Climate Change Vulnerability and Adaptation Challenges in the Caribbean SIDS: An Urban Morphological Approach*" reviews SIDS' vulnerability to sea-level rise. The research focuses on Negril, situated in Jamaica's west coast, a popular Caribbean tourist destination that contributes to about 5.5% to the national GDP. Dhar stresses how its densest and low-lying area of the Long Bay Resort has been losing beach at between 1–2 m/year, making its entire tourism industry highly vulnerable. After conducting elite interviews with planners, environmentalists and professionals along with observation and a GPS survey, this study identifies and maps the typology of Long Bay's built environment vulnerable to climate change by using urban morphology as an investigating tool. This chapter argues that Long Bay's current and proposed adaptation planning actions primarily consider the immediate and most crucial climatic impacts but overlooks more complex problems and local adaptive capacity. The author further argues that the lack of collaboration between development actors and locals in adaptation decision making process often increases the vulnerability and adaptation challenges of Negril.

Sea-level rise is considered as a major threat to low-lying SIDS. **Simonelli**, in her chapter entitled "*The Contentious Policies of Place and Space: The Maldives, Overpopulation, and Climate Concerns*", discusses this matter with reference to the loss of physical integrity in the Maldives. The research highlights how for the Maldives this threat has become an international issue when the former president went public with a plan to relocate the entire nation. However, the author contends that before climate processes will displace inhabitants on a large scale, other factors will come into play. The multi-level government structures of the Maldives provide several opportunities for contention over policy options and when it comes to continued population growth and development, and preferences differ at each level. Simonelli argues that no matter how terminal climate predictions are for some regions, potential responses cannot be taken out of the context of economic development and domestic political concerns.

The last chapter, by **Connell,** which focuses on Kiribati, a coral atoll state threatened by climate change, discusses the effect of this reality on the livelihoods of people living there. The author, in the chapter entitled "*The Margins Fade: Climate Change, Coral Islands and Distant Destinies?*" contends that in recent years economic growth has been constrained, which, coupled with rapid increase in population, led to considerable aid dependency and significant development challenges. Growing population pressure on resources led to the notion of 'migration with dignity': planned movement into better paid employment overseas with a consequent greater flow of remittances. I-Kiribati have grasped diverse employment opportunities overseas, including agricultural labour, employment on cruise ships or employment in the Australian tourism industry. While recent policy has favoured rural development on outer atolls, international migration in search of superior livelihoods and better economic well-being is unlikely to diminish. The author contends that meeting everyday needs remains more crucial than contemplating worsening environmental pressures. International migration has become more permanent, favouring New Zealand, while climate change is likely to increase the demand for migration and resettlement.

3 Conclusions

Climate change is adding a further layer of complexity to existing vulnerabilities and resiliences across SIDS, as shown in various chapters in this book. Many SIDS have managed to build and enhance their resilience, often by building on a set of norms and long-established practices that stem from a mix of tradition and innovation, spearheading mitigation and adaptation measures.

This is also true when it comes to the reaction of many SIDS to the Covid-19 pandemic. On the one hand, there is the increased isolation and the reduction of travel in and out of the islands have severely affected local economies (Kim, 2020), and generally impacted livelihoods by limiting education, healthcare and other life opportunities. On the other hand, many SIDS have responded (adapted) to this crisis by developing ad-hoc welfare and financial schemes to support their inhabitants (Nhamo et al., 2020), while others have adopted creative entrepreneurialism measures (Bishop et al., 2021), such as the case of the 'Welcome Stamp Visa' promoted by Barbados, or home gardening programmes in the Fiji, to foster post-pandemic resilience. Some of these programs may also deliver benefits that prove compatible with adaptation to climate change.

However, the promotion of such measures—employed necessarily urgently—comes at a cost, which is usually disproportionally higher for SIDS, given the indivisibility of their overhead costs. With luck and good planning, these pandemic responses won't detract from rising to meet the serious threat of climate change. Like climate change, the impacts from Covid-19 on island lives and economy have been various, with the relative remoteness and isolation protective of some, with a handful of Pacific nations still free of cases one year into the pandemic. Likewise, as argued by many authors in this volume, various social, cultural and economic factors specific to individual SIDS need to be taken into consideration, especially in the light of climate change impacts, as these are likely to shape and reshape the outcomes of such measures.

We believe that the scholarly research included in this book is relevant both to academia and to policy, as it provides crucial evidence from the perspective of SIDS, sometimes involving communities excluded from, or at the margins of, mainstream research.

References

Bishop, M., Bouhia, R., Carter, G., Corbett, J., Lindsay, C., Scobie, M., & Wilkinson, E. (2021). *Towards sustained development in Small Island Developing States: Why we need to reshape global governance*. ODI. https://www.odi.org/en/publications/towards-sustaineddevelopment-in-small-island-developing-states

Briguglio, L. P. (2016). Exposure to external shocks and economic resilience of countries: Evidence from global indicators. *Journal of Economic Studies, 43*(6), 1057–1068.

Kelman, I. (2020). Islands of vulnerability and resilience: Manufactured stereotypes? *Area, 52*(1), 6–13.

Kim, N. (2020). *How long will it take for LDCs and SIDS to recover from the impacts of COVID-19?* DESA Working Paper No. 170. New York: United Nations Department of Economic and Social Affairs.

Mercer, J., Dominey-Howes, D., Kelman, I., & Lloyd, K. (2007). The potential for combining indigenous and Western knowledge in reducing vulnerability to environmental hazards in Small Island Developing States. *Environmental Hazards, 7*(4), 245–256.

Moncada, S., Briguglio, L. P., Bambrick, H., & Kelman, I. (2018). Development and climate change in Small Island Developing States [guest editorial]. *International Journal of Climate Change Strategies and Management, 10*(2), 214–216.

Nhamo, G., Dube, K., & Chikodzi, D. (2020). Tourism economic stimulus packages as a response to COVID-19. In *Counting the cost of COVID-19 on the global tourism industry* (pp. 353–374). Springer.

Nunn, P. D. (2007). Holocene sea-level change and human response in Pacific Islands. *Earth and Environmental Science Transactions of the Royal Society of Edinburgh, 98*(1), 117–125.

Nunn, P. D., & Mimura, N. (1997). Vulnerability of South Pacific Island nations to sea-level rise. *Journal of Coastal Research, 24*(Fall), 133–151.

Robinson, S. A. (2020). Climate change adaptation in SIDS: A systematic review of the literature pre and post the IPCC Fifth Assessment Report. *Wiley Interdisciplinary Reviews: Climate Change* (p. e653).

Part I
Concepts and Dimensions

States of 'Knowing': Uncertainty, Ambiguity and Risk in SIDS Climate Change Impacts

Aideen Foley

Abstract This chapter considers the questions of not only *what* we know about climate change impacts for Small Island Developing States (SIDS) but also *how* we know what we know, and how *well* we know what we know. The term 'climate change risk' implies strong knowledge about the possible outcomes and their probabilities, and thus a firm basis for conventional, computation-based decision-making strategies. However, what we 'know' about climate change impacts is often uncertain—where knowledge about possible outcomes is strong but knowledge of their likelihood is weak—or ambiguous, where knowledge of possible outcomes is limited but probabilities can be assigned for those that are known. Therefore, this chapter presents a critical analysis of quantitative (e.g. climate models, meteorological datasets) and qualitative (e.g. historical photography, narrative accounts) approaches used to study climate change impacts, focusing on their applicability in island contexts, where smallness, boundedness and isolation may present challenges which are more pronounced than at broader scales. It explores how the inherent limitations of various data and methods, alongside the nonlinearity and unpredictability of the climate system, give rise to uncertainty and/or ambiguity, and in light of this, discusses how climate data can reasonably inform decision-making relating to climate change impacts for SIDS.

Keywords Climate change · Uncertainty · Climate change impacts · Small Island Developing States · Regional climate model · Islands · Sea level rise · Knowledge · Decision-making · Observations

A. Foley (✉)
Department of Geography, Birkbeck, University of London, London, UK
e-mail: a.foley@bbk.ac.uk

S. Moncada et al. (eds.), *Small Island Developing States*, The World of Small States 9, https://doi.org/10.1007/978-3-030-82774-8_2

1 Introduction

Climate change raises vital questions for our society, relating to our understanding of concepts such as uncertainty, risk, resilience-building, and societal fairness. These questions are of particular poignancy for Small Island Developing States (SIDS), as they are recognised as being highly vulnerable to the effects of climate change (Zellentin, 2015; Walshe et al., 2017), including sea level rise (e.g. Courchamp et al., 2014; Donner & Webber, 2014), while contributing minimally to global emissions (Betzold, 2015).

In everyday speech, 'uncertainty' and 'risk' are often taken as interchangeable concepts but, in the context of climate change assessment, there are fundamental distinctions between the two. Knight (1921) observes that the term 'risk' should only refer to measurable uncertainty, while 'uncertainty' should be restricted to non-quantitative uncertainty. Hubbard (2014) defines uncertainty as "the existence of more than one possibility" (Hubbard, 2014, p. 46), where the true outcome is not known, while risk is a state of uncertainty where some of the possibilities involve an undesirable outcome. Thus, one can have uncertainty without risk but not risk without uncertainty.

'Ambiguity' and 'uncertainty' are also frequently conflated, but Renn and Klinke (2015) argue that ambiguity is a state of ambivalence towards different and potentially divergent streams of thinking about a risk phenomenon, stemming from different interpretations of data or perspectives on what risks, boundaries and priorities are acceptable. Thus, while further scientific research might reduce uncertainty, by identifying and objectively characterising more of the possibilities associated with a risk phenomenon, it may not reduce ambiguity, which stems inherently from subjectivity. However, ambiguity may not necessarily hamstring decision-making, as stakeholders may be able to reach a consensus on what a desirable future scenario would be, even if they differ in why they prefer that outcome (Hull et al., 2002).

The term 'climate change risk', therefore, implies strong knowledge about the possible outcomes and their probabilities, and thus a firm basis for conventional, computation-based decision-making strategies. However, what we 'know' about climate change impacts is often uncertain and/or ambiguous. With less knowledge of possible outcomes, the basis for assigning probability becomes less firm, and alternative approaches to decision-making are required. Given that coastal and island communities are already tangibly encountering and responding to environmental change, it is important for all stakeholders in adaptation and disaster preparedness planning have a critical awareness of what kinds of knowledge exists to support decision-making and how different types of knowledge may contribute in adaptation planning.

Therefore, this chapter presents a critical analysis of quantitative (e.g. climate models, meteorological datasets) and qualitative (e.g. historical photography, narrative accounts) data sources used to study climate change impacts, focusing on their applicability in island contexts, where smallness, boundedness and isolation may present challenges not encountered at broader scales. It explores how the inherent limitations of various data and methods, alongside the nonlinearity and unpredictability of the climate system, give rise to uncertainty and/or ambiguity, and in light of this, discusses how climate data can reasonably inform decision-making relating to climate change impacts for SIDS.

2 Understanding the Past: Climate 'Data' in a Small Island Context

As inertia is an inherent characteristic of the climate system, some climate change impacts associated with current cumulative emissions may be slow to manifest, suggesting that we are likely already committed to some level of climate change impacts this century (Ramanathan & Feng, 2008; Hawkins, 2011). Understanding and anticipating these impacts requires consideration of the past, present and future, to assess how the complex relationship between climate, people and place changes over long timescales, and necessitates engagement with a wide range of data sources and types of knowledge.

The Intergovernmental Panel on Climate Change (IPCC) (Nurse et al., 2014) identifies the key drivers of climate change impact on small islands as variations in air and ocean temperatures; ocean chemistry; rainfall; wind strength and direction; sea levels and wave climate; and extremes events (e.g. tropical cyclones). Given that there are some overlaps between the types of data used to study these different influences, in this chapter data sources available with which to study these drivers and their impacts (summarised in Table 1) are discussed under the themes of:

1. Island climate (temperature, rainfall, wind, including extremes)
2. Oceans (temperature, chemistry and sea level)

2.1 *Island Climate*

When studying environments of the distant past, there are a number of natural datasets available in the form of speleothems (Shakun et al., 2007), corals (Cobb et al., 2001; von Reumont et al., 2016) and tree rings (Jonsson et al., 2002; Rozas et al., 2011). These climate 'proxies' can be used to estimate climate conditions prior to the modern era and, in some cases, even palaeo-climates. While past atmospheric CO_2 concentrations can be measured directly from the air bubbles trapped in ice cores such as those drilled in Antarctica and Greenland, climate parameters must be inferred (e.g. from isotopic fluctuations), making natural proxies an indirect source of climate data.

Table 1 A summary of key climate data sources with relevance for small islands

Type	Examples	Advantages	Limitations
Palaeo data	Tree rings, speleothems, sediments, coral.	Potential to reconstruct climate and ocean variability over geological timescales.	Site-specific records. Variable temporal resolution.
Traditional ecological knowledge	Oral traditions. Calendar systems based on weather/phenology.	Accumulated over very long timescales. Reflects biocultural diversity of place.	Ambiguity/conflicting viewpoints. Challenge of capturing/recording undocumented knowledge.
Historical narrative sources	Letters or diaries describing weather or weather-dependent phenomena, e.g. harvest. Reports of hazardous weather.	Diverse range of sources, potentially pre-dating instrumental observations at a site.	Unreliable narrators. Ambiguity/conflicting viewpoints. Temporal coverage may be discontinuous.
Historical observations and records	Instrumental observations. Ships logs. Records of phenological observations.	Potential to reconstruct climate and ocean variability before the modern era.	Human error (recording/transcribing). Sparse spatial and temporal coverage. Linking phenological responses to climate may be complex.
Pictorial sources	Aerial surveys Dated photographs of e.g. plants in flower	Potential to reconstruct climate and ocean variability before the modern era. Aerial surveys can complement modern satellite imagery.	Sparse spatial and temporal coverage. Linking phenological responses to climate may be complex.
Satellite data	Remotely sensed measurements and derived climate variables, e.g. CMORPH. Satellite altimetry and derived data products.	Comprehensive spatial coverage.	Short records. Mismatch between measured quantity and in situ variable, in some cases. Extreme precipitation may be under-sampled.

Given the specific environmental conditions required for the development of different types of proxy, these data will not be available for all island locations. Nevertheless, there are examples in the literature of proxy analysis undertaken on islands. Lake sediment from the crater lake on Isabel Island in the Eastern Pacific have been used to reconstruct rainfall anomalies for the period 1942–2006 (Romero-Viana et al., 2012), and tree rings have been used to reconstruct 1415–2010 summer precipitation in Cyprus (Touchan et al., 2014).

In terms of spatial resolution, natural proxies offer site-specific records of past climate, with data synthesis required to make wider inferences. For spatially heterogeneous climate variables such as precipitation, the extent to which a proxy record can be applied outside its local area may be limited (Haines & Olley, 2017). The finest temporal resolution reached by such proxies is annual or near-annual (e.g. tree rings, corals, ice cores). As such, human-made sources are still required to distinguish intra-annual detail.

The seventeenth century saw a surge in the invention of instruments that could be used to make meteorological measurements, but weather observation did not begin with the instrumental era. Rather, the instruments seek to make objective and scientific a heretofore more subjective practice. Many island communities possess Traditional Ecological Knowledge (TEK) of weather and climate, codified to varying degrees. Samoans possess a seasonal calendar system based on local environmental changes, identified via indirect phenological indicators such as the timing of the yam harvest, as opposed to direct weather and climate observations (Lefale, 2010), while Maldivians use a traditional weather forecasting system to determine the optimal times for fishing and inter-island travel (Hirsch, 2015). These examples of TEK also reflect the biocultural diversity of islands, and strong linkages between environment, culture and society (Hong et al., 2014).

Such data is mainly qualitative, transmitted from one generation to the next, often orally. Meleisea (1987) writes of historical knowledge in Samoa being shared only with trusted family members, and in certain locations in Indonesia, only older people now have detailed knowledge of the seasonal calendar systems, complicating efforts to officially document them (Hiwasaki et al., 2015). TEK can also be subject to significant ambiguity, with different interpretations of certain indicators emerging in different communities; in different Samoan communities, a proliferation of cockroaches is considered an indicator of oncoming gales, or of hot, clear conditions the following day, while some villages do not subscribe to the methodology as a weather indicator at all (Lefale, 2010), highlighting the importance of community validation of such knowledge if it is to be used in an empirical manner. Integrating this qualitative knowledge into decision-making structures in a culturally sensitive manner requires an appreciation that, as (Kvale, 1996, p. 239) writes: "Truth is constructed through a dialogue; valid knowledge claims emerge as conflicting interpretations and action possibilities are discussed and negotiated among the members of a community." For example, Hiwasaki et al. (2015) note the importance of not only integrating indigenous knowledge with science, but also "safeguarding" that knowledge which cannot be explained by natural science, with consideration of its social significance.

Letters and historical documents represent another important qualitative source of data relating to past climate. Such data can be direct or indirect. For instance, Berland et al. (2013) reconstruct precipitation variability in the Lesser Antilles using a combination of missionary, plantation and governmental papers, which include both direct descriptions of weather events and conditions, and indirect references to associated conditions such as crop yields or water availability. Photographs can also act as a phenological record, for instance, using dated photographs of plants in flower to estimate changes in flowering time (Miller-Rushing et al., 2006). Uncertainty can arise when linking weather-dependent phenomena to climate trends, as relationships between events such as flowering times and climatic conditions can be complex, with potential for contrasting responses to different seasonal climatic changes (Hart et al., 2014).

Instrumental observations, kept either by individuals or institutions, and systematic records of weather phenomena such as frost dates, are examples of direct quantitative historical climate data, and can aid in interpreting indirect observations. However, the use of different instruments over time, in different settings (e.g. screened vs. unscreened thermometers) can lead to biases (Böhm et al., 2009). The accuracy and calibration of early instruments can be a further source of uncertainty. In SIDS contexts, data sparsity is a recurring issue, particularly in peripheral areas (Campbell et al., 2011; Whan et al., 2014). For instance, in a study of temperature trends in Fiji, Kumar et al. (2013) note that all meteorological stations analysed were located on or near the coast, potentially overlooking differences between the coastal areas and the island interior. In island contexts the spatial gaps between observations can be large due to the isolation and fragmentation of islands (Rhee & Yang, 2018).

With their global coverage, satellite data overcomes issues of data sparsity, but records are still short. Satellite measurements of air temperature from the Microwave Sounding Unit begin in 1979 (Spencer & Christy, 1990), although there are discontinuities due to differences in satellites (Hurrell & Trenberth, 1998). Satellite data have been used to study the El Niño-Southern Oscillation (Yulaeva & Wallace, 1994) and tropical cyclones (Kidder et al., 2000), including hurricane prediction (Zhu et al., 2002). Satellites do not measure temperature directly. Surface temperature is derived from passive microwave measurements, but this measurement depends not only on temperature but also emissivity at the surface, which varies depending on e.g. soil composition and land cover (Mendelsohn et al., 2007). This lack of consistency between what is observed with various satellite sensors, and what we wish to measure on the ground necessitates some sort of model or algorithm required to process the satellite data, which may be a source of uncertainty (Foley et al., 2013b). An important part of the European Space Agency (ESA) Climate Change Initiative was algorithm intercomparison to identify the best available techniques for deriving climate variables from satellite measurements (Hollmann et al., 2013).

Precipitation events are frequently intermittent, leaving potential for satellites, which pass over a region a limited number of times each day, to miss them. For this reason, many data sets combine observations from multiple satellites (e.g. Climate Prediction Center morphing method (CMORPH): Joyce et al., 2004; Precipitation Estimation from Remotely Sensed Information using Artificial Neural Networks–Climate Data Record (PERSIANN-CDR): Ashouri et al., 2014). Precipitation data can be inferred from both microwave and infrared measurements, again using an algorithm to relate the measured variable to the in-situ precipitation variable.

2.2 Oceans

Although crustal movements and interannual to decadal sea-level variations can complicate identification of sea-level trend estimates (Douglas, 2001), the IPCC attributes high confidence both to the detection of global mean sea level rise and the attribution of that impact to climate change (Fig. 1). This global mean sea level rise combined with regional variability and vertical ground motion can result in significant 'total' or effective sea level rise at certain island locations, such as Funafuti (Becker et al., 2012).

On geological timescales, isotope analysis of corals and marine sediments can be used to reconstruct paleo sea-level changes. These approaches have been used in a number of island locations including Barbados (Schellmann et al., 2004), Haiti (Dumas et al., 2006) and Cape Verde (Zazo et al., 2010). Similar approaches have been used to reconstruct paleo ocean pH (Anagnostou et al., 2012).

The Permanent Service for Mean Sea Level (PSMSL) is the principal repository for tide gauge data (Woodworth & Player, 2003; Holgate et al., 2012). There are currently over 1500 tide gauge stations listed, with the earliest records dating to 1980 although average length of records is 10 years.[1] Around 2/3 of stations are included in the Revised Local Reference data set, for which sea level is measured relative to a known local land-based datum. Tide gauges without such a reference point are prone to calibration drift, making them less suitable for use in assessments of long-term change. The type of sensor utilised by the tide gauge can be a source of uncertainty, with bottom pressure sensors prone to instrumental drift compared with radar sensors (Míguez et al., 2012). Tectonic changes and local factors such as water extraction can also lead to bias in tide gauge data (Holgate, 2007).

Historical sources are being used to extend these records further into the past (Bradshaw et al., 2015). For example, Testut et al. (2010) estimated rate of relative sea level change over 135 years at Saint Paul Island using measurements on a tide staff from 1874, linking the historical measurement to the same datum as modern record through a combination of precise levelling at the site, and calibration using satellite altimetry and a Global Positioning System (GPS) buoy.

The satellite altimeter record is comprehensive and of high quality, but still short, commencing in the 1990s (Church & White, 2011; Ablain et al., 2015). These data have confirmed that sea level change is not spatially homogeneous, with some regions experiencing greater rates of change than the global mean (Cazenave & Nerem, 2004). Calibration phases between the different satellite altimeter missions, e.g. TOPEX/Poseidon (1992–2006), and Jason-1 (2001), Jason-2 (2008) and Jason-3 (2016), allow biases between the missions to be diagnosed, resulting in a highly accurate record (Nerem et al., 2010). It has been demonstrated that linking a new satellite to the mean sea level continuous record without a calibration phase would introduce significant uncertainty (Zawadzki & Ablain, 2016), highlighting the vital importance of forward planning in data collection activities.

[1] http://www.psmsl.org/data/obtaining/.

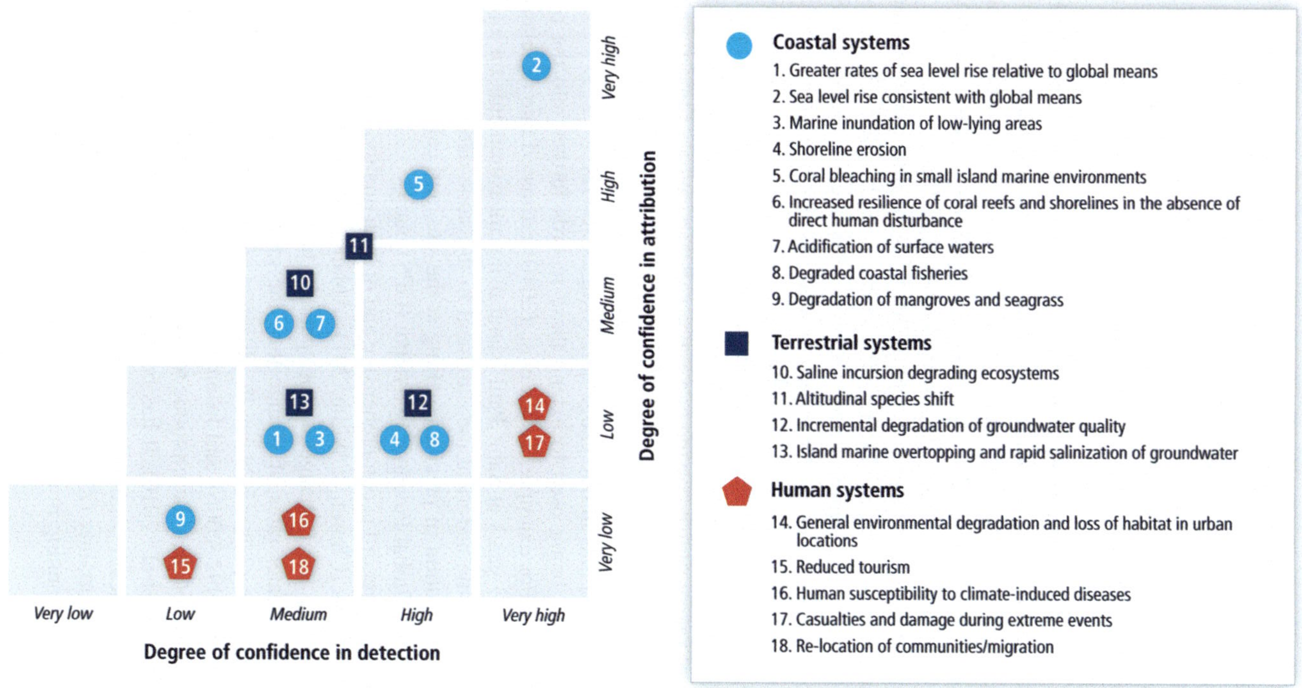

Fig. 1 A comparison of the degree of confidence in the detection of observed impacts of climate change on tropical small islands with the degree of confidence in attribution to climate change drivers at this time. For example, the blue symbol No. 2 (Coastal Systems) indicates there is very high confidence in both the detection of "sea level rise consistent with global means" and its attribution to climate change drivers; whereas the red symbol No. 17 (Human Systems) indicates that although confidence in detection of "casualties and damage during extreme events" is very high, there is at present low confidence in the attribution to climate change. It is important to note that low confidence in attribution frequently arises owing to the limited research available on small island environments.

Source: Figure 29-2 from Nurse et al. (2014), pp. 1613–1654. Reproduced with permission from Nurse, L.A., R.F. McLean, J. Agard, L.P. Briguglio, V. Duvat-Magnan, N. Pelesikoti, E. Tompkins, and A.Webb, 2014: Small islands. In: *Climate Change 2014: Impacts, Adaptation, and Vulnerability. Part B: Regional Aspects. Contribution of Working Group II to the Fifth Assessment Report of the Intergovernmental Panel on Climate Change* [Barros, V.R., C.B. Field, D.J. Dokken, M.D. Mastrandrea, K.J. Mach, T.E. Bilir, M. Chatterjee, K.L. Ebi, Y.O. Estrada, R.C. Genova, B. Girma, E.S. Kissel, A.N. Levy, S. MacCracken, P.R. Mastrandrea, and L.L.White (eds.)]. Cambridge University Press, Cambridge, United Kingdom and New York, NY, USA

Satellite altimetry observations can also be used to infer wave characteristics (e.g. European Space Agency's GlobWave project; Gavrikov et al., 2016), which can aid in studying shoreline change. However, while some features such as wave height are directly measured, other information must be derived from the measured quantities, using a model or algorithm to transform satellite proxy variables and arrive at in-situ variables. Other ocean parameters that can be remotely sensed using satellites include sea surface temperature (e.g. Alerskans et al., 2020), salinity, inferred from radiometer measurements (Lagerloef et al., 2008; Font et al., 2010; Kerr et al., 2010) and water quality variables, inferred from ocean colour (Garaba et al., 2014; Petus et al., 2016). Remote sensing data has also been used extensively to study changes in coral reefs (Purkis, 2018).

Several authors have used aerial imagery to study climate change impacts. This can be classed as another indirect, human-made data source. Identifying storm-generated features in such images can yield insights into the variability of change experienced over time. Historical photography can also be used to assess climate-driven changes in plant communities, by studying shifting tree lines; although in tropical ecosystems, usable photographs are scarce (Vellend et al., 2013). Availability of aerial photography in remote island locations is sometimes explained by a past strategic military significance. For example, in their study of shoreline change in Takapoto Atoll, Duvat and Pillet (2017) note that the availability of aerial photography from the late 1960s is linked to the location's role in French nuclear testing; similarly, the images used by Ford (2013) in his study of the shoreline change in the Marshall Islands dates back to World War II. Given that their original purpose was not the long-term monitoring of environmental change, it is to be expected that aerial image collections, where they exist for island locations, are fragmented, with limited temporal range. Satellite data has been used in many studies to supplement historic aerial imagery (Purkis et al., 2016).

Ships logs can potentially provide a rich source of instrumental data transecting oceans (Gergis et al., 2010; Brohan et al., 2012), providing insights into ocean climate. The Voluntary Observing Ship (VOS) data archive provides meteorological variables and visually observed wave characteristics for the period 1888–2015 (Grigorieva et al., 2017). The International Comprehensive Ocean-Atmosphere Data Set (ICOADS: Freeman et al., 2017), which include salinity, wave parameters and sea surface temperature data for the period 1662–2014, utilises observations from ships, buoys, coastal stations, and other marine platforms.

3 Anticipating the Future: Climate Models in Small Island Contexts

What we 'know' about past climate and current trends of change (Fig. 1) is grounded in the synthesis of a range of data sources: qualitative and quantitative, natural and human-made, direct and indirect observations. However, as we move towards the future and anticipate the impacts of climate change, much of our knowledge stems from climate modelling and, of particular relevance to SIDS, regional downscaling.

Misconceptions about what climate models are and what they seek to provide hamper their use as a decision-support tool (Weaver et al., 2013). Climate predictions and climate projections are commonly conflated, when each term has a specific meaning. A climate prediction is an estimate of the actual evolution of the climate system, e.g. at seasonal (Patt et al., 2005) or inter-annual scales. Modelling the climate system on these relatively short timescales is an 'initial value' problem: the greatest uncertainty stems from inaccurate knowledge of the current conditions, and improving this knowledge can reduce the uncertainty associated with the prediction.

A climate projection is an estimate of the response of the climate system to emission/concentration scenarios, or radiative forcing scenarios. Modelling the climate in this way becomes a 'boundary value' problem: uncertainty largely stems from the lack of knowledge about, and subsequent assumptions which must be made about conditions that influence the climate over the long term, such as changes in land use, greenhouse gases and aerosols, volcanic eruptions, and so forth. Several of these assumptions require postulation about future socioeconomic and technological developments, which is inherently unknowable knowledge, leading to irreducible uncertainty.

Arguably, the smallness and boundedness that frequently characterise SIDS pose technical challenges to climate modelling (Foley, 2017). Due to their small surface area and boundedness, grid cells corresponding to small islands are likely to be classed as ocean in a state-of-the-art Atmosphere-Ocean General Circulation Models (AOGCMs). For example, models used in the Coupled Model Intercomparison Project 5 (CMIP5) (Taylor et al., 2011) on average possess resolution of $\sim 1.85^{\circ} \times 2.25^{\circ}$, corresponding to over 200 km^2 at the Equator.[2] Jury and Bernard (2020) describe the Antilles islands as "'lost' between grid-points" in these models.

Regional downscaling can help to bridge this knowledge gap, using either dynamic (i.e. regional climate model) or statistical (using site-specific observational data to infer relationships between large-scale modelled variables and local climate parameters) techniques. Projects such as the Providing Regional Climates for Impacts Studies (PRECIS)-Caribbean initiative (Taylor et al. 2013) and Coordinated Regional Climate Downscaling Experiment (CORDEX) South Asia (Ghimire et al., 2015) provide a wealth of regional climate model data with particular relevance to small islands, at comparatively higher resolutions than those of General Circulation Models (GCMs), although smaller islands may still be missed (Centella-Artola et al., 2015). Additionally, as regional climate models are reliant on a parent GCM for their driving conditions (Foley et al., 2013a), the inability of a GCM to resolve island topography could still potentially impact the simulative skill of the Regional Climate Model (RCM) that does resolve the topography.

[2] https://portal.enes.org/data/enes-model-data/cmip5/resolution.

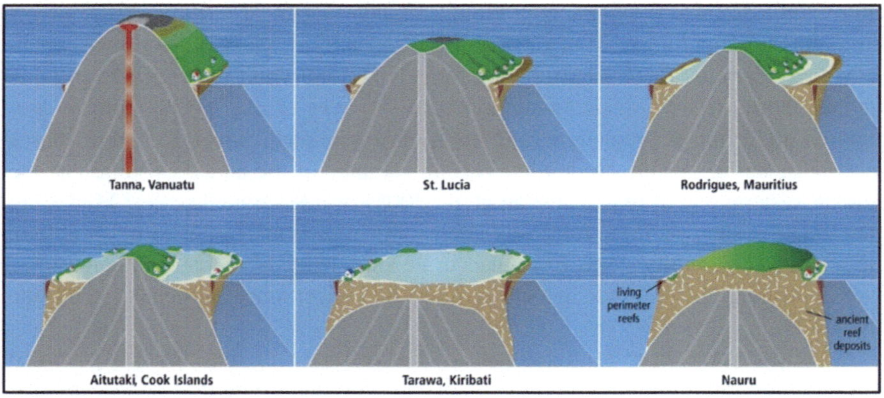

Fig. 2 Representative tropical island typologies. From top left: A young, active volcanic island (with altitudinal zonation) and limited living perimeter reefs (red zone at outer reef edge), through to an atoll (centre bottom), and raised limestone island (bottom right) dominated by ancient reef deposits (brown + white fleck). Atolls have limited, low-lying land areas but well developed reef/lagoon systems. Islands composed of continental rocks are not included in this figure. Source: Figure 29-1 from Nurse et al. (2014), pp. 1613–1654. Reproduced with permission from Nurse, L. A., R.F. McLean, J. Agard, L.P. Briguglio, V. Duvat-Magnan, N. Pelesikoti, E. Tompkins, and A. Webb, 2014: Small islands. In: *Climate Change 2014: Impacts, Adaptation, and Vulnerability. Part B: Regional Aspects. Contribution of Working Group II to the Fifth Assessment Report of the Intergovernmental Panel on Climate Change* [Barros, V.R., C.B. Field, D.J. Dokken, M.D. Mastrandrea, K.J. Mach, T.E. Bilir, M. Chatterjee, K.L. Ebi, Y.O. Estrada, R.C. Genova, B. Girma, E.S. Kissel, A.N. Levy, S. MacCracken, P.R. Mastrandrea, and L.L.White (eds.)]. Cambridge University Press, Cambridge, United Kingdom and New York, NY, USA

Small islands can possess significant orographic variation within a limited area (Fig. 2), creating a need for further, finer scale impact modelling (e.g. freshwater lens response to climate change; Holding & Allen, 2015). In this way, determining climate change impacts requires a series of inferences, with the range of possibilities at each stage (socio-economic scenarios, AOGCMs, RCMs, impacts models) each extending the envelope of uncertainty, resulting in a 'cascade of uncertainty' (Mitchell & Hulme, 1999).

The multi-model method is a pragmatic approach to regional climate modelling which acknowledges that all models are potentially lacking in skill and therefore it is unwise to rely on a single model. It also recognizes that all models represent a possible potential future and by utilizing output from many models, more of these potential futures are sampled. However, a significant issue with the multi-model approach is how to proceed when projections lack coherency with each other. For example, if one model projects an increase in rainfall under climate change and another model projects a decrease, how should decision-makers proceed when they are seeking to incorporate climate change into policy? One way to improve the multi-model approach would be to include some experiential knowledge about the individual models in the ensemble, to form a more coherent set of future projections and reduce the occurrence of contradictory scenarios which are of little use to the adaptation decision-making process (Foley et al., 2013a). Weighting model projections can be a subjective approach but it can be made more robust by using multiple diagnostics metrics and attempting to account for as much uncertainty as possible (Tebaldi & Knutti, 2007).

4 Planning for the Future: Informing Adaptation in SIDS

Adaptation will likely be required in order to prepare for negative effects by minimizing vulnerability, but also to maximize positive impacts, such as the potential to grow new crops, where they exist. There are various types of adaptation, including anticipatory (e.g. Smith et al., 1996; Linnenluecke et al., 2012; Kuruppu & Willie, 2015) or reactive adaptation, incremental or transformational (Kates et al., 2012) and autonomous (Allen, 2006) or planned (Moser & Ekstrom, 2010). Adaptation strategies can also be typified according to their approach to knowledge, as 'top-down' or 'bottom-up'.

Classical 'top-down' approaches often use modelled scenarios of future climate to determine impacts and suitable adaptation measures, and can be highly expert-dependant (Storbjörk, 2007). Such approaches are rooted in positivist research philosophy, which focuses on the data of experience, e.g. observations and experimental results. As Irwin (2010, p. 3) notes:

> Positivism expects a logical formula that explains the matter of the Earth. Complexity is read as a set of complicated causal stimuli that needs to be included in the model. The unknown and the uncertain are just the yet-to-be-discovered or better still, the yet-to-be-deduced.

However, knowledge is fallible and as such, certainty is impossible. Empirical or evidence-based knowledge about the climate system may have to be revised or rethought if further observations reveal previously unknown information; in a system undergoing change, there may be emergent phenomena that cannot be verified by experience. In modelling climate states that have not been experienced before, it cannot be stated with certainty that climate processes and feedbacks will behave as they are currently observed to behave. These uncertainties can be described an ontological, relating to the nature of the climate system itself, and epistemological, relating to gaps in our knowledge of the climate system (Foley, 2010).

'Bottom-up' approaches focus on assessing the current vulnerability and adaptive capacity of the community and tend to be highly participatory, focusing on the needs of at-risk populations. By exploring the process of adaptation and what makes it unique to a particular location and its inhabitants, factors that might constrain or render unviable certain adaptive measures are recognized and accounted for in the planning process (Smit & Wandel, 2006). 'Bottom-up' approaches focus on assessing current vulnerability and adaptive capacity. Approaches like this may be grounded in post-positivistic research philosophy, emphasizing the societal concerns associated with the scientific question of climate change (e.g. Rahmayanti, 2021), and applying qualitative methods to come to understanding through the subjective interpretations of individuals (Mohajan, 2018). However, a key limitation is that future climate change may be outside the scope of what has previously been experienced and may bring new challenges, such as sea level rise and ocean acidification. For these impacts, there may be no traditional adaptations, or the limits of existing adaptations may be unknown (Weir et al., 2017). As such, even within a bottom-up framework, scenarios of future climate may be beneficial for testing adaptation strategies.

The intersection between science and society is critical when considering climate change impacts, especially in the context of SIDS. The mathematical and physical representation of climate change cannot be pursued in isolation as there are human interests and societal concerns to address. As such, while improving the numerical representation of the climate is clearly beneficial to climate planning, a more useful approach would also attempt to address the consequences of uncertainty and ambiguity in the decision-making process. Useful strategies could include robust decision-making—which evaluates the conditions under which a particular decision would fail—to identify those that will endure under a range of scenarios (Lempert, 2013), and decision scaling, which identifies the climate states that favour a particular decision and assesses the probability of their occurrence using climate model data (Brown et al., 2012).

However, while decision-making can be instrumentally rational, weighing up costs and benefits, or procedurally rational, following consistent rules, emotions inevitably influence these processes and how decision-makers respond to uncertainty (Schwarz, 2000; Li et al., 2014). Adaptations that are reversible, flexible, soft, or 'no-regret' are one set of approaches to decision-making under uncertainty (Hallegatte, 2009), but there is a risk that pursuing these strategies exclusively delays more challenging transformative adaptation, e.g. relocation. Li et al. (2014) suggest affective construal as an approach that takes emphasis away from the impossible task of finding the 'right' solution under uncertain conditions and focuses instead on feelings associated with a choice, such as hope, loss, security or insecurity. Thresholds of uncertainty relating to data and climate change impacts might be similarly reformulated.

5 Outlook

Climate change presents a 'wicked problem' for coastal communities and especially SIDS, in which the intrinsic complexity and non-stationarity of the climate system and its interconnectedness with human society renders it difficult to arrive at an unambiguously 'correct' solution for the impacts of climate change (Moser et al., 2012). Improving the knowledge basis for decision-making can certainly help with determining which steps a community could helpfully take, but uncertainty will likely always be a feature of the decision-making process.

Given the uncertainties and limitations of all the data sources discussed, the optimal approach would be to use multiple datasets to identify robust signals of change. This is the approach taken by authors of the IPCC reports, with the type, amount and quality of evidence assessed alongside the levels of agreement between different evidence sources to arrive at confidence levels (Mastrandrea et al., 2010). The challenge to decision-making comes where there is high agreement but limited evidence, or robust evidence but low agreement. The former can be addressed with further research on small island environments, including data rescue activities (McGree et al., 2014; Whan et al., 2014) and citizen science approaches (Marshall et al., 2012). The latter is a case of using decision-making techniques that are robust under uncertainty. It is also important that decision-makers are fully aware of instances when lower confidence has been attached to a particular outcome due to an absence of research on small island environments, as limited or conflicting evidence of a particular impact cannot be taken as proof that there will be no impact.

Climate models provide valuable information about future climates. However, mismatches in scale can occur when tools designed for assessing climate impacts at regional scales are applied to small islands, necessitating local ground-truthing (Owen et al., 2016) and rigorous model evaluation (Foley & Kelman, 2018). While uncertainties will remain, given the natural variability of the climate system and the inherent unpredictability of anticipating the nature of anthropogenic intervention in the climate system in the future, climate models can be informative in the adaptation planning process, provided that uncertainty is communicated effectively with planners and decision-makers (Pidgeon & Fischhoff, 2011). Effective approaches might include tailoring the kinds of data provided, and the language or visualisations used to communicate them, to different contexts (Schroth et al., 2015).

Ultimately, although all information is potentially mistaken, it does not have to be viewed from the position of scepticism—that is, viewed with doubt. Theoretically, this means embracing pragmatism as a philosophical approach (Peirce, 1905) and acknowledging that while current knowledge may require revision as errors are corrected and new developments emerge, this does not prevent any progress being made. However, decision-making strategies may need to shift away from more conventional computation approaches (de Boer et al., 2010), to better support the work of adapting to uncertain and ambiguous climate change risks.

References

Ablain, M., Cazenave, A., Larnicol, G., Balmaseda, M., Cipollini, P., Faugère, Y., Fernandes, M. J., Henry, O., Johannessen, J. A., Knudsen, P., Andersen, O., Legeais, J., Meyssignac, B., Picot, N., Roca, M., Rudenko, S., Scharffenberg, M. G., Stammer, D., Timms, G., & Benveniste, J. (2015). Improved sea level record over the satellite altimetry era (1993–2010) from the Climate Change Initiative project. *Ocean Science, 11*(1), 67–82.

Alerskans, E., Høyer, J. L., Gentemann, C. L., Pedersen, L. T., Nielsen-Englyst, P., & Donlon, C. (2020). Construction of a climate data record of sea surface temperature from passive microwave measurements. *Remote Sensing of Environment, 236*, 111485.

Allen, K. M. (2006). Community-based disaster preparedness and climate adaptation: Local capacity-building in the Philippines. *Disasters, 30*(1), 81–101.

Anagnostou, E., Huang, K.-F., You, C.-F., Sikes, E. L., & Sherrell, R. M. (2012). Evaluation of boron isotope ratio as a pH proxy in the deep sea coral Desmophyllum dianthus: Evidence of physiological pH adjustment. *Earth and Planetary Science Letters, 349–350*, 251–260.

Ashouri, H., Hsu, K.-L., Sorooshian, S., Braithwaite, D. K., Knapp, K. R., Cecil, L. D., Nelson, B. R., & Prat, O. P. (2014). PERSIANN-CDR: Daily precipitation climate data record from multisatellite observations for hydrological and climate studies. *Bulletin of the American Meteorological Society, 96*(1), 69–83.

Becker, M., Meyssignac, B., Letetrel, C., Llovel, W., Cazenave, A., & Delcroix, T. (2012). Sea level variations at tropical Pacific islands since 1950. *Global and Planetary Change, 80–81*, 85–98.

Berland, A. J., Metcalfe, S. E., & Endfield, G. H. (2013). Documentary-derived chronologies of rainfall variability in Antigua, Lesser Antilles, 1770–1890. *Climate of the Past, 9*(3), 1331–1343.

Betzold, C. (2015). Adapting to climate change in small island developing states. *Climatic Change, 133*(3), 481–489.

Böhm, R., Jones, P. D., Hiebl, J., Frank, D., Brunetti, M., & Maugeri, M. (2009). The early instrumental warm-bias: A solution for long central European temperature series 1760–2007. *Climatic Change, 101*(1–2), 41–67.

Bradshaw, E., Rickards, L., & Aarup, T. (2015). Sea level data archaeology and the Global Sea Level Observing System (GLOSS). *GeoResJ, 6*, 9–16.

Brohan, P., Allan, R., Freeman, E., Wheeler, D., Wilkinson, C., & Williamson, F. (2012). Constraining the temperature history of the past millennium using early instrumental observations. *Climate of the Past, 8*(5), 1551–1563.

Brown, C., Ghile, Y., Laverty, M., & Li, K. (2012). Decision scaling: Linking bottom-up vulnerability analysis with climate projections in the water sector. *Water Resources Research, 48*(9), W09537.

Campbell, J. D., Taylor, M. A., Stephenson, T. S., Watson, R. A., & Whyte, F. S. (2011). Future climate of the Caribbean from a regional climate model. *International Journal of Climatology, 31*(12), 1866–1878.

Cazenave, A., & Nerem, R. S. (2004). Present-day sea level change: Observations and causes. *Reviews of Geophysics, 42*(3), RG3001.

Centella-Artola, A., Taylor, M. A., Bezanilla-Morlot, A., Martinez-Castro, D., Campbell, J. D., Stephenson, T. S., & Vichot, A. (2015). Assessing the effect of domain size over the Caribbean region using the PRECIS regional climate model. *Climate Dynamics, 44*(7–8), 1901–1918.

Church, J. A., & White, N. J. (2011). Sea-level rise from the late 19th to the early 21st century. *Surveys in Geophysics, 32*(4–5), 585–602.

Cobb, K. M., Charles, C. D., & Hunter, D. E. (2001). A central tropical Pacific coral demonstrates Pacific, Indian, and Atlantic decadal climate connections. *Geophysical Research Letters, 28* (11), 2209–2212.

Courchamp, F., Hoffmann, B. D., Russell, J. C., Leclerc, C., & Bellard, C. (2014). Climate change, sea-level rise, and conservation: Keeping island biodiversity afloat. *Trends in Ecology & Evolution, 29*(3), 127–130.

de Boer, J., Wardekker, J. A., & van der Sluijs, J. P. (2010). Frame-based guide to situated decision-making on climate change. *Global Environmental Change, 20*(3), 502–510.

Donner, S. D., & Webber, S. (2014). Obstacles to climate change adaptation decisions: A case study of sea-level rise and coastal protection measures in Kiribati. *Sustainability Science, 9*(3), 331–345.

Douglas, B. C. (2001). Chapter 3 Sea level change in the era of the recording tide gauge. In B. C. Douglas, M. S. Kearney, & S. P. Leatherman (Eds.), *International geophysics, sea level rise* (pp. 37–64). Academic Press. [online] Available from: http://www.sciencedirect.com/science/article/pii/S0074614201800061

Dumas, B., Hoang, C. T., & Raffy, J. (2006). Record of MIS 5 sea-level highstands based on U/Th dated coral terraces of Haiti. *Quaternary International, 145–146*, 106–118.

Duvat, V. K. E., & Pillet, V. (2017). Shoreline changes in reef islands of the Central Pacific: Takapoto Atoll, Northern Tuamotu, French Polynesia. *Geomorphology, 282*, 96–118.

Foley, A., Fealy, R., & Sweeney, J. (2013a). Model skill measures in probabilistic regional climate projections for Ireland. *Climate Research, 56*(1), 33–49.

Foley, A., & Kelman, I. (2018). EURO-CORDEX regional climate model simulation of precipitation on Scottish islands (1971–2000): Model performance and implications for decision-making in topographically complex regions. *International Journal of Climatology, 38*(2), 1087–1095.

Foley, A. M. (2010). Uncertainty in regional climate modelling: A review. *Progress in Physical Geography, 34*(5), 647–670.

Foley, A. M. (2017). Climate impact assessment and "islandness": Challenges and opportunities of knowledge production and decision-making for Small Island Developing States. *International Journal of Climate Change Strategies and Management.* [online] Retrieved February 7, 2018, from http://www.emeraldinsight.com/doi/10.1108/IJCCSM-06-2017-0142

Foley, A. M., Dalmonech, D., Friend, A. D., Aires, F., Archibald, A. T., Bartlein, P., Bopp, L., Chappellaz, J., Cox, P., Edwards, N. R., Feulner, G., Friedlingstein, P., Harrison, S. P., Hopcroft, P. O., Jones, C. D., Kolassa, J., Levine, J. G., Prentice, I. C., Pyle, J., Vázquez Riveiros, N., Wolff, E. W., & Zaehle, S. (2013b). Evaluation of biospheric components in Earth system models using modern and palaeo-observations: The state-of-the-art. *Biogeosciences, 10* (12), 8305–8328.

Font, J., Camps, A., Borges, A., Martin-Neira, M., Boutin, J., Reul, N., Kerr, Y. H., Hahne, A., & Mecklenburg, S. (2010). SMOS: The challenging sea surface salinity measurement from space. *Proceedings of the IEEE, 98*(5), 649–665.

Ford, M. (2013). Shoreline changes interpreted from multi-temporal aerial photographs and high resolution satellite images: Wotje Atoll, Marshall Islands. *Remote Sensing of Environment, 135*, 130–140.

Freeman, E., Woodruff, S. D., Worley, S. J., Lubker, S. J., Kent, E. C., Angel, W. E., Berry, D. I., Brohan, P., Eastman, R., Gates, L., Gloeden, W., Ji, Z., Lawrimore, J., Rayner, N. A., Rosenhagen, G., & Smith, S. R. (2017). ICOADS Release 3.0: A major update to the historical marine climate record. *International Journal of Climatology, 37*(5), 2211–2232.

Garaba, S., Badewien, T., Braun, A., Schulz, A.-C., & Zielinski, O. (2014). Using ocean colour remote sensing products to estimate turbidity at the Wadden Sea time series station Spiekeroog. *Journal of the European Optical Society-Rapid publications, 9*.

Gavrikov, A. V., Krinitsky, M. A., & Grigorieva, V. G. (2016). Modification of Globwave satellite altimetry database for sea wave field diagnostics. *Oceanology, 56*(2), 301–306.

Gergis, J., Brohan, P., & Allan, R. (2010). The weather of the First Fleet voyage to Botany Bay, 1787–1788. *Weather, 65*(12), 315–319.

Ghimire, S., Choudhary, A., & Dimri, A. P. (2015). Assessment of the performance of CORDEX-South Asia experiments for monsoonal precipitation over the Himalayan region during present climate: Part I. *Climate Dynamics*, 1–24.

Grigorieva, V. G., Gulev, S. K., & Gavrikov, A. V. (2017). Global historical archive of wind waves based on Voluntary Observing Ship data. *Oceanology, 57*(2), 229–231.

Haines, H. A., & Olley, J. M. (2017). The implications of regional variations in rainfall for reconstructing rainfall patterns using tree rings. *Hydrological Processes, 31*(16), 2951–2960.

Hallegatte, S. (2009). Strategies to adapt to an uncertain climate change. *Global Environmental Change, 19*(2), 240–247.

Hart, R., Salick, J., Ranjitkar, S., & Xu, J. (2014). Herbarium specimens show contrasting phenological responses to Himalayan climate. *Proceedings of the National Academy of Sciences, 111*(29), 10615–10619.

Hawkins, E. (2011). Our evolving climate: Communicating the effects of climate variability. *Weather, 66*(7), 175–179.

Hirsch, E. (2015). "It won't be any good to have democracy if we don't have a country": Climate change and the politics of synecdoche in the Maldives. *Global Environmental Change, 35*, 190–198.

Hiwasaki, L., Luna, E., & Syamsidik & Marçal, J.A. (2015). Local and indigenous knowledge on climate-related hazards of coastal and small island communities in Southeast Asia. *Climatic Change, 128*(1–2), 35–56.

Holding, S., & Allen, D. M. (2015). From days to decades: Numerical modelling of freshwater lens response to climate change stressors on small low-lying islands. *Hydrology and Earth System Sciences, 19*(2), 933–949.

Holgate, S. J. (2007). On the decadal rates of sea level change during the twentieth century. *Geophysical Research Letters, 34*(1), L01602.

Holgate, S. J., Matthews, A., Woodworth, P. L., Rickards, L. J., Tamisiea, M. E., Bradshaw, E., Foden, P. R., Gordon, K. M., Jevrejeva, S., & Pugh, J. (2012). New data systems and products at the permanent service for mean sea level. *Journal of Coastal Research*, 493–504.

Hollmann, R., Merchant, C. J., Saunders, R., Downy, C., Buchwitz, M., Cazenave, A., Chuvieco, E., Defourny, P., de Leeuw, G., Forsberg, R., Holzer-Popp, T., Paul, F., Sandven, S., Sathyendranath, S., van Roozendael, M., & Wagner, W. (2013). The ESA climate change initiative: Satellite data records for essential climate variables. *Bulletin of the American Meteorological Society, 94*(10), 1541–1552.

Hong, S.-K., Maffi, L., Oviedo, G., Mastuda, H., & Kim, J.-E. (2014). Development and vision of island biocultural diversity initiative. In *Biocultural landscapes* (pp. 207–214). Springer. [online] Retrieved February 16, 2018, from https://link.springer.com/chapter/10.1007/978-94-017-8941-7_14

Hubbard, D. W. (2014). *How to measure anything: Finding the value of intangibles in business* (3rd ed.). John Wiley & Sons, Inc..

Hull, R. B., Robertson, D. P., Richert, D., Seekamp, E., & Buhyoff, G. J. (2002). *Assumptions about ecological scale and nature knowing best hiding in environmental decisions*. [online] Retrieved February 14, 2018, from https://www.nrs.fs.fed.us/pubs/3983

Hurrell, J. W., & Trenberth, K. E. (1998). Difficulties in obtaining reliable temperature trends: Reconciling the surface and satellite microwave sounding unit records. *Journal of Climate, 11* (5), 945–967.

Irwin, R. (2010). Climate change and Heidegger's philosophy of science. *Essays in Philosophy, 11* (1), 16–30.

Jonsson, S., Gunnarson, B., & Criado, C. (2002). Drought is the major limiting factor for tree-ring growth of high-altitude Canary Island pines on Tenerife. *Geografiska Annaler: Series A, Physical Geography, 84*(1), 51–71.

Joyce, R. J., Janowiak, J. E., Arkin, P. A., & Xie, P. (2004). CMORPH: A method that produces global precipitation estimates from passive microwave and infrared data at high spatial and temporal resolution. *Journal of Hydrometeorology, 5*(3), 487–503.

Jury, M. R., & Bernard, D. (2020). Climate trends in the East Antilles Islands. *International Journal of Climatology, 40*(1), 36–51.

Kates, R. W., Travis, W. R., & Wilbanks, T. J. (2012). Transformational adaptation when incremental adaptations to climate change are insufficient. *Proceedings of the National Academy of Sciences, 109*(19), 7156–7161.

Kerr, Y. H., Waldteufel, P., Wigneron, J. P., Delwart, S., Cabot, F., Boutin, J., Escorihuela, M. J., Font, J., Reul, N., Gruhier, C., Juglea, S. E., Drinkwater, M. R., Hahne, A., Martin-Neira, M., & Mecklenburg, S. (2010). The SMOS mission: New tool for monitoring key elements of the global water cycle. *Proceedings of the IEEE, 98*(5), 666–687.

Kidder, S. Q., Goldberg, M. D., Zehr, R. M., DeMaria, M., Purdom, J. F. W., Velden, C. S., Grody, N. C., & Kusselson, S. J. (2000). Satellite analysis of tropical cyclones using the Advanced Microwave Sounding Unit (AMSU). *Bulletin of the American Meteorological Society, 81*(6), 1241–1260.

Knight, F. H. (1921). *Risk, uncertainty and profit*. Hart, Schaffner & Marx.

Kumar, R., Stephens, M., & Weir, T. (2013). Temperature trends in Fiji: A clear signal of climate change. *The South Pacific Journal of Natural and Applied Sciences, 31*(1), 27–38.

Kuruppu, N., & Willie, R. (2015). Barriers to reducing climate enhanced disaster risks in Least Developed Country-Small Islands through anticipatory adaptation. *Weather and Climate Extremes, 7*, 72–83.

Kvale, S. (1996). *Interviews: An introduction to qualitative research interviewing*. Sage Publications.

Lagerloef, G., Colomb, F. R., Le Vine, D., Wentz, F., Yueh, S., Ruf, C., Lilly, J., Gunn, J., Chao, Y., Decharon, A., Feldman, G., & Swift, C. (2008). The Aquarius/SAC-D mission: Designed to meet the salinity remote-sensing challenge. *Oceanography, 21*(1), 68–81.

Lefale, P. F. (2010). Stormy weather today: Traditional ecological knowledge of weather and climate. The Samoa experience. *Climatic Change, 100*(2), 317–335.

Lempert, R. (2013). Scenarios that illuminate vulnerabilities and robust responses. *Climatic Change, 117*(4), 627–646.

Li, Y., Ashkanasy, N. M., & Ahlstrom, D. (2014). The rationality of emotions: A hybrid process model of decision-making under uncertainty. *Asia Pacific Journal of Management, 31*(1), 293–308.

Linnenluecke, M. K., Griffiths, A., & Winn, M. (2012). Extreme weather events and the critical importance of anticipatory adaptation and organizational resilience in responding to impacts. *Business Strategy and the Environment, 21*(1), 17–32.

Marshall, N. J., Kleine, D. A., & Dean, A. J. (2012). CoralWatch: Education, monitoring, and sustainability through citizen science. *Frontiers in Ecology and the Environment, 10*(6), 332–334.

Mastrandrea, M. D., Field, C. B., Stocker, T. F., Edenhofer, O., Ebi, K. L., Frame, D. J., Held, H., Kriegler, E., Mach, K. J., Matschoss, P. R., Plattner, G.-K., Yohe, G. W., & Zwiers, F. W. (2010). *Guidance note for lead authors of the IPCC fifth assessment report on consistent treatment of uncertainties*. Intergovernmental Panel on Climate Change (IPCC).

McGree, S., Whan, K., Jones, D., Alexander, L. V., Imielska, A., Diamond, H., Ene, E., Finaulahi, S., Inape, K., Jacklick, L., Kumar, R., Laurent, V., Malala, H., Malsale, P., Moniz, T., Ngemaes, M., Peltier, A., Porteous, A., Pulehetoa-Mitiepo, R., Seuseu, S., Skilling, E., Tahani, L., Teimitsi, F., Toorua, U., & Vaiimene, M. (2014). An updated assessment of trends and variability in total and extreme rainfall in the western Pacific. *International Journal of Climatology, 34*(8), 2775–2791.

Meleisea, M. (Ed.). (1987). *Lagaga: A short history of Western Samoa*. University of the South Pacific.

Mendelsohn, R., Kurukulasuriya, P., Basist, A., Kogan, F., & Williams, C. (2007). Climate analysis with satellite versus weather station data. *Climatic Change, 81*(1), 71–83.

Míguez, B. M., Testut, L., & Wöppelmann, G. (2012). Performance of modern tide gauges: Towards mm-level accuracy. *Scientia Marina, 76*(S1), 221–228.

Miller-Rushing, A. J., Primack, R. B., Primack, D., & Mukunda, S. (2006). Photographs and herbarium specimens as tools to document phenological changes in response to global warming. *American Journal of Botany, 93*(11), 1667–1674.

Mitchell, T. D., & Hulme, M. (1999). Predicting regional climate change: Living with uncertainty. *Progress in Physical Geography: Earth and Environment, 23*(1), 57–78.

Mohajan, H. K. (2018). Qualitative research methodology in social sciences and related subjects. *Journal of Economic Development, Environment and People, 7*(1), 23–48.

Moser, S. C., & Ekstrom, J. A. (2010). A framework to diagnose barriers to climate change adaptation. *Proceedings of the National Academy of Sciences, 107*(51), 22026–22031.

Moser, S. C., Williams, S. J., & Boesch, D. F. (2012). Wicked challenges at land's end: Managing coastal vulnerability under climate change. *Annual Review of Environment and Resources, 37*(1), 51–78.

Nerem, R. S., Chambers, D. P., Choe, C., & Mitchum, G. T. (2010). Estimating mean sea level change from the TOPEX and Jason altimeter missions. *Marine Geodesy, 33*(sup1), 435–446.

Nurse, L. A., McLean, R. F., Agard, J., Briguglio, L. P., Duvat-Magnan, V., Pelesikoti, N., Tompkins, E., & Webb, A. (2014). Small islands. In V. R. Barros, C. B. Field, D. J. Dokken, M. D. Mastrandrea, K. J. Mach, T. E. Bilir, M. Chatterjee, K. L. Ebi, Y. O. Estrada, R. C. Genova, B. Girma, E. S. Kissel, A. N. Levy, S. MacCracken, P. R. Mastrandrea, & L. L. White (Eds.), *Climate Change 2014: Impacts, adaptation, and vulnerability. Part B: Regional aspects. Contribution of Working Group II to the Fifth Assessment Report of the Intergovernmental Panel on Climate Change* (pp. 1613–1654). Cambridge University Press.

Owen, S. D., Kench, P. S., & Ford, M. (2016). Improving understanding of the spatial dimensions of biophysical change in atoll island countries and implications for island communities: A Marshall Islands' case study. *Applied Geography, 72*, 55–64.

Patt, A., Suarez, P., & Gwata, C. (2005). Effects of seasonal climate forecasts and participatory workshops among subsistence farmers in Zimbabwe. *Proceedings of the National Academy of Sciences of the United States of America, 102*(35), 12623–12628.

Peirce, C. S. (1905). What pragmatism is. *The Monist*, 161–181.

Petus, C., Devlin, M., Thompson, A., McKenzie, L., Teixeira da Silva, E., Collier, C., Tracey, D., & Martin, K. (2016). Estimating the exposure of coral reefs and seagrass meadows to land-sourced contaminants in river flood plumes of the great barrier reef: Validating a simple satellite risk framework with environmental data. *Remote Sensing, 8*(3), 210.

Pidgeon, N., & Fischhoff, B. (2011). The role of social and decision sciences in communicating uncertain climate risks. *Nature Climate Change, 1*(1), 35–41.

Purkis, S. J. (2018). Remote sensing tropical coral reefs: The view from above. *Annual Review of Marine Science, 10*(1), 149–168.

Purkis, S. J., Gardiner, R., Johnston, M. W., & Sheppard, C. R. C. (2016). A half-century of coastline change in Diego Garcia – The largest atoll island in the Chagos. *Geomorphology, 261*, 282–298.

Rahmayanti, K. P. (2021). Promote collaborative governance? Review of disaster risk reduction strategy in Jakarta. In *IOP conference series: Earth and environmental science* (Vol. 724, No. 1, p. 012044). IOP Publishing.

Ramanathan, V., & Feng, Y. (2008). On avoiding dangerous anthropogenic interference with the climate system: Formidable challenges ahead. *Proceedings of the National Academy of Sciences, 105*(38), 14245–14250.

Renn, O., & Klinke, A. (2015). Risk governance and resilience: New approaches to cope with uncertainty and ambiguity. In U. Fra.Paleo (Ed.), *Risk governance* (pp. 19–41). Springer. [online] Available from: http://link.springer.com/10.1007/978-94-017-9328-5_2

Rhee, J., & Yang, H. (2018). Drought prediction for areas with sparse monitoring networks: A case study for Fiji. *Water, 10*(6), 788.

Romero-Viana, L., Kienel, U., & Sachse, D. (2012). Lipid biomarker signatures in a hypersaline lake on Isabel Island (Eastern Pacific) as a proxy for past rainfall anomaly (1942–2006AD). *Palaeogeography, Palaeoclimatology, Palaeoecology, 350–352*, 49–61.

Rozas, V., Pérez-de-Lis, G., García-González, I., & Arévalo, J. R. (2011). Contrasting effects of wildfire and climate on radial growth of *Pinus canariensis* on windward and leeward slopes on Tenerife, Canary Islands. *Trees, 25*(5), 895–905.

Schellmann, G., Radtke, U., Potter, E.-K., Esat, T. M., & McCulloch, M. T. (2004). Comparison of ESR and TIMS U/Th dating of marine isotope stage (MIS) 5e, 5c, and 5a coral from Barbados—Implications for palaeo sea-level changes in the Caribbean. *Quaternary International, 120*(1), 41–50.

Schroth, O., Pond, E., & Sheppard, S. R. J. (2015). Evaluating presentation formats of local climate change in community planning with regard to process and outcomes. *Landscape and Urban Planning, 142*, 147–158.

Schwarz, N. (2000). Emotion, cognition, and decision making. *Cognition and Emotion, 14*(4), 433–440.

Shakun, J. D., Burns, S. J., Fleitmann, D., Kramers, J., Matter, A., & Al-Subary, A. (2007). A high-resolution, absolute-dated deglacial speleothem record of Indian Ocean climate from Socotra Island. *Yemen Earth and Planetary Science Letters, 259*(3), 442–456.

Smit, B., & Wandel, J. (2006). Adaptation, adaptive capacity and vulnerability. *Global Environmental Change, 16*(3), 282–292.

Smith, J. B., Ragland, S. E., & Pitts, G. J. (1996). A process for evaluating anticipatory adaptation measures for climate change. In *Climate change vulnerability and adaptation in Asia and the Pacific* (pp. 229–238). Springer. [online] Retrieved January 16, 2018 from https://link.springer.com/chapter/10.1007/978-94-017-1053-4_21

Spencer, R. W., & Christy, J. R. (1990). Precise monitoring of global temperature trends from satellites. *Science, 247*(4950), 1558–1562.

Storbjörk, S. (2007). Governing climate adaptation in the local arena: Challenges of risk management and planning in Sweden. *Local Environment, 12*(5), 457–469.

Taylor, K. E., Stouffer, R. J., & Meehl, G. A. (2011). An overview of CMIP5 and the experiment design. *Bulletin of the American Meteorological Society, 93*(4), 485–498.

Taylor, M. A., Centella, A., Charlery, J., Bezanilla, A., Campbell, J., Borrajero, I., Stephenson, T., & Nurmohamed, R. (2013). The Precis Caribbean Story: Lessons and legacies. *Bulletin of the American Meteorological Society, 94*(7), 1065–1073.

Tebaldi, C., & Knutti, R. (2007). The use of the multi-model ensemble in probabilistic climate projections. *Philosophical Transactions of the Royal Society A: Mathematical, Physical and Engineering Sciences, 365*(1857), 2053–2075.

Testut, L., Miguez, B. M., Wöppelmann, G., Tiphaneau, P., Pouvreau, N., & Karpytchev M. (2010). Sea level at Saint Paul Island, southern Indian Ocean, from 1874 to the present *Journal of Geophysical Research: Oceans, 115*(C12), C12028.

Touchan, R., Christou, A. K., & Meko, D. M. (2014). Six centuries of May–July precipitation in Cyprus from tree rings. *Climate Dynamics, 43*(12), 3281–3292.

Vellend, M., Brown, C. D., Kharouba, H. M., McCune, J. L., & Myers-Smith, I. H. (2013). Historical ecology: Using unconventional data sources to test for effects of global environmental change. *American Journal of Botany, 100*(7), 1294–1305.

von Reumont, J., Hetzinger, S., Garbe-Schönberg, D., Manfrino, C., & Dullo, W.-C. (2016). Impact of warming events on reef-scale temperature variability as captured in two Little Cayman coral Sr/Ca records. *Geochemistry, Geophysics, Geosystems, 17*(3), 846–857.

Walshe, R. A., Seng, D. C., Bumpus, A., & Auffray, J. (2017). Perceptions of adaptation, resilience and climate knowledge in the Pacific: The cases of Samoa, Fiji and Vanuatu. *International Journal of Climate Change Strategies and Management.* [online] Available from: http://www.emeraldinsight.com/doi/full/10.1108/IJCCSM-03-2017-0060

Weaver, C. P., Lempert, R. J., Brown, C., Hall, J. A., Revell, D., & Sarewitz, D. (2013). Improving the contribution of climate model information to decision making: The value and demands of robust decision frameworks. *Wiley Interdisciplinary Reviews: Climate Change, 4*(1), 39–60.

Weir, T., Dovey, L., & Orcherton, D. (2017). Social and cultural issues raised by climate change in Pacific Island countries: An overview. *Regional Environmental Change, 17*(4), 1017–1028.

Whan, K., Alexander, L. V., Imielska, A., McGree, S., Jones, D., Ene, E., Finaulahi, S., Inape, K., Jacklick, L., Kumar, R., Laurent, V., Malala, H., Malsale, P., Pulehetoa-Mitiepo, R., Ngemaes, M., Peltier, A., Porteous, A., Seuseu, S., Skilling, E., Tahani, L., Toorua, U., & Vaiimene, M. (2014). Trends and variability of temperature extremes in the tropical Western Pacific. *International Journal of Climatology, 34*(8), 2585–2603.

Woodworth, P. L., & Player, R. (2003). The permanent service for mean sea level: An update to the 21st century. *Journal of Coastal Research, 19*(2), 287–295.

Yulaeva, E., & Wallace, J. M. (1994). The signature of ENSO in global temperature and precipitation fields derived from the microwave sounding unit. *Journal of Climate, 7*(11), 1719–1736

Zawadzki, L., & Ablain, M. (2016). Accuracy of the mean sea level continuous record with future altimetric missions: Jason-3 vs. Sentinel-3a. *Ocean Science, 12*(1), 9–18.

Zazo, C., Goy, J. L., Hillaire-Marcel, C., Dabrio, C. J., González-Delgado, J. A., Cabero, A., Bardají, T., Ghaleb, B., & Soler, V. (2010). Sea level changes during the last and present interglacials in Sal Island (Cape Verde archipelago). *Global and Planetary Change, 72*(4), 302–317.

Zellentin, A. (2015). Climate justice, small island developing states & cultural loss. *Climatic Change, 133*(3), 491–498.

Zhu, T., Zhang, D.-L., & Weng, F. (2002). Impact of the advanced microwave sounding unit measurements on hurricane prediction. *Monthly Weather Review, 130*(10), 2416–2432.

Climate and Development Research in Small Island Developing States: The Benefits of a Political Ecology Approach

Heather Brown, Emma L. Tompkins, Malcolm Hudson, Kate Schreckenberg, and Jack Corbett

Abstract Political ecology has been widely applied to analyse processes of agricultural development, most notably where there are complex relationships between ecological, political and economic factors. Political ecology explores how the impacts of environmental change are felt unequally by economies and societies. Small island developing states, which often produce low levels of greenhouse gas emissions, yet are on the frontline of climate change impacts, demonstrate the unequal nature of the impact of environmental change. The unique vulnerabilities of small island developing states have been documented in numerous international environmental agreements. However, there is an absence of both political ecology in island studies and islands in political ecology. Here we make the case for adopting a political ecology approach when studying adaptations to environmental and climate change in small island developing states. We focus on several universal themes prevalent in islands research, notably: sovereignty, migration, disaster risk reduction and natural resource management trade-offs. This chapter also explores what political ecology can bring to the subject of climate and development in small islands, and concludes that a political ecology approach to sustainable development in small islands can support a reconceptualisation of the challenges faced, as well as reshape perceptions of adaptive capacity, and opportunities for future adaptation.

H. Brown (✉) · E. L. Tompkins · M. Hudson
School of Geography and Environmental Science, University of Southampton, Southampton, UK
e-mail: H.L.Brown@soton.ac.uk; E.L.Tompkins@soton.ac.uk; mdh@soton.ac.uk

K. Schreckenberg
Department of Geography, King's College London, London, UK
e-mail: kate.schreckenberg@kcl.ac.uk

J. Corbett
Politics & International Relations, Faculty of Social Sciences, University of Southampton, Southampton, UK
e-mail: J.Corbett@soton.ac.uk

S. Moncada et al. (eds.), *Small Island Developing States*, The World of Small States 9, https://doi.org/10.1007/978-3-030-82774-8_3

Keywords Political ecology · Small island developing states · Climate change adaptation · Environmental change · Disaster risk reduction · Equality · Climate change mitigation · Sovereignty

1 Introduction

Islands have always faced developmental challenges, in part due to smallness (e.g. limited land area, limited freshwater, limited access to foreign exchange, small domestic markets), isolation and remoteness (Chasek, 2010; Connell, 2013). These challenges are well documented, and experience has shown the varied pathways that islands have chosen to address them. Broadly islands have relied on combinations of overseas remittances and aid, exploitation of natural resources and development of niche services (e.g. from offshore finance, online gaming and shipping registration to philately and leasing internet domain names) (Connell, 2013).

Evidence from the Intergovernmental Panel on Climate Change (IPCC) increasingly points to the potentially damaging impacts that islands will experience as a result of climate change (Nurse et al., 2014). These impacts could include coastal erosion from large transboundary oceanic waves, health impacts from invasive species and transcontinental dust clouds, and the spread of aquatic pathogens (Nurse et al., 2014). Islands and island people have historically adapted to many different changes (social, economic, political and environmental) with varying degrees of success; however, the IPCC suggests that even though islands are taking action to adapt to current and future climate impacts this may not be sufficient given the magnitude of expected changes (Kates et al., 2012; Nurse et al., 2014).

While significant research has been undertaken on islands that consider: island ecology, island politics, climate change adaptation and economic development in islands (some examples include Gorman, 1979; Barnett & Campbell, 2010; Corbett & Ng Shiu, 2014; Jayaraman et al., 2016; Barnett, 2020), not all of this explicitly takes into account the interface between climate change, ecology, politics and economy. As with other climatic hotspots, this has resulted in studies that focus either on the changing environment or the politics of decision-making, but fewer studies that focus on how power relationships affect the ways in which different people use the environment, and how power and the environment affect economic drivers of environmental change. We believe that the lack of explicit focus on political ecology in much island research may hinder a complete understanding of macro and micro drivers of future development under climate change on islands.

In this chapter, we argue for the application of an explicitly political ecology approach to the study of environmental change in Small Island Developing States (SIDS). The political ecology approach is unique because it combines an empirical interest in explaining how and why different people access and distribute resources with normative concerns about equality and justice. The political ecology approach emerged in the 1970s, but it suffers from a double absence: few political ecology scholars study SIDS, and so there is very little research on SIDS in political ecology. The absence is surprising, both because questions about equality and justice are central to debates about the impact of climate change but also because SIDS are widely viewed as the proverbial canaries in the global climate coal mine. Echoing this, SIDS have been extremely successful at highlighting their unique vulnerability, and by extension, their acute sense of injustice, at the international level (Corbett et al., 2019). But there is little explicit research on how environmental change will alter existing patterns of power *within* SIDS (although much of the existing research that we canvas here makes this claim implicitly). An explicitly political ecology approach, therefore, provides the theoretical justification and analytic framework for scholars seeking to link empirical work on the impact of environmental change on SIDS with normative concerns about equality and justice within affected communities.

We substantiate this claim by focusing on some universal themes prevalent in island research. We explore the new understandings that applying a political ecology approach can bring to the subject of climate and development in islands. We focus specifically on sovereignty, migration, disaster risk reduction and natural resource management trade-offs. We conclude with a review of what can be learned from adopting a political ecology framework of analysis for the future development of islands in a changing climate.

2 Why Political Ecology Research on Climate and Development in Islands?

Political ecology is concerned with equality and justice in the way resources are (re)distributed as a consequence of environmental change. The approach draws our attention to politics, people and power when thinking about environment, climate and development problems—making it ideal for considering island development in the context of a changing climate, changing climate policy and changing climate institutions. Political ecology allows us to identify the upheavals in peoples' lives caused by social, economic or environmental change, and specifically the impacts of unequal access to resources which can be the result of power imbalances (Batterbury, 2015). It does this by adopting a multi-scale approach to reveal how power is manifested in human-environment relations (Robbins, 2004). In the context of a changing climate driving multi-level geophysical and social-economic processes, political ecology focuses on the relationships between the environment and vulnerable societal groups from global to local scales, particularly how global political, climatic and environmental pressures can impact on marginalised communities.

One strength of a political ecology approach is the ability to identify justice and equity issues associated with indigenous systems of resource management during the process of incorporation into the global economy (Blaikie & Brookfield, 1987). A political ecology approach has also enabled an analysis of environmental issues related to the impacts of international markets, social inequalities, and large-scale political conflicts (Paulson et al., 2003). Originating in the 1970s (Peet & Watts, 1996), political ecology emerged in research circles from 'green politics', an ideology that aims to create an ecologically stable society rooted in social justice, democracy, nonviolence and environmental rights, i.e. research into politics, environment and policy, which previously operated independently of each other (Somma, 1993). At the time, this was a marked change in our understanding of human-environment relationships. The nascent political ecology movement was able to draw on a variety of developing ideas about the role of society in relation to environmental degradation, risk and global trade. Instead of seeing environmental change as the result of ecological dynamics, the role of social risk and social action was recognised (Nygren & Rikoon, 2008). Instead of seeing risk as being generated by environmental problems, the role of society in shaping risk was made clear and evolved into the political ecology of hazards literature (Wisner et al., 2004).

In the 1980s, the emphasis in political ecology shifted from ecology to politics (Nightingale, 2003; Walker, 2005). Ecological processes, including processes of climate impacts and adaptations, have at times been omitted from more recent studies of power relationships, while sophisticated insights have been made into the power relations of resource access and control (Neumann, 2004). This has culminated in relatively little attention being placed on what natural scientists have to say about environmental change (Nygren & Rikoon, 2008). An explicit political ecology approach brings normative concerns about equality and justice to the forefront of empirical studies that seek to explain human-environmental interactions (Schubert, 2005). Specifically, political ecology brings explicit attention to normative concerns in the study of environmental change.

Contemporary scholars engaged in political ecology research have studied in a variety of contexts, e.g. extractive economies and mining, international land grabs for food security and timber, the impact of protected areas on livelihoods, food politics, urban environmental dynamics and the impact of climate change on local vulnerabilities (for example see: Adams & Hutton, 2007; Allen, 2013; Wolford et al., 2013; Taylor, 2014; Moragues-Faus & Marsden, 2017; Lanzas, 2020). Despite this broad reach of political ecology, the explicit application of political ecology to consider current issues, notably climate adaptation and mitigation, poverty, environment and development on islands is less prevalent. In part, this may be due to a current focus on understanding climate impacts in islands from a physical impacts perspective, and understanding adaptive processes from a social perspective. A political ecology approach draws attention to the fact that climate change will alter how resources are allocated and which groups in society will bear the burden of environmental degradation and climate impacts (Schubert, 2005). Ultimately, there is a double absence: of political ecology in SIDS and SIDS in political ecology. So, bringing political ecology to SIDS is significant and important for both research communities.

3 Vulnerability, Climate and Development on Islands

The vulnerability of islands is undoubtedly multi-scalar, with global to local influences shaping the future on SIDS (Ghosal, 2016). These vulnerabilities have reduced the ability of SIDS to trade and follow conventional development trajectories, evolving from agricultural economies to production economies (Connell & Corbett, 2016). SIDS come in multiple forms, e.g. low-lying coral atolls and volcanic islands, which tend to have very different physical characteristics, such as steep slopes and narrow coastal fringes (Barnett & Campbell, 2010), however, all SIDS have limited land area. Small size leads to small economies of scale, small resource bases, and often, competing land-use practices. Communities often live in close proximity to the sea and are at risk of flooding and saltwater intrusion (Barnett & Campbell, 2010). For example, roughly 75% of Samoa's population (193,000 in 2015 (World Bank, 2015b)) live on or near the coast (Daly et al., 2010).

Isolated and remote locations mean transport between SIDS and to the rest of the world is time-consuming and expensive. Reduced connectivity of SIDS constrains participation in international trade (Connell, 2013). To improve international trade, some countries have taken extreme measures, for example, Samoa and Tokelau moved across the dateline in 2011 and 2012 respectively, to increase their ability to trade with China, New Zealand and Australia (BBC News, 2011). However, development vulnerabilities have shifted from economic issues such as the intrinsic lack of resources and constraints of distance to weaknesses in politics and governance (Pelling & Uitto, 2001; Connell, 2013). At the national and local level, economic growth has been limited in many SIDS leading to high rates of unemployment and poverty at local scales (Abbott & Pollard, 2004).

Islands are highly exposed to hazards, partly due to physical characteristics such as small size, remoteness, isolation, lack of connectivity and proximity to hazards. Many SIDS lie in the tropics and hence are exposed to weather hazards such as El Niño events and annual tropical cyclones. Cyclone damages can reverse years of development (Barnett & Campbell, 2010). Climate change will affect the frequency and magnitude of fast- and slow-onset hazards in some regions leading to more frequent or more intense natural hazards over time (Schipper & Pelling, 2006; Nurse et al., 2014). For many SIDS, climate change is expected to bring increased temperatures and changing rainfall patterns over the next century (Schipper & Pelling, 2006). The number of intense cyclones could increase in some areas, suggesting a strong possibility of increased risks of more persistent and devastating tropical cyclones in a warmer world (Nurse et al., 2014). Vulnerability to these hazards is also increasing due to rising poverty, a growing global population, and other underlying development issues (Schipper & Pelling, 2006).

Islands are not only affected socially and economically by their physical characteristics, but they are also affected politically. Within island states developing and implementing policies is a challenge due to "*weak short-term governments, under-equipped and inexperienced bureaucracies, scarce resources and limited access to information*" (Connell, 2013, p. 53). At the global scale, it is less clear whether islands have power or not. Payne (2004) suggests the smallness of small island states leads to vulnerabilities rather than opportunities in global politics. Lack of economic and political power and authority arguably make it difficult for small island states to exert an effective presence in global negotiations, for example, on trade or climate change (Connell, 2013). To overcome this lack of visibility in global debates, 43 island states formed the Alliance of Small Island States (AOSIS) in 1990. AOSIS has made some progress in championing local and national island needs on the international political stage (Pelling & Uitto, 2001; Baldacchino & Kelman, 2014), for example in pushing for an agreement to limit global warming to 1.5 °C. Global climate change negotiations may be one area where small islands have used their vulnerability to their advantage, and their active participation in international organisations has brought them benefits and recognition on the international stage (Corbett et al., 2019).

Given these physical, economic and political characteristics of many SIDS, we argue that political ecology advances a closer link between natural, social and political sciences allowing better planning for climate change adaptations, identification of environmental risk management solutions, and implementation of local societal changes in response to global environmental and political change (Maguigad et al., 2015). An explicitly political ecology approach provides the theoretical justification and analytical framework for scholars seeking to link empirical work on the impact of environmental change on SIDS with normative concerns about equality and justice within affected communities. We argue that there is clear interest in the analyses of current and future sustainability in island social-ecological systems (Baldacchino, 2006; Banos-González et al., 2016), and endeavour to contribute to scholarship in this area by applying a political ecology lens to the question of how to address climate and development in islands. Specifically, we interrogate four key policy areas—island sovereignty and climate justice; population movement; climate and disaster risk management; and natural resources management—highlighting gaps for additional research.

4 Sovereignty and Climate (in)Justice

Sovereignty is vital for SIDS to retain rights over their ocean resources, have political representation in international forums, and to control the direction of development, allowing them to keep their citizenship and culture. Sovereignty refers to the authority of a state to govern itself (Philpott, 2011) and is a concept that drives the choices of many political leaders. Climate change is creating complicated sovereignty issues for islands, as sea level rise, flooding and erosion are affecting the accessible land area, and changes in sea surface temperature are affecting ocean resources. As a result, current sovereignty issues relating to climate change arise from legal rights over land and ocean resources, and recognition of islands as a political entity in processes of '*deterritorialisation*'—i.e. legal rights to land and sea assets after island inhabitants have been forced to leave due to climate or environmental change (Connell & Corbett, 2016).

Limited land available on SIDS means that surrounding waters are particularly important for food security, transport, trade and economic development (Briguglio, 1995). While oceans play a key role in everyone's lives, oceans, coastal and marine environments are an integral part of island lifestyle (Hau'ofa, 1995; D'Arcy, 2008). All coastal countries have a designated area of the ocean that is their Exclusive Economic Zone (EEZ), where they have jurisdiction over the exploration and exploitation of marine resources (OECD, 2001). The rights provided by an EEZ include those for fishing and deep-sea mineral mining. Tuvalu's EEZ, for example, is 27,000 times the size of its land area. The Republic of Kiribati has the 13th largest EEZ on Earth. In total, SIDS are the custodians of 30% of the 50 largest EEZs on the planet (Jumeau, 2013). One example of the economic benefit provided by EEZs is the licence fees paid by foreign fishing fleets to SIDS to obtain the right to fish their seas (Connell, 2013). SIDS have a small land area, but with consideration of their EEZ, they can be considered as large ocean states (Jumeau, 2013). EEZs, therefore, give SIDS economic power, and also greater sovereign territory. Climate change has the potential to affect the resources in the EEZs (Barnett, 2020). For example, rising sea surface temperatures can change the composition of marine life in an area and affect the quality of the marine environment (Paice & Chambers, 2016). As pressure grows on ocean resources, SIDS will face the challenge of managing their EEZs to enable long-term, sustainable exploitation of their ocean and coastal resources, while at the same time acknowledging that a changing climate may give rise to new environmental challenges in the oceans that SIDS may not have the resources to manage (Garcia & Rosenberg, 2010). A political ecology approach allows the broader perspectives of justice, environmental and climate impacts, adaptations, and the politics of ocean resource management to be carefully balanced.

For low-lying SIDS, maintaining sovereignty at a time of climate change poses novel sovereignty issues relating to the mobility of people. Due to rising sea levels, there is speculation that, without adaptation, some islands may become uninhabitable, leaving island populations without a home country (Kittel, 2014). Several small island nations, including the Maldives, are currently starting to evaluate the options to relocate parts of their populations to different areas (Yamamoto & Esteban, 2010). However, it is unclear what the sovereignty implications under international law, specifically under the United Nations Convention on the Law of the Sea (UNCLOS), are for how SIDS would exist and govern following the physical disappearance of islands due to sea level rise. At present, the law regarding economic use of an island's EEZ after island inhabitants have been forced to leave due to sea level rise is not well defined and leaves low-lying small island communities at risk of losing their statehood (Yamamoto & Esteban, 2010; Kittel, 2014). International discussions of 'deterritorialisation' have started (Connell & Corbett, 2016), to allow inhabitants' identity and culture to be preserved (Kittel, 2014). A complimentary term to a 'deterritorialised' state, is a 'government-in-exile', although the latter is based on the premise that sea level will eventually return to its current level and islands will be re-inhabited by the descendants of the 'exiled' inhabitants (Yamamoto & Esteban, 2010).

Past cases of forced relocation due to loss of territory, such as the relocation of Banaban islanders to Rabi Island in Fiji, do not offer encouraging evidence that sovereignty and political identity has been maintained. Phosphate mining on Banaba by colonial enterprises in the 1940s led to the reduced habitability of Banaba and caused the migration of the entire island population (Tabucanon, 2012). Economic wealth and livelihoods, cultural land, land titles and traditional systems of land tenure were lost (Maude, 1946), with the island now jointly governed by Kiribati and Fiji (Tabucanon, 2012; Stoutenburg, 2015).

A political ecology approach to the issue of deterritorialisation and forced relocation would require explicit consideration of the physical environmental changes, at both local and global scale, and the global political factors, and the risk and justice implications of these interactions. Ignoring these factors risks relying on existing decision-making frameworks and systems that disadvantage SIDS. A political ecology approach makes justice an explicit component of analysis. Therefore, as climate impacts continue to create sovereignty challenges for SIDS, political ecology requires us to consider the political, environmental and justice issues, i.e. special statuses for relocated groups, thinking about dual nationality, land rights, and the maintenance of discrete, self-governing communities, if populations are internationally relocated in the future due to environmental pressures (Tabucanon, 2012). In the context of a changing climate, with possible changes to land area, land access, and rights to a home, sovereignty is increasingly important as both a political and an environmental issue.

5 Migration for a Better Future?

Migration—voluntary and involuntary—emigration and immigration are said to embody the socio-economic and demographic history of islands (King, 2009). The smallest Pacific island countries depend on labour migration, remittances, aid and sustaining urban bureaucracy, i.e. 'MIRAB' economies (Bertram & Watters, 1985). Migration, in particular, has become an important development strategy for small islands (Luthria, 2008; Gibson & McKenzie, 2014; Craven, 2015). Migrant remittances (and aid) do not just supplement local incomes but act as the foundations for the modern economy (Bertram, 1986; Cook & Kirkpatrick, 1998). In Samoa, for example, personal remittances equate to 20% of Samoa's annual GDP (World Bank, 2015a). Migration is often used by islanders where mainstream models of economic development have failed, sometimes due to lack of natural resources, industrialisation and job creation (Connell & Corbett, 2016).

A growing body of research on migration reveals that it is driven by multiple factors; the most often stated drivers by individuals are the search for opportunities of better employment and education, medical treatment, family reunion, or to escape warfare (Guan & McElroy, 2012). However, it is increasingly evident that environmental change is an indirect driver of migration. Families seek better economic opportunities after losing homes or livelihoods in extreme events, or where land is slowly lost, for example, through erosion or salinisation (Kelman, 2017). A political ecology approach to migration requires reflection on all the drivers of migration, direct and indirect, and on the justice implications of this.

There are multiple scales of benefits and costs of migration that are experienced according to the levels of vulnerability and political context. Migrants send home remittances, yet migrants are often the strongest, most economically active member of their household. Losing key household members to migration can affect the quality of the local workforce and social structures at both the national and local level (Connell, 2013). Remittances earned through migration have been seen to contribute to development (e.g. Luthria, 2008). However, many studies of migration fail to account for the multifaceted nature of vulnerability and the longer-term effects of climate change (Eriksen & O'Brien, 2007; Brown, 2011). For example, in Lamen Bay, Vanuatu, migration was a cause and consequence of local vulnerabilities and development failures. Islanders who had migrated due to lack of economic opportunities found on their return home, a lack of local opportunities to invest their savings. A positive outcome of this was an investment in building weather-resilient housing. However, habitats cleared for this housing damaged the environment and put water resources under stress (Craven, 2015). Further, returning migrants felt less willing to act for the community and took part in less collective decision-making and management of community resources. Political ecology requires us to consider the multiple scales at which politics, environment and justice interact. By using a multi-scale and interdisciplinary approach, the research in Lamen Bay revealed unexpected social and environmental impacts of migration that may have otherwise been missed by a single-scaled or disciplinary approach.

In SIDS, discussions around adapting to climate change and managing the environment could be used as ways to encourage and facilitate internal migration which may be a good opportunity for island development (Connell & Corbett, 2016). Using an interdisciplinary approach that considers environment, politics and social issues, Kothari (2014) examines the resettlement policies in the Maldives. While not termed 'political ecology' Kothari applies these ideas in relation to a government proposal to merge a population dispersed over 200 islands onto 10–15 islands. It is evident that environmental discourses are being mobilised to re-introduce previously unpopular resettlement and migration policies. Kothari's interdisciplinary research that considered politics, ecology and justice has generated findings that indicate possible positive outcomes from migration. This may not have been identified without a specific focus on justice.

6 Disaster Risk Reduction in a Changing Climate

Islands, amongst some other notable examples, are considered to be at the forefront of climate impacts (Connell, 2015; Klöck & Nunn, 2019). Although there are few large-scale studies of climate adaptation practice (Nurse et al., 2014), there are an increasing number of small-scale case studies looking at past adaptive practices (Tompkins, 2005; Kelman et al., 2015; Buggy & McNamara, 2016; Walshe et al., 2017). Climate change impacts, rising sea level and global average temperatures are projected to increase the frequency of severe natural hazards like cyclones (Knutson et al., 2019; Zhang et al., 2020) that disproportionately affect the development of low-income countries, low-lying countries and SIDS (UNFCCC, 2005). Climate justice, as embodied in the Paris Agreement (UNFCCC, 2015), acknowledges that low-income countries, low-lying countries and SIDS need financial support to adapt to climate change impacts.

Despite repeated calls since the 1990s by the IPCC for more research on climate impacts and adaptations in islands (e.g. Nurse et al., 2014), there remains limited evidence of the spectrum and prevalence of types of adaptations to hazards and their effectiveness. From the small-scale case studies that exist, there is some evidence that social structures have supported some effective disaster risk reduction practices over time (Mercer et al., 2010). Following the 2004 Indian Ocean tsunami, disaster and development research explored the role of scientific knowledge and indigenous institutions in reducing risk, improving response and recovery, and adapting to long-term climatic change (Rumbach & Foley, 2014). In some SIDS, indigenous institutions play important roles in Disaster Risk Reduction (DRR). For example, Rumbach and Foley (2014) identified a number of different areas in which the *fa'a Samoa*, the culture of Samoa, and its attendant institutions helped to save lives and speed recovery following the 2009 tsunami in American Samoa. Characteristics important for effective DRR related particularly to decision-making, with designated responsibilities for all DRR activities, mobilising local action and facilitating communication with external actors as well as providing a system that enabled vulnerable people to hold decision-makers accountable (Rumbach & Foley, 2014). A political ecology approach here could advance our understanding of climate change adaptation by considering the effectiveness of past DRR practices in the context of a changing climate. Many small-scale adaptation studies evaluate the distributional consequences of adaptations without necessarily taking into account their effectiveness under different environmental or climatic impacts. Both perspectives matter. Both perspectives are considered in political ecology.

The Barbados Programme of Action (BPoA) (United Nations, 1994) noted that natural disasters are of special concern to SIDS and called for assistance in establishing and/or strengthening national and regional institutional mechanisms and policies designed to reduce the impacts of natural disasters, improve disaster preparedness and integrate natural disaster considerations in development planning. In the same year, the World Conference on Disaster Reduction held in Yokohama, Japan, expressed particular concern at the high vulnerability of SIDS, as they are the least equipped to mitigate disasters. SIDS continue to be recognised as being particularly vulnerable to disasters (Nurse et al., 2014; UNISDR, 2015; Shultz et al., 2020).

Countries that are already vulnerable to hazards are least able to manage risks and will suffer the most (Popke et al., 2016; Shultz et al., 2020). As we have seen when SIDS group together, they can be influential in climate discussions, however individually, SIDS are small and often insignificant actors on an international scale with little influence (Briguglio, 1995). It is often argued that SIDS have barely contributed to global greenhouse gas emissions, yet will be the first to be impacted by climate impacts, and by mitigation strategies adopted by other countries, e.g. in relation to international tourism (Hoad, 2016). The impacts of climate change have been recognised by the Malé Declaration as a breach of human rights. In November 2007, representatives of SIDS signed the Malé Declaration on the Human Dimension of Global Climate Change (CIEL, 2011). The Declaration notes that the environment provides the infrastructure for human civilization, and that the impacts of climate change pose the most immediate, fundamental and far-reaching threat to the environment, as well as to individuals and communities around the planet (CIEL, 2011).

Many islands have received aid funding that has largely focused on economic growth, sometimes at the expense of the needs of marginalised people. Disaster risk reduction agendas on SIDS that are driven by external agencies and the decisions of external experts, rather than by local needs, have been largely recognised as ineffective (Barnett, 2008; Buggy & McNamara, 2016). These decisions have often related to 'hard' engineered responses but have been recognised as not contributing to long-term sustainable development.

Ultimately, there remains a lack of political ecology research in this area. There is a need to understand the impacts of alternative governance approaches to environmental and climate change management, for example, by exploring 'soft' adaptation approaches (Nunn, 2009). A political ecology approach involves taking a more socially based view of adaptation that while recognising the physical environmental changes, also focuses on underlying vulnerabilities, and takes into consideration the indirect impacts that climate change and variability will have across social, political, environmental and economic spheres for local communities (Buggy & McNamara, 2016). An example of a socially based view of adaptation can be seen in the Caribbean where local knowledge and participation is being integrated into climate change adaptation programmes (Mercer et al., 2012), reducing the divide between local and external knowledge and power. 'Soft' community-based adaptation has been encouraged to reduce power imbalances between top-down aid donors and beneficiary communities. In Yadua Island, Fiji, community-based management approaches are encouraged to reduce community reliance on the government implemented disaster risk reduction (Martin et al., 2018).

'Soft' community-based adaptation approaches do not always work, often due to uneven power dynamics within committees and individuals governing the projects. Ten out of 34 community-based climate change adaptation projects on Pele Island, Vanuatu, failed as committees have broken down or were deemed ineffective

(Buggy & McNamara, 2016). The governance and leadership environment on Pele Island enabled particular groups and individuals to take over projects and as a result, restrict the voice and participation of other community members. In consequence, project benefits were captured by powerful people (Mansuri & Rao, 2013; Buggy & McNamara, 2016). On Pele Island, the adaptation projects further worsened power imbalances between community members, causing increased division and conflict that may have significant impacts on the future ability of communities to adapt to climate change. A political ecology approach is as important, therefore, for exploring power dynamics within communities as it is for understanding power imbalances between local and global scales.

7 Natural Resource Management Trade-offs in a Changing Climate

Political ecology can highlight trade-offs between economic development and land use sustainability, for example in relation to tourism by identifying the multi-scale tensions that exist between society, ecology, environment and economics (Connell, 2018). In some SIDS, e.g. the Maldives, tourism has been widely encouraged by government policy for economic development (Scheyvens, 2011). Tourism supports the Gross Domestic Product (GDP) of some SIDS; in 2014, 18% of Samoa's GDP was provided by tourism (World Bank, 2017). Yet, tourism can exert pressure on the natural environment and livelihoods of local people by placing significant demands on natural resources such as marine protected areas, coastal fisheries, marine and terrestrial biodiversity, ecosystems and ecosystem services (Chen et al., 2017), and water (Lu et al., 2013). As well as adding stress to the natural environment and creating competing objectives for land use, tourism has also led to people being displaced and excluded from culturally or economically important natural resources (Towner & Milne, 2017).

Political ecology considers how people are affected by land use conflicts. Given the importance of both natural resources and tourism to SIDS, political ecology offers a useful lens to explore political pressures at the national scale and the highly uneven power relations between land developers and land residents. In the Maldives, political ecologists are arguing that greater attention needs to be paid to issues of power, including social and political dimensions of tourism development, to alleviate negative social and environmental impacts of tourism (Scheyvens, 2011). Connell (2018), following Brown et al. (2001) argues that making trade-offs between conservation and development on islands is tricky, and this remains one of the great challenges of political ecology.

Trade-offs also exist between top-down and local/ traditional governance systems of natural resource management on SIDS. There are many cases where the use of traditional knowledge and traditional institutions that exhibit flexible and customary land tenure have delivered sustainable self-management of resources in the past. For example, in Simbo, Solomon Islands, loosely organised governance structures that draw on traditional knowledge (McMillen et al., 2014), such as the practice of customary tenure can spontaneously adapt to changing social-ecological conditions (Lauer et al., 2013). Evidence from across Oceania reveals evidence of natural-resource dependent people understanding the natural cycles of a whitebait run and using this knowledge to manage their food source (Jenkins et al., 2018). Lauer (2018), however, emphasises the importance of both customary and top-down governance, highlighting how traditional land tenure systems in the Solomon Islands support sustainable land management, while top-down approaches and formal courts ensure social cohesion and reduce conflict.

Top-down pressures can enhance or hinder traditional institutions and practices. Local traditions have been maintained through years of forest resource management in Timor-Leste while under pressures from various political transitions including the development of regional trade links, Portuguese political control, Indonesian occupation, and political independence. However, a successful trade in beeswax and sandalwood led to depletion of forest resources, which play a critical role in traditional practices (Yoder, 2011). Political ecology offers an important understanding of sustainable customary traditions and the pressures that may break them. Under a changing climate, political ecology requires us to consider the multi-scalar institutions and environmental and climate changes. For islands where customary land and community-based management systems predominate, some level of top-down coordination may be necessary to ensure collective resilience. Hence, a political ecology approach which considers the relationship between multi-scalar institutions and their effect on the environment and human resilience can make an important contribution to promoting the resilience of islands.

8 Conclusions: Towards a Political Ecology of Islands

Islands deserve increased attention from the international political ecology community due to their unique development challenges, expected future climate injustices and opportunities for innovative adaptation (Kueffer & Kinney, 2017). Equally, island studies scholars could benefit from adopting an explicit political ecology approach. With a focus on some universal themes prevalent in island research, we have explored what political ecology can bring to the subject of climate and development in islands. The political ecology approach is unique because it combines an empirical interest in explaining how and why different peoples access and distribute resources with normative concerns about equality and justice, while at the same time taking into account the physical dimensions of environmental and climatic change. SIDS have been successful at highlighting that climate change will impact them unequally relative to other states at the international level, reinforcing the need for an effective political ecology approach to analyse climate and development issues in SIDS.

There is little explicit research on how environmental change will alter existing patterns of power within SIDS (although plenty of implied analyses, many of which are referenced). Political ecology offers an approach for examining social structures in their global and historical settings and their role in navigating future climate change adaptations. Political ecology is well suited to study issues of climate adaptation and climate justice as it considers environmental change, uneven political power, and uneven social impacts. The key strength of a political ecology approach is that it focuses attention on how the impacts of environmental change are felt unequally by economies and societies. SIDS have been successful at highlighting this at an international level, with the recognition of the unique vulnerabilities of SIDS documented in numerous international agreements illustrating the need for climate justice. However, the same research approach has been less prevalent *within* SIDS.

Without an explicit political ecology approach to the study of SIDS development, a full picture of the shocks, stresses, and opportunities on SIDS will remain absent. Political ecology research is methodologically plural, albeit there is an emphasis on identifying and giving voice to the most marginalised members of society. Marginalised peoples tend to be the most vulnerable to environmental change. The normative dimension of a political ecology approach is also important for both natural resource scientists and social scientists. Indeed, we posit that it may be able to tie together interdisciplinary teams of researchers around common concerns where other approaches cannot. It also allows us to examine problems at multiple scales: global-local, national-regional, urban-rural, and between islands.

Political ecology is not without its problems: it continues to be under-theorized and its definition contested. But, at its best, it offers an approach for examining social structures in their global and historical contexts to explain environmental change. Adding SIDS will greatly strengthen political ecology research as these cases have been conspicuously absent from the existing literature. Indeed, SIDS are most likely cases for this approach: they experience an acute sense of injustice in the face of anthropogenic climate change due to their having contributed little to a global problem that threatens their existence. As such, we argue we cannot do global political ecology research without paying attention to their condition.

References

Abbott, D., & Pollard, S. (2004). *Hardship and poverty in the Pacific: Strengthening poverty analysis and strategies in the Pacific*. Asian Development Bank.

Adams, W. M., & Hutton, J. (2007). People, parks and poverty: Political ecology and biodiversity conservation. *Conservation and Society, 5*(2), 147–183.

Allen, M. G. (2013). Melanesia's violent environments: Towards a political ecology of conflict in the western Pacific. *Geoforum, 44*, 152–161.

Baldacchino, G. (2006). Islands, island studies, island studies journal. *Island Studies Journal, 1*(1), 3–18.

Baldacchino, G., & Kelman, I. (2014). Critiquing the pursuit of island sustainability. *Shima: The International Journal of Research into Island Cultures, 8*(2), 1–21.

Banos-González, I., Martínez-Fernández, J., & Esteve, M. Á. (2015). Tools for sustainability assessment in island socio-ecological systems: An application to the Canary Islands. *Island Studies Journal, 11*(1), 9–34.

Barnett, J. (2008). The effect of aid on capacity to adapt to climate change: Insights from Niue. *Political Science, 60*(1), 31–45.

Barnett, J. (2020). Climate change and food security in the Pacific Islands. In J. Connell & K. Lowitt (Eds.), *Food security in small island states* (pp. 25–38). Springer.

Barnett, J., & Campbell, J. (2010). *Climate change and small island states. Power, knowledge and the south Pacific*. Earthscan.

Batterbury, S. (2015). Doing political ecology inside and outside the academy. In R. L. Bryant (Ed.), *The international handbook of political ecology* (pp. 27–43). Edward Elgar.

Bertram, G. (1986). "Sustainable development" in Pacific micro-economies. *World Development, 14*(7), 809–822.

Bertram, L. G., & Watters, R. F. (1985). The MIRAB economy and south Pacific microstates. *Pacific Viewpoint, 26*(3), 497–512.

Blaikie, P., & Brookfield, H. (1987). *Land degradation and society*. Routledge.

Briguglio, L. (1995). Small island developing states and their economic vulnerabilities. *World Development, 23*(9), 1615–1632.

British Broadcasting Corporation (BBC) News. (2011). *Samoa and Tokelau skip a day for dateline change, BBC News*. [online] Retrieved March 26, 2018, from http://www.bbc.co.uk/news/world-asia-16351377

Brown, K. (2011). Sustainable adaptation: An oxymoron? *Climate and Development, 3*(1), 21–31.

Brown, K., Adger, W. N., Tompkins, E., Bacon, P., Shim, D., & Young, K. (2001). Trade-off analysis for marine protected area management. *Ecological Economics, 37*(3), 417–434.

Buggy, L., & McNamara, K. E. (2016). The need to reinterpret "community" for climate change adaptation: A case study of Pele Island, Vanuatu. *Climate and Development, 8*(3), 270–280.

Chasek, P. S. (2010). In S. Halapua, T. Kabutaulaka, & G. M. White (Eds.), *Confronting environmental treaty implementation challenges in the Pacific Islands*. East West Center.

Chen, H., Liu, W., & Hsieh, C. (2017). Integrating ecosystem services and eco-security to assess sustainable development in Liuqiu Island. *Sustainability, 9*(6), 1002.

Connell, J. (2013). *Islands at risk? Environments, economies and contemporary change*. Edward Elgar.

Connell, J. (2015). Climate change, tectonic change, and changing livelihoods in the western Pacific. *The Contemporary Pacific, 27*(1), 1–36.

Connell, J. (2018). Islands: Balancing development and sustainability? *Environmental Conservation, 45*(2), 111–124.

Connell, J., & Corbett, J. (2016). Deterritorialisation: Reconceptualising development in the Pacific Islands. *Global Society, 30*(4), 583–604.

Cook, P., & Kirkpatrick, C. (1998). Labor market adjustment in small open economies: The case of Micronesia. *World Development, 26*(5), 845–855.

Corbett, J., & Ng Shiu, R. (2014). Leadership succession and the high drama of political conduct: Corruption stories from Samoa. *Pacific Affairs, 87*(4), 743–763.

Corbett, J., Xu, Y., & Weller, P. (2019). Norm entrepreneurship and diffusion "from below" in international organisations: How the competent performance of vulnerability generates functional benefits for small states. *Review of International Studies, 45*(4), 647–648.

Craven, L. K. (2015). Migration-affected change and vulnerability in rural Vanuatu. *Asia Pacific Viewpoint, 56*(2), 223–236.

D'Arcy, P. (2008). *The people of the Sea: Environment, identity and history in Oceania*. University of Hawai'i Press.

Daly, M., Poutasi, N., Nelson, F., & Kohlhase, J. (2010). Reducing the climate vulnerability of coastal communities in Samoa. *Journal of International Development, 22*(1), 265–281.

Eriksen, S., & O'Brien, K. (2007). Vulnerability, poverty and the need for sustainable adaptation measures. *Climate Policy, 7*(4), 242–261.

Garcia, S. M., & Rosenberg, A. A. (2010). Food security and marine capture fisheries: Characteristics, trends, drivers and future perspectives. *Philosophical Transactions of the Royal Society B: Biological Sciences*, 2869–2880.

Ghosal, R. D. (2016). Climate change in the Caribbean: A multi-scalar account of context and inequality climate change in the Caribbean. *Journal of Critical Thought and Praxis, 5*(2), 1–20.

Gibson, J., & McKenzie, D. (2014). The development impact of a best practice seasonal worker policy. *The Review of Economics and Statistics, 96*(2), 229–243.

Gorman, M. (1979). *Island ecology* (1st ed.). Springer.

Guan, J., & McElroy, J. L. (2012). The determinants of migration in small islands. *Shima: The International Journal of Research into Island Cultures, 7*(1), 80–95.

Hau'ofa, E. (1995). Our sea of islands. In E. Waddell, V. Naidu, & E. Hau'ofa (Eds.), *A new oceania: Rediscovering our sea of islands* (pp. 39–48). University of the South Pacific.

Hoad, D. (2016). The 2015 Paris Climate Agreement: Outcomes and their impacts on small island states. *Island Studies Journal, 11*(1), 315–320.

Jayaraman, T. K., Choong, C.-K., & Chand, P. (2016). Do foreign aid and remittance inflows hurt competitiveness of exports of pacific island countries? An empirical study of Fiji. *Journal of Economic Development, 41*(2), 111–125.

Jenkins, A., Horwitz, P., & Arabena, K. (2018). My island home: Place-based integration of conservation and public health in Oceania. *Environmental Conservation, 45*(2), 125–136.

Jumeau, R. (2013). *Small island developing states, large ocean states, Expert Group Meeting on Oceans, Seas and Sustainable Development: Implementation and follow-up to Rio + 20*. United Nations. [online] Retrieved June 15, 2020, from https://sustainabledevelopment.un.org/content/documents/1772Ambassador Jumeau_EGM Oceans FINAL.pdf

Kates, R. W., Travis, W. R., & Wilbanks, T. J. (2012). Transformational adaptation when incremental adaptations to climate change are insufficient. *Proceedings of the National Academy of Sciences, 109*(19), 7156–7161.

Kelman, I. (2017). How can island communities deal with environmental hazards and hazard drivers, including climate change? *Environmental Conservation, 44*(3), 244–253.

Kelman, I., Lewis, J., Gaillard, J. C., & Mercer, J. (2015). Island contributions to disaster research. *Global Environment, 8*(1), 16–37.

King, R. (2009). Geography, islands and migration in an era of global mobility. *Island Studies Journal, 4*(1), 53–84.

Kittel, J. (2014). The global "disappearing act": How island states can maintain statehood in the face of disappearing territory. *Michigan State Law Review, 2014*(4), 1207–1250.

Klöck, C., & Nunn, P. D. (2019). Adaptation to climate change in small island developing states: A systematic literature review of academic research. *The Journal of Environment & Development, 28*(2), 196–218.

Knutson, T., Camargo, S. J., Chan, J. C. L., Emanuel, K., Ho, C. H., Kossin, J., Mohapatra, M., Satoh, M., Sugi, M., Walsh, K., & Wu, L. (2019). Tropical cyclones and climate change assessment. *Bulletin of the American Meteorological Society, 100*(10), 1987–2007.

Kothari, U. (2014). Political discourses of climate change and migration: Resettlement policies in the Maldives. *The Geographical Journal, 180*(2), 130–140.

Kueffer, C., & Kinney, K. (2017). What is the importance of islands to environmental conservation? *Environmental Conservation, 44*(4), 311–322.

Lanzas, G. (2020). From water abundance to water scarcity: The case of the Chontalpa, Mexico. *Journal of Political Ecology, 27*(1), 263–278.

Lauer, M. (2018). Governing uncertainty: Resilience, dwelling, and flexible resource management in Oceania. *Conservation and Society, 14*(1), 34–47.

Lauer, M., Albert, S., Aswani, S., Halpern, B. S., Campanella, L., & La Rose, D. (2013). Globalization, Pacific Islands, and the paradox of resilience. *Global Environmental Change, 23*(1), 40–50.

Lu, F., Valdivia, G., & Wolford, W. (2013). Social dimensions of 'nature at risk' in the Galapagos Islands, Ecuador. *Conservation and Society, 11*(1), 83–95.

Luthria, M. (2008). Seasonal migration for development? Evaluating New Zealand's RSE program. *Pacific Economic Bulletin, 23*(3), 165–170.

Maguigad, V., King, D., & Cottrell, A. (2015). Political ecology approach to island tourism planning and climate change adaptation: A methodological exploration. *Urban Island Studies, 1*(1), 133–151.

Mansuri, G., & Rao, V. (2013). *Localising development: Does participation work?* World Bank.

Martin, P. C. M., Nunn, P., Leon, J., & Tindale, N. (2018). Responding to multiple climate-linked stressors in a remote island context: The example of Yadua Island, Fiji. *Climate Risk Management, 21*, 7–15.

Maude, H. E. (1946). *Memorandum, the future of the Banaban population of Ocean Island: With special relation to their land and funds*. University of Adelaide.

McMillen, H. L., Ticktin, T., Friedlander, A., Jupiter, S. D., Thaman, R., Campbell, J., Veitayaki, J., Giambelluca, T., Nihmei, S., Rupeni, E., Apis-Overhoff, L., Aalbersberg, W., & Orcherton, D. F. (2014). Small islands, valuable insights: Systems of customary resource use and resilience to climate change in the Pacific. *Ecology and Society, 19*(4), 44.

Mercer, J., Kelman, I., Alfthan, B., Kurvits, T., Mercer, J., Kelman, I., Alfthan, B., & Kurvits, T. (2012). Ecosystem-based adaptation to climate change in Caribbean small island developing states: Integrating local and external knowledge. *Sustainability, 4*(8), 1908–1932.

Mercer, J., Kelman, I., Taranis, L., & Suchet-Pearson, J. (2010). Framework for integrating indigenous and scientific knowledge for disaster risk reduction. *Disasters, 34*(1), 214–239.

Moragues-Faus, A., & Marsden, T. (2017). The political ecology of food: Carving "spaces of possibility" in a new research agenda. *Journal of Rural Studies, 55*, 275–288.

Neumann, R. P. (2004). Moral and discursive geographies in the war for biodiversity in Africa. *Political Geography, 23*, 813–837.

Nightingale, A. (2003). Nature-society and development: Social, cultural and ecological change in Nepal. *Geoforum, 34*(4), 525–540.

Nunn, P. D. (2009). Responding to the challenges of climate change in the Pacific Islands: Management and technological imperatives. *Climate Research, 40*(2–3), 211–231.

Nurse, L. A., McLean, R. F., Agard Trinidad, J., Briguglio, L., Duvat-Magnan, V., Pelesikoti, N., Tompkins, E., & Webb, A. (2014). Small islands. In V. R. Barros, C. B. Field, D. J. Dokken, M. D. Mastrandrea, K. J. Mach, T. E. Bilir, M. Chatterjee, K. L. Ebi, Y. O. Estrada, R. C. Genova, B. Girma, E. S. Kissel, A. N. Levy, S. MacCracken, P. R. Mastrandrea, & L. L. White (Eds.), *Climate Change 2014: Impacts, adaptation, and vulnerability. Part B: Regional aspects. Contribution of Working Group II to the Fifth Assessment Report of the Intergovernmental Panel on Climate Change* (pp. 1613–1654). Cambridge University Press.

Nygren, A., & Rikoon, S. (2008). Political ecology revisited: Integration of politics and ecology does matter. *Society and Natural Resources, 21*(9), 767–782.

Organisation for Economic Co-operation and Development (OECD). (2001). *Exclusive economic zone (EEZ), glossary of statistical terms*. [online] Retrieved January 22, 2020, from http://stats.oecd.org/glossary/detail.asp?ID=884

Paice, R., & Chambers, J. (2016). *Climate change impacts on coastal ecosystems*. Gold Coast, Australia: National Climate Change Adaptation Research Facility. [online] Retrieved June 15, 2020, from https://coastadapt.com.au/sites/default/files/factsheets/T312_9_Coastal_Ecosystems.pdf

Paulson, S., Gezon, L. L., & Watts, M. (2003). Locating the political in political ecology: An introduction. *Human Organization, 62*(3), 205–217.

Payne, A. (2004). Small states in the global politics of development. *The Round Table: The Commonwealth Journal of International Affairs, 93*(376), 623–635.

Peet, R., & Watts, M. (1996). *Liberation ecologies: Environment, development, social movements*. Routledge.

Pelling, M., & Uitto, J. I. (2001). Small island developing states: Natural disaster vulnerability and global change. *Environmental Hazards, 3*(2), 49–62.

Philpott, D. (2011). Sovereignty. In G. Klosko (Ed.), *The Oxford handbook of the history of political philosophy*. Oxford University Press.

Popke, J., Curtis, S., & Gamble, D. W. (2016). A social justice framing of climate change discourse and policy: Adaptation, resilience and vulnerability in a Jamaican agricultural landscape. *Geoforum, 73*, 70–80.

Robbins, P. (2004). *Political ecology: A critical introduction, critical introductions to geography*. Wiley-Blackwell.

Rumbach, A., & Foley, D. (2014). Indigenous institutions and their role in disaster risk reduction and resilience: Evidence from the 2009 tsunami in American Samoa. *Ecology and Society, 19* (1), 19.

Scheyvens, R. (2011). The challenge of sustainable tourism development in the Maldives: Understanding the social and political dimensions of sustainability. *Asia Pacific Viewpoint, 52*(2), 148–164.

Schipper, L., & Pelling, M. (2006). Disaster risk, climate change and international development: Scope for, and challenges to, integration. *Disasters, 30*(1), 19–38.

Schubert, J. (2005). *Political ecology in development research, an introductory overview and annotated bibliography*. National Centre for Competence in Research (NCCR) North-South.

Shultz, J. M., Sands, D. E., Kossin, J. P., & Galea, S. (2020). Double environmental injustice - Climate change, Hurricane Dorian, and the Bahamas. *New England Journal of Medicine, 382* (1), 1–2.

Somma, M. (1993). Theory building in political ecology. *Biology and social life, 32*(3), 371–385.

Stoutenburg, J. (2015). *Disappearing island states in international law*. BRILL.

Tabucanon, G. M. (2012). The Banaban resettlement: Implications for Pacific environmental migration. *Pacific Studies, 35*(3), 343.

Taylor, M. (2014). *The political ecology of climate change adaptation: Livelihoods, agrarian change and the conflicts of development* (1st ed.). Routledge/Earthscan.

The Centre for International Environmental Law (CIEL). (2011). *Climate change and human rights: A primer*. Geneva.

Tompkins, E. L. (2005). Planning for climate change in small islands: Insights from national hurricane preparedness in the Cayman Islands. *Global Environmental Change, 15*(2), 139–149.

Towner, N., & Milne, S. (2017). Sustainable surfing tourism development in the Mentawai Islands, Indonesia: Local stakeholder perspectives. *Tourism Planning and Development, 14*, 503–526.

United Nations. (1994). *Barbados Plan of Action, sustainable development knowledge platform*. [online] Retrieved January 22, 2020, from https://sustainabledevelopment.un.org/conferences/bpoa1994

United Nations Framework Convention on Climate Change (UNFCCC). (2005). *Climate change, small island developing States*. Bonn, Germany. [online] Retrieved June 15, 2020, from http://unfccc.int/resource/docs/publications/cc_sids.pdf

United Nations Framework Convention on Climate Change (UNFCCC). (2015). The Paris Agreement. In *21st Conference of the Parties*. Paris: United Nations, p. 25. [online] Retrieved June 15, 2020, from https://unfccc.int/files/essential_background/convention/application/pdf/english_paris_agreement.pdf

United Nations Office for Disaster Risk Reduction (UNISDR). (2015). *Sendai Framework for Disaster Risk Reduction 2015 - 2030, Third World Conference on Disaster Risk Reduction*. Sendai, Japan. [online] Retrieved June 15, 2020, from https://www.preventionweb.net/files/43291_sendaiframeworkfordrren.pdf

Walker, P. A. (2005). Political ecology: Where is the ecology? *Progress in Human Geography, 29* (1), 73–82.

Walshe, R. A., Chang Seng, D., Bumpus, A., & Auffray, J. (2017). Perceptions of adaptation, resilience and climate knowledge in the Pacific. *International Journal of Climate Change Strategies and Management, 20*(2), 303–322.

Wisner, B., Blaikie, P., Cannon, T., & Davis, I. (2004). *At risk: Natural hazard's, people's vulnerability and disasters* (2nd ed.). Routledge.

Wolford, W., Borras, S. M., Hall, R., Scoones, I., & White, B. (2013). Governing global land deals: The role of the state in the rush for land. *Development and Change, 44*(2), 189–210.

World Bank. (2015a). *Personal remittances, received (% of GDP), Data.* [online] Retrieved March 2, 2017, from http://data.worldbank.org/indicator/BX.TRF.PWKR.DT.GD.ZS

World Bank. (2015b). *Population, total, Data.* [online] Retrieved March 2, 2017, from http://data.worldbank.org/indicator/SP.POP.TOTL

World Bank. (2017). *International tourism, receipts (current US$), Data.* [online] Retrieved January 22, 2020, from http://data.worldbank.org/indicator/ST.INT.RCPT.CD

Yamamoto, L., & Esteban, M. (2010). Vanishing island states and sovereignty. *Ocean and Coastal Management, 53*(1), 1–9.

Yoder, L. S. M. (2011). Political ecologies of wood and wax: Sandalwood and beeswax as symbols and shapers of customary authority in the Oecusse enclave, Timor. *Journal of Political Ecology, 18*(1), 11–24.

Zhang, G., Murakami, H., Knutson, T. R., Mizuta, R., & Yoshida, K. (2020). Tropical cyclone motion in a changing climate. *Science Advances, 6*(17), 7.

Part II
Sectors

Community Participation, Situated Knowledge and Climate Change (Mal-) Adaptation in Rural Island Communities: Evidence from Artificial Shoreline-Protection Structures in Fiji

Michael Fink, Carola Klöck, Isoa Korovulavula, and Patrick D. Nunn

Abstract Small Island Developing States like Fiji are climate change hotspots. Adaptation to climate change is thus paramount. Research has underlined the importance of indigenous or local knowledge and community participation for island communities to successfully adapt to the effects of a changing climate, such as sea-level rise and shoreline change. Yet, indigenous knowledge and community participation are not enough. We here point to the need to combine indigenous and scientific knowledges.

We use the example of seawalls in rural Fiji communities to illustrate our argument. Although seawalls are very popular throughout the Fiji archipelago (and beyond), they are largely ineffective and unsustainable solutions to a long-term problem. Particularly in rural locations, seawalls fail to reduce shoreline erosion and groundwater salinization, or to protect infrastructure and settlements from flooding. Although the decision-making process is participatory and bottom-up, and although local knowledge inputs to decision-making may be considerable, integration of local and scientific knowledge to create adaptive, situated knowledge and to build climate resilient communities is generally lacking.

Successful climate change adaptation requires informed investigation of the local context, the drivers of change, and local inhabitants' awareness of the consequences

---

M. Fink (✉)
Institute of Geography, University of Hamburg, Hamburg, Germany
e-mail: michael.fink@uni-hamburg.de

C. Klöck
Centre for International Research, Sciences Po Paris, Paris, France
e-mail: carola.kloeck@sciencespo.fr

I. Korovulavula
Institute of Applied Sciences, University of the South Pacific (USP), Suva, Fiji
e-mail: isoa.korovulavula@usp.ac.fj

P. D. Nunn
School of Law and Society, University of the Sunshine Coast, Sippy Downs, QLD, Australia
e-mail: pnunn@usc.edu.au

S. Moncada et al. (eds.), *Small Island Developing States*, The World of Small States 9, https://doi.org/10.1007/978-3-030-82774-8_4

of different response measures. To create such situated knowledge through community participation, scientific information on climate change as well as the advantages and disadvantages of various coping strategies must be effectively communicated to community decision-makers and integrated with existing local cultural knowledge. Real empowerment requires appropriately skilled persons with both a scientific understanding of climate change combined with a sense of locality and a vested interest in the long-term security of its inhabitants.

Keywords Climate change adaptation · Community based adaptation · Community participation · Maladaptation · Local/indigenous knowledge · Seawalls · Shoreline-protection · Sea-level rise · Pacific islands · Fiji

1 Introduction

Fiji and other Small Island Developing States (SIDS) are popularly regarded as climate-change hotspots (Nunn & Kumar, 2018). SIDS are particularly vulnerable to the adverse effects of climate change for two reasons. First, their geography—long coast lines compared to land mass and concentration of infrastructure and populations along the often narrow coastal strip—makes them highly exposed to climate change impacts such as sea-level rise. Second, structural economic disadvantages suggest SIDS have low adaptive capacities, meaning a lack of abilities to cope with damage and only limited resources for transformative learning and adaptation to changes in climate in the long term (Nurse et al., 2014; Briguglio, 2016). Adaptation to climate change is thus urgent, yet effective and sustainable adaptation solutions have not proved readily forthcoming (Nunn, 2009). Scholars have hence called for "greater community involvement in the development of adaptation strategies as well as the incorporation of traditional knowledge from communities in these strategies to increase the likelihood of their uptake" (Mercer et al., 2007; see also e.g. McNaught et al., 2014; Janif et al., 2016, p. 7; McNamara et al., 2020).

Although the benefits of increasing Community Participation (CP) and integration of local/indigenous knowledge into adaptation interventions are well documented, we argue that CP and local knowledge alone do not automatically lead to effective adaptation and might even result in maladaptive solutions. For communities to choose effective and sustainable adaptation solutions, they often also need more awareness of western scientific insights into climate change—in particular likely future shoreline changes—technical capacity and support. Assistance should not be limited to funding but should facilitate a two-way exchange of knowledge, sharing of information, and learning from each other. For successful adaptation, local and western knowledge systems need to be integrated and ideally, action knowledge should be co-produced.

We use the case of seawalls in different locations in Fiji (see Fig. 1) to illustrate our argument. Artificial shoreline protection, mostly through seawalls, is common

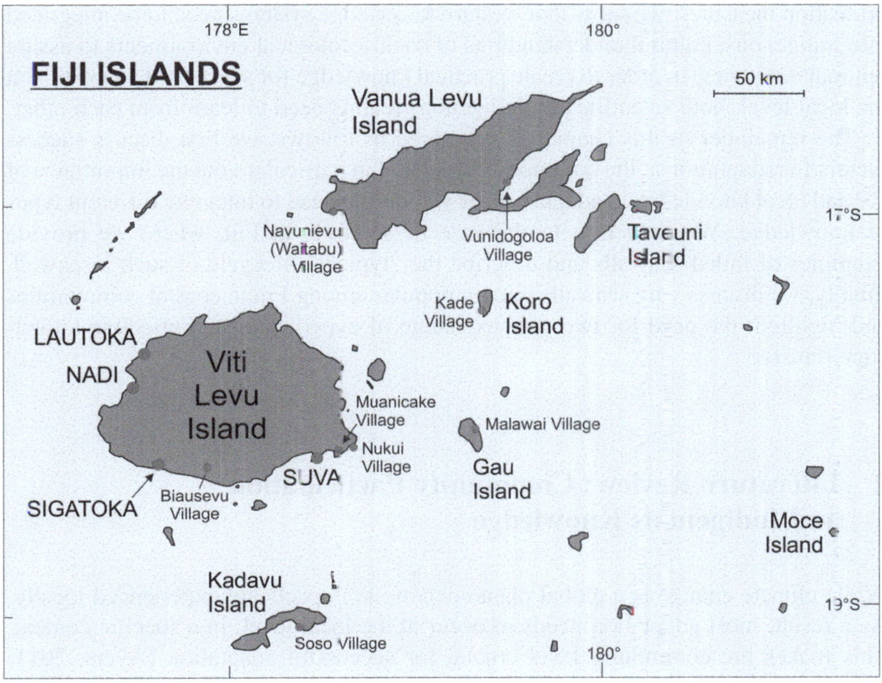

Fig. 1 Map of the Fiji Islands showing places mentioned in the text

throughout the Fiji archipelago, as it is in many other island groups elsewhere (Kench, 2012; Betzold & Mohamed, 2017). Many coastal communities actively demand seawalls be constructed or repaired and sometimes secure government funding for their construction. Yet, despite their popularity, seawalls are largely ineffective and unsustainable (Nunn, 2013). Particularly in rural locations, while sometimes providing a short-lived 'fix', seawalls fail to reduce shoreline erosion and groundwater salinization, or to protect low-lying coastal infrastructure and settlements from flooding.

Although communities across the Pacific islands, including Fiji, are dealing with and responding to the effects of climate change, interventions and initiatives at community level are rarely fully documented, monitored or evaluated. To better understand how Pacific Islanders adapt to climate change, what interventions work or do not work, we need to document, evaluate and communicate experiences and lessons learned (e.g. Hay & Mimura, 2006; Remling & Veitayaki, 2016; McNamara et al., 2020), a parallel process to dissemination of knowledge about Locally-Managed Marine Areas (LMMAs) in Fiji (Veitayaki et al., 2003; Aalbersberg et al., 2005).

In this chapter, we address this knowledge gap and seek in particular to document and critique responses to shoreline change in Fijian communities, with a focus on the role of different types of knowledge. Based on the case of artificial shoreline protection measures, we show that western knowledge systems need to be integrated into indigenous, cultural understandings of socio-ecological environments to assure optimal outcomes. In order to create practical knowledge for successful adaptation at the local level, both scientific and indigenous realms need to learn from each other.

The remainder of this chapter is structured as follows: we first discuss success factors for adaptation at the community level and in particular note the importance of CP and local knowledge in adaptation, as well as the need to integrate different types of knowledge. We then turn to the case of seawalls in Fiji, where we provide examples of failed seawalls and describe the "typical" lifecycle of such a seawall. Finally, we discuss why seawalls remain popular among Fijian coastal communities and highlight the need for two-way exchange of experiences and effective knowledge transfer.

2 Literature Review: Community Participation and Indigenous Knowledge

While climate change is a global phenomenon, its impacts are experienced locally. As a result, most adaptation needs to occur at the local level, in a specific context. This makes the community level crucial for successful adaptation (Ayers, 2011; Fenton et al., 2014; Mimura et al., 2014). The community or village level is even more important in geographically remote and fragmented areas such as Fiji and other Pacific Island nations, where national-level strategies and policy do not necessarily reach more remote rural locations (Nunn et al., 2014; Korovulavula et al., 2019; Nunn & Kumar, 2019).

The concept of Community-Based Adaptation (CBA) is based on the idea that successful adaptation must be deeply rooted at the community level. CBA is a bottom-up approach to dealing with climate change, a "community-led process, based on communities' priorities, needs, knowledge, and capacities, which should empower people to plan for and cope with the impacts of climate change" (Reid et al., 2009, p. 13). CBA acknowledges that climate change is only one among many livelihood stressors and therefore seeks to integrate livelihood benefits and poverty reduction with efforts to reduce vulnerability to climate change (Reid et al., 2009; Fenton et al., 2014).

Adaptation scholars agree that CP is central to, even a precondition for, successful adaptation in such contexts, for two main reasons: first, CP empowers the local population and helps ensure that adaptation projects reflect the priorities, preferences and values of the affected people. Second, CP also ensures that local or indigenous knowledge feeds effectively into the design of the community response to climate change (e.g. Kumar, 2002; Reid et al., 2009).

CP is closely linked to concepts such as empowerment, ownership and appropriateness of measures; and takes dissimilarities within and between communities into account. Development research has long promoted participatory processes such that policies and interventions balance internal and external contributions and address asymmetrical power relations between the local community and external actors (Cooke & Kothari, 2001). These lessons also apply to Climate Change Adaptation (CCA) (Maclellan, 2011; Kelman, 2017). CP means involving the community from the beginning of any adaptation intervention, that is, from the planning through to the implementation phases (McNamara, 2013). Successful adaptation starts with ascertaining the needs and perceptions of the community: "it is vital to *ask* local communities what support they require, rather than *tell* them" (Maclellan, 2011, p. 25; emphasis in original). CP thus ensures that the final intervention reflects the community's priorities and preferences and is culturally and socially appropriate (Reid et al., 2009; Kelman, 2017). Through such involvement, it is the community itself that owns the adaptation process: "Adaptation is not something that can be done *to* a community. It is something that needs to be done *by* a community, determined by its own needs and values" (Barnett, 2008, p. 45; our emphasis).

A second pillar of CP is local knowledge, also referred to as indigenous, traditional, or ecological knowledge (Lauer, 2017). While local knowledge is often contrasted with western science, research has increasingly questioned to what extent these two types of knowledge can be neatly divided, not least because the two have interacted for centuries (Dove, 2006). Scholars increasingly explore the dynamic and contested nature of indigenous knowledge (Lauer, 2017, p. 337). We here use local knowledge to refer to knowledge that is passed on from generation to generation within communities, as opposed to western knowledge that is acquired in schools, universities and other formal institutions of learning.

It is well documented that Pacific Island societies are quite resilient; most have multimillennial histories of successfully dealing with environmental change. Over generations, island people have developed ways to monitor their environment and to both anticipate and respond to sudden and gradual changes (e.g. Bridges & McClatchey, 2009; McMillen et al., 2014; McNamara & Prasad, 2014; Johnston, 2015; Janif et al., 2016). To what extent these "reservoirs of ancient islander wisdom" (Lauer, 2017, p. 337) translate into higher adaptive capacity and/or more effective adaptation to climate change, however, is unclear.

On the one hand, experiences and traditional coping mechanisms enhance adaptive capacity and provide valuable insights into potential adaptation interventions (McNamara & Prasad, 2014; Janif et al., 2016). On the other hand, past experiences and traditional coping mechanisms may no longer be applicable under a changing climate and potentially even lead to maladaptive solutions. Additionally, while local populations do recognise changes in their environment, they do not necessarily link them to global climate change (Fazey et al., 2011; Macintosh, 2013). Therefore, the incorporation of western knowledge on the ground sounds promising for CCA. While this might lead to contradictions and uncertainties and hinder action and adaptation it might also be incorporated into the manifold local strategies for adaptive transformation. In case local knowledge is helpful for CCA, we refer to it as situated knowledge.

Studies across the Pacific document a limited awareness of global climate change, especially in rural areas, and that some environmental changes are uncritically misattributed to this. Irreversible climate change and reversible climate variability are often confused and environmental changes attributed to local practices such as deforestation rather than global climate change. Lack of awareness and misperception about climate change, its causes and consequences help explain why the Pacific Island countries are not as prepared for climate change as they might be expected to be after several decades of direct external support for CCA (Nunn, 2009; Lata & Nunn, 2012; Betzold, 2015; Janif et al., 2016; Weir et al., 2017).

While "indigenous strategies alone cannot be expected to successfully address the problems that have arisen as a result of environmental processes" (Mercer et al., 2007, p. 253), it is abundantly clear that western scientific knowledge alone is not sufficient either to address these effectively and sustainably. Instead, indigenous and western knowledges need to complement and learn from each other. There should be constant "collaboration and exchange" (Mercer et al., 2007, p. 253). Yet, such a process is not without challenges for it demands not only mutual respect (for the other's knowledge preference) but also acceptance of the dynamic and place-based nature of all knowledge types (Lauer, 2017). Ideally, such situated consensus knowledge should be co-produced through inclusive dialogue between stakeholders at all levels. In this sense, the social process of mutual knowledge exchange around sharing best practices can be a bond within communities, a bridge between communities and a link between power and institutions (McMillen et al., 2014).

As a starting point, western knowledge needs to be translated into the community's preferred vernacular language and knowledge co-production generated in a culturally-appropriate setting that is respectful of the community's way of reaching decisions. Such tasks require persons who are both culturally sensitive and academically skilled. Therefore, in the long run, an academic education of local residents seems promising. Academics who combine scientific and cultural/indigenous knowledge without losing contact with their home society can act as mediators and pioneers of change and adaptation (WBGU, 2011). In order to mainstream agency towards sustainable management of commons, rules and norms should be generated and governed by the communities. Promising strategies for robust institutions are CP on local levels as they need to be perceived as trustworthy, fair, and legitimised to restrict access and create incentives (Ostrom et al., 1999). Effective governing can be supported by monitoring resource use and rates of change (ideally by the communities themselves) as well as dialogue in multi-scale, complex and layered robust institution networks. (Dietz et al., 2003).

3 Seawalls as Popular Yet Maladaptive Adaptations

Seawalls illustrate the importance of integrating and co-producing knowledge. In the next section, we first outline how seawalls in rural coastal Fiji communities have become the default response to erosion and flooding yet mostly fail to achieve their purpose. In Box 1 we then present some examples of failed or failing seawalls across Fiji, before finally sketching the typical lifecycle of a seawall in Fiji.

3.1 Responding to Shoreline Changes in Fiji

Although coastal communities in Fiji adapted to changing sea levels and coastlines well before the twentieth century, the response/s to such changes has changed. Pre-colonial knowledge about artificial shoreline protection hardly existed in Fiji (Mimura & Nunn, 1998). Until about 1960, Fijians do not generally appear to have built sea defences or other structures, but instead found other ways to live with changing coastlines, typically through relocation of vulnerable settlements; an exception is shown in Fig. 2, a sea defence on Kadavu from around 1920.[1]

Prior to 1960, coastal change was likely less pronounced than today for several reasons. First, it seems that coastal population densities were significantly lower and coastal mangrove forests were commonly more extensive (Mimura & Nunn, 1998). Second, the rate of sea-level rise around Fiji was significantly less than the 5.5 mm/year[2] which has prevailed since 1972. Finally, before the advent of brick-built houses in Fiji, settlements were designed to be easier to move and relocation appears to have been a common response to coastal change (Fujieda & Kobayashi, 2013); an anecdotal example from Moce Island recalls that the 'second' settlement (Kororua) was relocated to the 'third' settlement (Korotolu) after a whale carcass was washed ashore and began to smell. Three of the Fijian communities studied in a recent paper had stories about relocating following ocean flooding, going back as early as 1868 (Janif et al., 2016).

Since 1960 and probably increasingly since Fiji became independent in 1970, coastal population densities have increased and human activities in the coastal zone have become more exposed to the ocean as a result of mangrove clearance and an acceleration in the rate of sea-level rise, which has in turn amplified the impacts of extreme wave events; large waves can reach farther inland than they once did. As a result, shoreline change, erosion and flooding are today widespread across Fiji (and similar Pacific island groups) and are likely to have been caused primarily by sea-level rise.

Seawalls have become the default response to shoreline change and erosion in Fiji, where the number of seawalls has increased exponentially since the mid-twentieth century. Yet, seawalls and other ways of artificially protecting the shoreline have many pitfalls (Cooper & Pilkey, 2012). Shorelines are highly dynamic systems that change naturally. Erosion is hence not necessarily a problem,

[1]It is possible that this inference is wrong, based on an absence of evidence rather than evidence of absence. Recent research in the islands of Micronesia shows that coastal management involving the construction of hard structures has continued for at least 1000 years (Nunn et al., 2017).

[2]Data is for Suva from the Permanent Service for Mean Sea Level.

Fig. 2 A photo of Soso Village, Kadavu Island, Fiji from the 1920s showing shoreline protection consisting of sticks. Note the artificial island offshore (photo courtesy of Dr Rod Ewins)

but rather part of "a natural process that creates, modifies and destroys coastal landforms through linked processes of erosion, transport and deposition" (Cooper & McKenna, 2008, p. 316). Any interference with the shoreline, such as through seawalls, necessarily alters natural processes and often has unintended consequences such as increasing, rather than decreasing, erosion rates. To minimise these consequences, "natural environmental processes must be completely understood prior to any coastal modification" (Brayshaw & Lemckert, 2012, p. 12). This includes understanding present conditions like tidal currents and seasonal patterns as well as predicted sea-level rise, but also understanding how the envisaged shoreline modification might alter these present conditions.

Most coastal locations in the developing world lack local-level data on shoreline processes and predicted sea-level rise, as well as the capacity and resources to collect such data. Additionally, the local population is often unaware of the consequences of building a seawall, or of alternative forms and designs of 'hard' shoreline stabilisation and softer interventions. In some iconic highly visible locations like the cities of Lautoka, Nadi and Suva (the nation's capital), the central government has modified the shoreline at a large scale, often with the assistance of bilateral aid donors. Critically, because these seawalls protect highly-valued infrastructure of national significance, they are well-designed, well-maintained and accomplish what they were intended to accomplish (Fig. 3). In contrast to these successful examples of seawalls, the great majority of seawalls in rural locations have failed to achieve what was intended. In addition, many have worsened pre-existing problems and/or created unanticipated problems for coastal dwellers. Accounts of badly built and ineffective seawalls abound for Fiji (Mimura & Nunn, 1998; Hay & Mimura, 2006;

Fig. 3 Views of seawalls in Fiji. (**a**) Part of the seawall along the Suva Peninsula, part of Fiji's capital city. This well-maintained seawall is likely to have inspired numerous residents of rural settlement about the efficacy of hard structures for coastal protection (photo by Anamaiasa Vunituraga, used with permission). (**b**) Newly-built (December 2015) seawall at Kumi Village

Piggott-McKellar et al., 2020) as well as other islands in the Pacific and elsewhere (e.g. Sinane et al., 2010; Kench, 2012; Betzold & Mohamed, 2017).

For different locations, Box 1 shows examples of largely ineffective seawalls (see also Fig. 1). These mini case studies and our discussion draws on decades of fieldwork and research experience that the authors have from across the Fiji islands. The cases come from Fiji's two main islands, Viti Levu and Vanua Levu, and show that even in less remote areas, rural communities typically fail to construct proper seawalls, as opposed to the well-maintained seawalls in urban locations like Suva and Nadi.

Box 1: Examples of Seawalls in Fiji

At **Navunievu (Waitabu) village** in Bua, a moderately peripheral location, the community's experience with coastal change reportedly began in the 1950s, a decade or so after the colonial government had ordered the mangrove forest fringing the shoreline to be cleared, supposedly so that the number of disease-carrying insects infecting the community would fall and its health would improve: a commonly-reported action by colonial medical officers in Fiji. Removal of the mangrove fringe (which has since partly been replanted) exposed the edge of the sand flat on which the village is built, causing its fringe to be eroded and allowing large waves to wash into the heart of the village. In response to these problems, the community used their own resources to build a seawall across the front of the village in the 1970s; the timing may be significant as an acceleration in the rate of sea-level rise around this time may have amplified the observed problems. This seawall collapsed but the problems persisted so in the 1990s another seawall was constructed. This too collapsed, since which time the community has endured the effects of continued shoreline erosion and flooding (Fig. 4). In acknowledgement of the ongoing sea-level rise, some new houses in Navunievu have been constructed slightly upslope and inland, an iterative adaptation strategy also adopted autonomously by the inland floodplain community of Biausevu where newly-married couples are required to build their family houses upslope away from the danger zone of flooding (Campbell et al., 2014). Today the

(continued)

Fig. 3 (continued) (Tailevu) is a solid impermeable barrier that will indeed protect the community from sea-level rise for decades. Yet it is also likely to alter surrounding water and sediment movements; the effects of scour at the base of the seawall are already visible. This seawall was built with the assistance of the Government of Japan (photo by Government of Fiji's Media Center, used with permission). (**c**) Seawall at Kade Village (Koro Island) showing the effects of scour in 2001 associated with the construction of a vertical impermeable structure. Local people complain that the amount of food obtainable from nearshore areas (at low tide) has fallen drastically since seawall construction about 1995 (photo by Patrick Nunn). Note that Severe Tropical Cyclone Winston (February 2016) fundamentally altered the coastline of Koro Island

Box 1 (continued)
Navunievu community has gained new hope from their involvement in an intentional mangrove replanting scheme. Yet there is also a persistent belief—among some residents—that sea level will not continue to rise but will (soon) start falling so that there is no need to consider relocating the entire community upslope, a sentiment similar to that expressed by some in Muanicake Village in the increasingly flood-prone Rewa Delta (Lata & Nunn, 2012).

An example from the fringe of the Rewa Delta is that of **Nukui Village**, which has experienced the effects of coastal change, especially during tropical cyclones, for at least several decades (Mimura & Nunn, 1998). Originally on a promontory, the sea has been steadily encroaching on three sides of Nukui, washing away coastal vegetation and forcing some dwellings to be relocated. The first defensive structures the inhabitants recall consisted of a kind of stick and log framework filled in with soil and refuse, probably similar to that shown in Fig. 2. The first seawall was constructed with the help of 'foreign' engineers in the early 1960s; beachrock and reef rock were cemented together with the help of iron rods. This seawall did not last long and, having been rebuilt a few times, was completely destroyed during Tropical Cyclone Bebe in 1972; it is now buried by sand. Some years later, a Japanese engineer came and worked with the villagers to build a similar seawall that proved stronger as its foundations went into the beachrock below the sand cover. Despite having been repaired/rebuilt several times since, a seawall still protects the main part of Nukui from wave attack although, as it has been rendered less effective over the last few years as a result of sea-level rise, the Fiji Red Cross announced in 2017 a project to raise this wall (Fig. 5).

Hailed as Fiji's first climate change induced settlement relocation, the village of Vunidogoloa in Cakaudrove was relocated in January 2014 from an exposed coastal location to a new one 2 km inland (McNamara & Combes, 2015; Gharbaoui & Blocher, 2016). Perhaps because sea-level rise here was amplified by (tectonic) subsidence, the relative pace of inundation at Vunidogoloa was greater than at most other coastal sites in Fiji. For decades, residents of the 'old' village had attempted to adapt *in situ*, both by raising houses on stilts and building seawalls that eventually proved ineffective, even exacerbating problems. Reportedly the first seawall at Vunidogoloa was built in the 1960s—its remains lay underwater 60 m offshore in 2012—a second seawall was built in the late 1990s and repeatedly repaired (Fig. 6). By 2012, it had become clear this seawall was ineffective—high water simply went around its edges to enter the village—and also unduly prolonged the effects of flooding by stopping seawater readily draining off the village land once the tide fell.

Fig. 4 The shoreline at Navunievu (Waitabu) Village, Bua, in December 2015, showing the (recently-extended) mangrove fringe, the remains of the two seawalls and the erosion of the edge of the coastal flat on which the village is built (photo by Patrick Nunn)

3.2 The Life Cycle of a Seawall in Fiji

A typical scenario for a rural coastal location experiencing shoreline erosion and increasingly-frequent coastal flooding is that the local community discusses the need to respond, discussions that almost invariably end with a decision to either remain unresponsive or to construct a seawall. In Fijian villages, these discussions take place in many different settings, including in the household, during informal gatherings happening daily, religious meetings, in official village meetings that foster custom and tradition, or in official committees. It is these committees that can apply for government assistance via the elected village headman (*turaga-ni-koro*). Although age, sex, and social status might affect an individual's level of influence, our observations are that everyone seems able to participate in the decision-making processes to an adequate degree. Usually in the Pacific Island communities, social bonds are very strong, inclusive and a majority of community members identify with and legitimize decisions even when they may contradict personal opinions (Fink, 2016).

In general, the discussions immediately revolve around building a seawall; other options are not typically discussed. The possibility of relocation in particular is rarely aired, largely because many coastal-dwelling Fijians—like people elsewhere

Fig. 5 Views of Nukui Village in the Rewa Delta in February 2015 (all photos by Jeff Tan, used with permission). (**a**) The seawall protecting the main part of Nukui leaks and is regularly overtopped by high waves, flooding parts of the village. To maintain this seawall, villagers gather rocks and cement is provided by non-governmental organisations (NGOs) (Caritas and Fiji Red Cross). (**b**) Not all parts of Nukui are protected by seawalls. The part adjoining the mouth of the Rewa distributary floods twice daily during high tide. (**c**) On Turaga Island, along the coast from

in today's world—view shorelines as fixed and their right to occupy them immutable. While villages traditionally were built further inland and occasionally moved (as the earlier example from Moce Island indicated) relocation is no longer as easily feasible as houses today are often concrete, solid buildings that cannot easily be rebuilt elsewhere. Land tenure systems and the inseparable connection between land and identity further complicate relocation (Farbotko et al., 2018). Seawalls seem thus an efficacious solution that can protect property and land. After a decision to build a seawall has been taken, the community discusses the best location, building material, funding, and actual construction of the seawall. With regard to where the seawall might best be placed, communities often reach a compromise decision that has to acknowledge issues like boat access, which may require gaps to be left in the structure, or outlay cost, which may see seawalls not extended along less-populated parts of an exposed shoreline. The design of the seawall is rarely debated; seawalls are like house walls, it is often said, that should be vertical and impermeable (for maximum protection).

In contrast, there is some variation in terms of the material used. Communities discuss how to acquire the raw materials needed for the seawall and then how to have it constructed. In many more peripheral communities where little cash is available, seawalls are often made from boulders or reef rock (sometimes levered off living reef), stacked vertically and veneered with cement. In richer rural communities, the rocks may all be cemented together to give a stronger structure. But what are often regarded as the 'best' seawalls are ones that utilize funds from outside the community, often through government schemes that will split costs 50:50 or wholly funded by bilateral donors (Fig. 3). The beneficiary community will provide (free) labour. Finally, we note that although seawall construction may be completed quickly, it is more often completed in stages, typically because of a discontinuous supply of raw materials or (free community) labour, or because the money needed for the purchase of the former is insufficient. Incomplete seawalls are seen in many parts of Fiji yet most are eventually completed as planned and often opened with great fanfare by a local dignitary. Because the seawall is constructed by the community, villagers are proud of it.

For many such (vertical impermeable) seawalls as described, the most immediate effects are the scouring of the foreshore—an effect of replacing a sloping permeable (sandy) shoreline with a vertical impermeable one—and the associated loss of useful nearshore bioproductivity. Scouring removes the soft sediment cover of the rocky platform fringing the shoreline resulting in a loss of habitat for burrowing organisms, including many species of edible crustacean and mollusc.

The seawall itself normally does not survive for long. Most seawalls in rural Fiji coastal settlements remain in place for 18–24 months before they show signs of collapsing. The reasons for this are many, the most common being (1) undermining

---

Fig. 5 (continued) Nukui, a seawall has failed to halt shoreline erosion or groundwater salinization, marked by the fallen coconut trees

of the base of the seawall by wave scour, (2) ponding of water (from wave splash-over) on the landwards side of the seawall, and (3) inability of the structure (either because of its structure or composition) to withstand extreme waves, such as occur during tropical cyclones, occasionally tsunamis (see Figs. 4, 5, and 6).

Seawall collapse, wholly or partly, increases the exposure of the (soft) shoreline to wave erosion and flooding, so many communities make it a priority to repair the seawall. The examples in Box 1 show how seawalls have collapsed and been repaired, in some locations repeatedly. Repairing/maintenance is easier for communities that have surplus cash to quickly purchase the materials needed but more difficult for (more peripheral) communities that do not. Yet even when repaired, most seawalls continue to collapse periodically; the enthusiasm for repair (and the associated cost of this) often dwindles quite quickly, which has given rise to a situation where Fiji's rural coasts—like many in the tropical Pacific islands—are littered with the remains of collapsed seawalls, as the examples in Box 1 also show. These examples also highlight how an affected community may sometimes decide simply to accommodate the effects of sea-level rise; sometimes communities will plant vegetation or dump refuse along the eroding coast; sometimes they will implement set-back, sometimes staggered or even complete relocation. More often than not, affected communities register a request for outside assistance from the national government and/or foreign donors, and await an intervention that may never eventuate.

4 Lack of Knowledge Transfers and Climate Change Maladaptation

If seawalls are shown to be ineffective and have repeatedly failed to protect Fijian communities from erosion and flooding—as the examples in Box 1 clearly indicate—why do they remain popular? Although adaptation decision-making in Fijian communities is participatory and incorporates local knowledge, it often fails to integrate local and western knowledge. Government interaction remains at the subnational level and is often limited to financial support from the government to the community. There is little monitoring and evaluation. Knowledge and experiences on both failures and best practice are rarely shared, let alone co-produced, although there are some exceptions, such as the Gau Island initiative (Lomani Gau) presented in Box 2.

The importance of adaptation being designed and implemented at the community level appears paramount to ensure CCA is successful. In Fiji, most adaptation happens at the community level; it is both participatory and builds on local or indigenous knowledge. As outlined earlier, Fijian communities have a number of contexts in which community issues are discussed and debated, and it is the communities that decide, through these discussions, on a particular course of action. Many communities also have high levels of local knowledge; they are intimately

Fig. 6 Views of the seawall at old Vunidogoloa Village (Cakaudrove) in May 2012 (photos by Simione Sorowale, used with permission). In January 2014, Vunidogoloa became the first coastal settlement in Fiji to be relocated inland explicitly because of climate change. (**a**) The 'useless' seawall that seawater went over and around every high tide, flooding the village (and excavating tree roots), and leading to prolonged periods of standing water. This seawall was constructed from reef rocks stacked vertically with a topping of cement; collapse of the stack is visible here. (**b**) Close-up of seawall structure showing the reef-rock stack and the uneven veneer of cement. Gaps in the seawall like that shown on the left are sometimes deliberate to allow ready access yet also funnel water through to the area behind. Each high tide, seawater would erode the edge of the coastal plain and enter the village. Note the house on wooden piles.

familiar with the environment in which they live and with which they interact on a daily basis. They are hence also aware of changes in the shoreline and their effects on food production (marine and terrestrial), though they may not necessarily connect these observed changes to global climate change and sea-level rise. While adaptive capacity at the community level is therefore potentially very high in Fiji, this does not necessarily lead to sustainable and effective adaptation interventions.

At the community level in Fiji, there remains a widespread conviction that seawalls are an adequate solution to shoreline change. Notably, it seems that successful seawalls, like those that are well-maintained in highly visible urban locations, serve as an example for communities in more rural locations which seek to emulate these successful seawalls. Living in a 'sea of islands' (Hau'ofa, 1993), Pacific Islanders have always been highly mobile and in constant exchange with neighbouring islands. With globalisation this connectedness and mobility increased drastically. Many Fijians travel overseas and report their impressions via modern communication technologies or once they get home and urge their village to implement development projects that are seen as modern and efficient, such as seawall construction (Fink, 2016).

A belief in the efficacy of a proposed solution (like a seawall) appears to be more important than the efficacy itself; in other words, it seems more important that people support a particular adaptive intervention than that it is scientifically adjudged to be optimally effective and sustainable. Yet, as the case of seawalls shows, adaptation strategies cannot be easily transferred to the local context. This applies not only to seawalls, which are demonstrably ineffective and unsustainable in almost all but a few urban contexts, but also to other interventions like river-channel dredging or upland reforestation intended to reduce the frequency and magnitude of lowland/coastal flooding (Nunn, 2013).

The Government of Fiji supports communities in the construction of seawalls. The trust that most rural Fijians have in seawalls, a trust built on uncritical observation and emulation of others' seawalls, appears to have persuaded the Government of Fiji, especially at subnational level, that their construction should continue to be supported. Government support and interaction occurs mainly at the subnational level through providing/matching funds and/or sometimes through engaging engineers to help design/construct hard shoreline structures. In Fiji, both provincial councils (for *iTaukei* people in a rural area) and district offices (for all people in a rural area) are in regular contact with most readily-accessible communities, albeit more often for agricultural, commercial, infrastructural or health-linked issues than for those affecting environmental sustainability. A formal request for assistance will normally be processed at this subnational level and funds may be disbursed for purpose without any reference to the central government.

Government support for rural communities is thus limited to implementation, but does not often include technical support with regard to feasibility or efficacy of the selected intervention. Not least because interactions mostly remain at the subnational level, there is no formalised process of sharing information. Knowledge is neither transferred from local to national levels nor the other way round. Information on best practices or typical mistakes do not reach villages contemplating their response to shoreline change, nor are alternatives to seawalls—even different designs or locations of such seawalls—usually presented or discussed.

Box 2 A Successful Adaptation Initiative: Lomani Gau

A good example that successful coastal protection is possible even in peripheral communities is the *Lomani Gau* (caring for Gau [Island]) initiative that has raised awareness among subsistence communities on Gau Island (see Fig. 1) about the need to manage the resource base sustainably (Remling & Veitayaki, 2016; Medina Hidalgo et al., 2021). The 16 villages and settlements try to evolve best practices and exchange information in a participatory way. Each village implemented locally managed marine areas and installed wardens to monitor fish occurrence and shoreline changes. The overarching initiative is led by a committee of locals accepted by all communities. The committee further consults with religious and chiefly bodies such that decisions are made in a culturally acceptable way, which further strengthens their legitimacy. On top of that, the committee constantly exchanges information with a native Gau Islander who initiated the project and pioneered the change. He graduated with a PhD in the sustainable communal management of marine resources and now works for the University of the South Pacific in Suva, a few hours by boat from Gau Island. As some villagers are highly mobile, the exchange of information does not only take place via telecommunication but also through personal visits. Furthermore, together with external experts from NGOs and the university, the academic from Gau Island frequently visits his home island and organises workshops on ecosystem management, focusing on marine protection. These experts explicitly discourage the construction of hard shoreline structures and favour mangrove conservation and replanting and other environmentally-friendly methods of reducing the exposure of coastal settlements (Veitayaki & Holland, 2015). In a nutshell, science teaches the villagers to trust, relearn and re-establish traditional adaptive capacities underpinned by scientific findings on optimisation. Though some communities still build and repair seawalls, such structures become integrated into more holistic approaches. For example, in 2017 the villagers of Malawai repaired their seawall to reclaim land and plant and tend mangroves in front that successfully gather soft sediments, sap wave energy and stabilize the wall (Fig. 7). The settlement of Naovuka investigates and reports positive effects of nonvertical, concrete breakwaters some distance offshore. After decades of building seawalls in Fiji, the example shows that even among more subsistence-focused remote communities, ‘hard’, artificial, or modern solutions have at least the same acceptance as ‘soft’, natural, or traditional adaptation.

The success of the initiative can be explained by (1) the existence of a far-sighted pioneer of change combined with the (2) installation of a regional institution—the Lomani Gau initiative—which builds on existing local actors and institutions and seeks meaningful dialogue with national and international bodies.

Fig. 7 Rehabilitated coastal protection in Malawai (photo by Joeli Veitayaki, used with permission)

5 Conclusion

CP is key to the success of CCA in Fiji and other Pacific Island countries. Rural communities in many poorer (developing) countries, especially archipelagic ones like Fiji where ocean gaps amplify geographical peripherality, are effectively beyond the influence of central government when it comes to environmental decision-making. In the future, it is likely that such communities will continue to make their own decisions, especially when deciding how to respond to slow-onset climate-driven stressors like sea-level rise and temperature rise. While such local decision-making is bottom-up and participatory and based on local knowledge, it often does not make use of western scientific insights. If community-level decision-makers in countries like Fiji are to make the best-informed choices on how to respond to climate changes, they need to be empowered. Empowerment of this kind requires not only an intimate knowledge of the local environment (with which the community routinely interacts) and traditional ways of coping with environmental adversity, but also of science-based predictions of likely future climate and other changes.

It is clear that interactions with rural communities should be on their terms, not dictated by outsiders: terms that extend to both the language and cultural context of engagement (Nunn, 2009). Interventions should not privilege western scientific approaches and should rather focus on enhancing (not replacing) traditional coping capacity (Dumaru, 2010). There is ample evidence to show that the cultural resilience which enabled Fijians to occupy mid-ocean islands for more than 3000 years is of value in contemporary Fiji. Traditional resource management practices and interpretations of climate-ocean phenomena, for example, while somewhat attenuated over the last few decades, nevertheless comprise a *corpus* of knowledge that should be acknowledged and centrally incorporated into community-scale adaptation planning (Johnston, 2015; Janif et al., 2016). Yet, western adaptive solutions are preferred by many external donors, and their efficacy is hardly questioned by rural Fiji communities. That these solutions invariably sideline Fijian science and ways of understanding the natural environment is seldom reflected; western knowledge and indigenous knowledge are not integrated (Nunn et al., 2020).

Seawalls illustrate this need to combine western and local knowledge systems. Seawalls are a popular but ineffective response to shoreline change in Fiji (and elsewhere in rural communities in the Pacific). While decision-making is participatory and bottom-up, and while local knowledge inputs to decision-making may be considerable, integration of local and scientific knowledge to create adaptive, situated knowledge and to build climate resilient communities is generally lacking. Yet the example of Lomani Gau shows how community decision-making can effectively and sustainably be empowered to make informed environmental choices that will enable community livelihoods to be sustained, even enhanced, in the face of future climate change challenges. Situated knowledge emerges from robust institutional networks together with individual pioneers of change and bridge-builders—e.g. Pacific islanders who have a western academic education can mediate between western and indigenous knowledge systems.

Acknowledgements Some of the research into climate-change adaptation in Fiji reported in this chapter was funded by the Australian Research Council (ARC) through Linkage Grant LP160100941 (to PN) and the Asia-Pacific Network for Global Change Research (APN) through grant CRRP2015-FP02 (to PN and IK).

References

Aalbersberg, W., Tawake, A., & Parras, T. (2005). Village by village: Recovering Fiji's coastal fisheries. In G. Mock & P. Whitten (Eds.), *World Resources 2005: The wealth of the poor: Managing ecosystems to fight poverty* (pp. 144–152). World Resources Institute.

Ayers, J. (2011). Resolving the adaptation paradox: Exploring the potential for deliberative adaptation policy-making in Bangladesh. *Global Environmental Politics, 11*(1), 62–88.

Barnett, J. (2008). The effect of aid on capacity to adapt to climate change: Insights from Niue. *Political Science, 60*(1), 31–45.

Betzold, C. (2015). Adapting to climate change in small island developing states. *Climatic Change, 133*(3), 481–489. https://doi.org/10.1007/s10584-015-1408-0

Betzold, C., & Mohamed, I. (2017). Seawalls as a response to coastal erosion and flooding: A case study from Grande Comore, Comoros (West Indian Ocean). *Regional Environmental Change, 17*(4), 1077–1087. https://doi.org/10.1007/s10113-016-1044-x

Brayshaw, S., & Lemckert, C. (2012). Pitfalls of shoreline stabilisation – Tweed River Mouth, Gold Coast, Australia. In J. A. G. Cooper & O. H. Pilkey (Eds.), *Pitfalls of shoreline stabilization: Selected case studies* (pp. 1–13). Springer.

Bridges, K. W., & McClatchey, W. C. (2009). Living on the margin: Ethnoecological insights from Marshall Islanders at Rongelap atoll. *Global Environmental Change - Human and Policy Dimensions, 19*(2), 140–146. https://doi.org/10.1016/j.gloenvcha.2009.01.009

Briguglio, L. P. (2016). Exposure to external shocks and economic resilience of countries: Evidence from global indicators. *Journal of Economic Studies, 43*(6), 1057–1078. https://doi.org/10.1108/JES-12-2014-0203

Campbell, J., Bedford, R., & Bedford, R. (2014). Migration and climate change in Oceania. In E. Piguet & F. Laczko (Eds.), *People on the move in a changing climate: The regional impact of environmental change on migration*. Springer.

Cooke, B., & Kothari, U. (2001). The case for participation as Tyranny. In B. Cooke & U. Kothari (Eds.), *Participation: The new Tyranny?* (pp. 1–15). Zed Books.

Cooper, J. A. G., & McKenna, J. (2008). Working with natural processes: The challenge for coastal protection strategies. *The Geographical Journal, 174*(4), 315–331. https://doi.org/10.1111/j.1475-4959.2008.00302.x

Cooper, J. A. G., & Pilkey, O. H. (Eds.). (2012). *Pitfalls of shoreline stabilization: Selected case studies* (Vol. 3). Springer.

Dietz, T., Ostrom, E., & Stern, P. C. (2003). The struggle to govern the commons. *Science, 302* (5652), 1907–1912. https://doi.org/10.1126/science.1091015

Dove, M. R. (2006). Indigenous people and environmental politics. *Annual Review of Anthropology, 35*, 191–208. https://doi.org/10.1146/annurev.anthro.35.081705.123235

Dumaru, P. (2010). Community-based adaptation: Enhancing community adaptive capacity in Druadrua Island, Fiji. *Wiley Interdisciplinary Reviews-Climate Change, 1*(5), 751–763. https://doi.org/10.1002/wcc.65

Farbotko, C., McMichael, C., Dun, O., Ransan-Cooper, H., McNamara, K. E., & Thornton, F. (2018). Transformative mobilities in the Pacific: Promoting adaptation and development in a changing climate. *Asia and the Pacific Policy Studies*, 1–15. https://doi.org/10.1002/app5.254

Fazey, I., Pettorelli, N., Kenter, J., Wagatora, D., & Schuett, D. (2011). Maladaptive trajectories of change in Makira, Solomon Islands. *Global Environmental Change, 21*(4), 1275–1289. https://doi.org/10.1016/j.gloenvcha.2011.07.006

Fenton, A., Gallagherd, D., Wright, H., Huq, S., & Nyandigag, C. (2014). Up-scaling finance for community-based adaptation. *Climate and Development, 6*(4), 388–397. https://doi.org/10.1080/17565529.2014.953902

Fink, M. (2016). *Soziale Sicherung im Spannungsfeld gesellschaftlicher Transformation. Eine partizipative Studie in Küstendörfern der Fidschi-Inseln* (Vol. 17). Göttingen: Universitätsverlag Göttingen.

Fujieda, A., & Kobayashi, H. (2013). The potential of Fijian Traditional Housing to cope with natural disasters in rural Fiji. *Journal of Disaster Research, 8*, 18–27

Gharbaoui, D., & Blocher, J. (2016). The reason land matters: Relocation as adaptation to climate change in Fiji Islands. In A. Milan, B. Schraven, K. Warner, & N. Cascone (Eds.), *Migration, risk management and climate change: Evidence and policy responses* (pp. 149–173). Springer.

Hau'ofa, E. (1993). Our sea of Islands. In E. Waddell, V. Naidu, & E. Hau'ofa (Eds.), *A new oceania. Rediscovering our sea of Islands* (pp. 2–16). The University of the South Pacific.

Hay, J. E., & Mimura, N. (2006). Supporting climate change vulnerability and adaptation assessments in the Asia-Pacific region: An example of sustainability science. *Sustainability Science, 1* (1), 23–35. https://doi.org/10.1007/s11625-006-0011-8

Janif, S. Z., Nunn, P. D., Geraghty, P., Aalbersberg, W., Thomas, F. R., & Camailakeba, M. (2016). Value of traditional oral narratives in building climate-change resilience: Insights from rural communities in Fiji. *Ecology and Society, 21*(2), 7. https://doi.org/10.5751/ES-08100-210207

Johnston, I. (2015). Traditional warning signs of cyclones on remote islands in Fiji and Tonga. *Environmental Hazards, 14*(3), 210–223. https://doi.org/10.1080/17477891.2015.1046156

Kelman, I. (2017). How can island communities deal with environmental hazards and hazard drivers, including climate change? *Environmental Conservation, 44*(3), 244–253.

Kench, P. (2012). Compromising reef island shoreline dynamics: Legacies of the engineering paradigm in the Maldives. In J. A. G. Cooper & O. H. Pilkey (Eds.), *Pitfalls of shoreline stabilization: Selected case studies* (pp. 165–186). Springer.

Korovulavula, I., Nunn, P. D., Kumar, R., & Fong, T. (2019). Peripherality as key to understanding opportunities and needs for effective and sustainable climate-change adaptation: A case study from Viti Levu Island, Fiji. *Climate and Development*. https://doi.org/10.1080/17565529.2019.1701972

Kumar, S. (2002). *Methods for community participation. A complete guide for practitioners*. Vistaar.

Lata, S., & Nunn, P. D. (2012). Misperceptions of climate-change risk as barriers to climate-change adaptation: A case study from the Rewa Delta, Fiji. *Climatic Change, 110*, 169–186.

Lauer, M. (2017). Changing understandings of local knowledge in island environments. *Environmental Conservation, 44*(4), 336–347. https://doi.org/10.1017/S0376892917000303

Macintosh, A. (2013). Coastal climate hazards and urban planning: How planning responses can lead to maladaptation. *Mitigation and Adaptation Strategies for Global Change, 18*(7), 1035–1055. https://doi.org/10.1007/s11027-012-9406-2

Maclellan, N. (2011). *Turning the tide: Improving access to climate financing for the Pacific Islands*. Lowy Institute for International Policy.

McMillen, H. L., Ticktin, T., Friedlander, A., Jupiter, S. D., Thaman, R., Campbell, J., Veitayaki, J., Giambelluca, T., Nihmei, S., Rupeni, E., Apis-Overhoff, L., Aalbersberg, W., & Orcherton, D. F. (2014). Small islands, valuable insights: Systems of customary resource use and resilience to climate change in the Pacific. *Ecology and Society, 19*(4), 44. https://doi.org/10.5751/ES-06937-190444

McNamara, K. E. (2013). Taking stock of community-based climate-change adaptation projects in the Pacific. *Asia Pacific Viewpoint, 54*(3), 398–405. https://doi.org/10.1111/apv.12033

McNamara, K. E., Clissold, R., Westoby, R., Piggott-McKellar, A., Kumar, R., Clarke, T., Namoumou, F., Areki, F., Joseph, E., Warrick, O., & Nunn, P. D. (2020). An assessment of community-based adaptation initiatives in the Pacific Islands. *Nature Climate Change, 10*, 628–639.

McNamara, K. E., & Combes, H. J. D. (2015). Planning for community relocations due to climate change in Fiji. *International Journal of Disaster Risk Science, 6*, 315–319. https://doi.org/10.1007/s13753-015-0065-2

McNamara, K. E., & Prasad, S. S. (2014). Coping with extreme weather: Communities in Fiji and Vanuatu share their experiences and knowledge. *Climatic Change, 123*, 121–132. https://doi.org/10.1007/s10584-013-1047-2

McNaught, R., Warrick, O., & Cooper, A. (2014). Communicating climate change for adaptation in rural communities: A Pacific study. *Regional Environmental Change, 14*, 1491–1503.

Medina Hidalgo, D., Nunn, P. D., Beazley, H., Sovinasalevu, J. S., & Veitayaki, J. (2021). Climate change adaptation planning in remote contexts: Insights from community-based natural resource management and rural development initiatives in the Pacific Islands. *Climate and Development*. https://doi.org/10.1080/17565529.2020.1867046

Mercer, J., Dominey-Howes, D., Kelman, I., & Lloyd, K. (2007). The potential for combining indigenous and western knowledge in reducing vulnerability to environmental hazards in small island developing states. *Environmental Hazards, 7*, 245–256. https://doi.org/10.1016/j.envhaz.2006.11.001

Mimura, N., & Nunn, P. D. (1998). Trends of beach erosion and shoreline protection in rural Fiji. *Journal of Coastal Research, 14*(1), 37–46.

Mimura, N., Pulwarty, R. S., Duc, D. M., Elshinnawy, I., Redsteer, M. H., Huang, H.-Q., Nkem, J. N., & Rodriguez, R. A. S. (2014). Adaptation planning and implementation. In C. B. Field, V. R. Barros, D. J. Dokken, K. J. Mach, M. D. Mastrandrea, T. E. Bilir, M. Chatterjee, K. L. Ebi, Y. O. Estrada, R. C. Genova, B. Girma, E. S. Kissel, A. N. Levy, S. MacCracken, P. R. Mastrandrea, & L. L. White (Eds.), *Climate Change 2014: Impacts, adaptation, and vulnerability. Part A: Global and sectoral aspects. Contribution of Working Group II to the Fifth Assessment Report of the Intergovernmental Panel on Climate Change* (pp. 869–898). Cambridge University Press.

Nunn, P. D. (2009). Responding to the challenges of climate change in the Pacific Islands: Management and technological imperatives. *Climate Research, 40*, 211–231. https://doi.org/10.3354/cr00806

Nunn, P. D. (2013). The end of the Pacific? Effects of sea level rise on Pacific Island livelihoods. *Singapore Journal of Tropical Geography, 34*(2), 143–171. https://doi.org/10.1111/sjtg.12021

Nunn, P. D., Aalbersberg, W., Lata, S., & Gwilliam, M. (2014). Beyond the core: Community governance for climate-change adaptation in peripheral parts of Pacific Island countries. *Regional Environmental Change 14*(1), 221–235. https://doi.org/10.1007/s10113-013-0486-7

Nunn, P. D., & Kumar, R. (2018). Understanding climate-human interactions in Small Island Developing States (SIDS): Implications for future livelihood sustainability. *International Journal of Climate Change Strategies and Management, 10*(2), 245–271. https://doi.org/10.1108/IJCCSM-01-2017-0012

Nunn, P. D., & Kumar, R. (2019). Measuring peripherality as a proxy for autonomous community coping capacity: A case study from Bua Province, Fiji Islands, for improving climate change adaptation. *Social Sciences, 8*(225).

Nunn, P. D., McLean, R. F., Dean, A., Fong, T., Iese, V., Katonivualiku, M., Klöck, C., Korovulavula, I., Kumar, R., & Tabe, T. (2020). Adaptation to climate change: Contemporary challenges and perspectives. In L. Kumar (Ed.), *Climate change and impacts in the Pacific* (pp. 499–524). Springer.

Nunn, P. D., Runman, J., Falanruw, M., & Kumar, R. (2017). Culturally grounded responses to coastal change on islands in the Federated States of Micronesia, northwest Pacific Ocean. *Regional Environmental Change, 17*(4), 959–971. https://doi.org/10.1007/s10113-016-0950-2

Nurse, L. A., McLean, R. F., Agard, J., Briguglio, L. P., Duvat-Magnan, V., Pelesikoti, N., Tompkins, E., & Webb, A. (2014). Small Islands. In V. R. Barros, C. B. Field, D. J. Dokken, M. D. Mastrandrea, K. J. Mach, T. E. Bilir, M. Chatterjee, K. L. Ebi, Y. O. Estrada, R. C. Genova, B. Girma, E. S. Kissel, A. N. Levy, S. MacCracken, P. R. Mastrandrea, & L. L. White (Eds.), *Climate Change 2014: Impacts, adaptation, and vulnerability. Part B: Regional aspects. Contribution of Working Group II to the Fifth Assessment Report of the Intergovernmental Panel on Climate Change* (pp. 1613–1654). Cambridge University Press.

Ostrom, E., Burger, J., Field, C. B., Norgaard, R. B., & Policansky, D. (1999). Revisiting the commons: Local lessons, global challenges. *Science, 284*(5412), 278–282. https://doi.org/10.1126/science.284.5412.278

Piggott-McKellar, A., Nunn, P. D., McNamara, K. E., & Sekinini, S. T. (2020). Dam(n) seawalls: A case of climate change maladaptation in Fiji. In W. Leal-Filho (Ed.), *Managing climate change adaptation in the Pacific Region* (pp. 69–84). Cham.

Reid, H., Alam, M., Berger, R., Cannon, T., Huq, S., & Milligan, A. (2009). Community-based adaptation to climate change: An overview. *Participatory Learning and Action, 60*, 11–33

Remling, E., & Veitayaki, J. (2016). Community-based Action in Fiji's Gau Island: A model for the Pacific? *International Journal of Climate Change Strategies and Management, 8*(3), 375–398. https://doi.org/10.1108/IJCCSM-07-2015-0101

Sinane, K., David, G., Pennober, G., & Troadec, R. (2010). Fragilisation et modification des formations littorales meubles sur l'île d'Anjouan (Comores) : Quand l'érosion d'origine anthropique se conjugue au changement climatique. VertigO - la revue électronique en sciences de l'environnement, *10*(3).

Veitayaki, J., & Holland, E. (2015). Lessons from Lomani Gau Project, Fiji: A local community's response to climate change. In T. Cook & P. Rudiak-Gould (Eds.), *Appropriating climate change: Pacific reception of a scientific prophecy*. De Gruyter.

Veitayaki, J., Tawake, A., Aalbersberg, W., Rupeni, E., & Tabunakawai, K. (2003). Mainstreaming resource conservation: The Fiji locally-managed marine area network and national policy development. In H. Jaireth & D. Smyth (Eds.), *Innovative governance, indigenous people, local communities and protected areas* (pp. 105–124). Ane Books.

Weir, T., Dovey, L., & Orcherton, D. (2017). Social and cultural issues raised by climate change in Pacific Island countries: An overview. *Regional Environmental Change, 17*(4), 1017–1028. https://doi.org/10.1007/s10113-016-1012-5

Wissenschaftlicher Beirat der Bundesregierung Globale Umweltveränderungen (WBGU). (2011). *World in transition: A social contract for sustainability*. WBGU.

Widening the Scope of Disaster Preparedness in the Caribbean: Building Resilience Through Improving Climate Information

Denyse S. Dookie and Daniel E. Osgood

Abstract Despite the frequency and severity of disasters in the insular Caribbean, and in particular the impacts of hydrological/meteorological ('hydromet') events, there are relatively few insights into the necessary relevance and scope of disaster preparedness. Such plans should not be limited to evacuation plans, assembling disaster toolkits, and preparing emergency response, but rather widened to include a deeper and more updated awareness of impending disasters and response options, and the timely communication of appropriate climate information. Not only could this facilitate building comprehensive disaster risk resilience but also holistic adaptation strategies by highlighting local vulnerabilities and policy gaps, given the specific challenge of climate change and variability for Caribbean SIDS.

This chapter reviews global- and Caribbean-relevant literature regarding disaster risk management and reduction, as well as ongoing strategies and policies. It advocates for a wider appreciation of disaster preparedness strategies, including the utility of enhanced climate information, storm forecasting and hydromet service delivery, as well as improved coordination and communication between and amongst weather, disaster and other national and international agencies and the local public. It also offers a brief review of the detrimental impacts of Tropical Storm Erika in Dominica in August 2015 towards a contextual understanding of disaster risk management lessons learnt and the need for encouraging the improvement and use of climate information within the Caribbean. Such a specific focus on storm disaster preparedness, and in particular that preparedness matters, could assist in channelling more attention to disaster preparedness and resilience strategies, assisting disaster policy directives as well as resource allocation planning within the Caribbean and also other small island states.

D. S. Dookie (✉)
Grantham Research Institute on Climate Change and the Environment, London School of Economics and Political Science, London, UK
e-mail: d.dookie@lse.ac.uk

D. E. Osgood
Financial Instruments Sector Team, International Research Institute for Climate and Society (IRI), Columbia University, New York, NY, USA
e-mail: deo@iri.columbia.edu

S. Moncada et al. (eds.), *Small Island Developing States*, The World of Small States 9, https://doi.org/10.1007/978-3-030-82774-8_5

Keywords Caribbean · Disasters · Climate information · Resilience · Disaster risk reduction · Disaster risk management · Climate change · Tropical storm · Preparedness · Humanitarian assistance · Small Island Developing States

1 The Caribbean: Inherent Vulnerability and Disaster Risk

It is commonly known that, like other small island regions, the countries and territories of the Caribbean are particularly vulnerable to external shocks (Briguglio, 1995, 2014; Thomas et al., 2020), including international economic and business busts, and local environmental shocks such as climatic changes, variability and increases in the frequency and magnitude of natural hazards. For Caribbean Small Island Developing States (SIDS), their location and characteristics such as small sizes of economies and population, dependence on notably climate-dependent primary industries such as agriculture and tourism, evolving economic and governance systems, levels of development and environmental degradation, as well as typical residence along a coastline or otherwise low-lying area, often offers limited resilience in times of a shock. Understanding this complex and multi-layered nature of vulnerability, as well as its quantification and comparison across countries, within a holistic context (for example, with particular attention to women and children), has been a developing research area within the region towards reducing vulnerability and promoting resilience (for example, see UNECLAC, 2011).

Together with this inherent vulnerability, the increasing exposure of Caribbean societies and ecological systems to a variety of natural hazards give rise to significant disaster risk. Such risk may be especially emphasised when there are multiple and/or different types of hazard events, and/or repeated events over time. As shown in Table 1, since the start of EM-DAT[1] records in 1900 to end-2019, there have been 608 natural hazard-based disasters (encompassing hydrological, meteorological, climatological, biological and geophysical hazard events) within the insular Caribbean region (in addition to 132 technological disasters). According to the EM-DAT database, over the years these disasters have affected or injured about 56.4 million people (of note, the 2019 regional population is about 42.8 million[2]), resulted in 301,939 deaths, and led to approximately US$172.6 billion in total damages (at 2019 constant prices). It is of note that the majority of these impacts have happened since 1990, arguably a good benchmark of impact data reporting and availability.

---

[1]Guha-Sapir et al. (2020). It should be noted that EM-DAT defines a 'natural' disaster as the following: climatological (e.g. drought, wildfire), geophysical (earthquake, mass movements, volcanic activity), hydrological (flood, landslide), meteorological (extreme temperature, storm), biological (epidemic, insect infestation, animal epidemic), or extra-terrestrial (impact, space weather).EM-DAT includes all disasters from 1900 until present, which fit at least one of the following criteria: 10 or more people dead; 100 or more people affected; declaration of a state of emergency; call for international assistance.

[2]World Bank Development Indicators, using 2019 data.

Table 1 Disasters in the Caribbean, 1900–2019

Disaster group	Disaster type	Number of events	Total deaths	Total affected	Total damages (mil US$ - cur)	Total damages (mil US$ - con)
Hydrological	Flood	155	6382	7,507,035	992.7	1685.4
	Landslide	7	443	2435		
Meteorological	Storm	355	31,858	35,874,544	128,171.7	159,567.9
Climatological	Drought	29		8,331,762	283.6	433.3
	Wildfire	5			1.0	3.1
Biological	Epidemic	31	7628	766,198		
Geophysical	Earthquake	15	223,988	3,837,599	8104.0	10,924.6
	Mass movement (dry)	1	40			
	Volcanic activity	10	31,599	110,403	8.0	12.7
Natural hazard-based disaster sub-total		608	301,938	56,429,976	137,561.1	172,637.1
Technological	Industrial accident	8	59	5290	22.4	50.6
	Miscellaneous accident	24	657	524,518	50.3	89.3
	Transport accident	100	6435	2549		
All disasters total		740	309,089	56,962,333	137,633.8	172,777.[illegible]
Sum, All disasters (since 1990)		497	247,328	46,160,685	133,241.4	152,024.9
Sum, Natural hazard-based (since 1990)		397	241,609	46,155,271	133,184.9	151,929.3
Sum, Hydro-met-clim (since 1990)		261	11,403	41,630,806	125,152.0	142,497.0

Source: Data compilation from EM-DAT database, www.emdat.be. Database version: 15th June 2020

Notes: Data compiled for 1900 to end-2019 for Caribbean countries including: Anguilla, Antigua and Barbuda, Aruba, The Bahamas, Barbados, Belize, Cayman Islands, Cuba, Curaçao, Dominica, Dominican Republic, Grenada, Guadeloupe, Haiti, Jamaica, Martinique, Montserrat, Netherland Antilles (thereafter Bonaire, Sint Eustatius and Saba), Puerto Rico, Saint Barthèlemy, Saint Kitts and Nevis, Saint Lucia, Saint Martin (French), Saint Vincent and the Grenadines, Sint Maarten (Dutch), Trinidad and Tobago, Turks and Caicos Islands, British Virgin Islands, US Virgin Islands. *Total Damages* values expressed in millions of US dollars in current prices in the year of event (cur) as well as constant 2019 prices (con) as calculated using CPI data as per EM-DAT database

While major events within the region include earthquakes (such as the devastating magnitude 7.0 earthquake in Haiti in 2010) and volcanic eruptions (such as in Montserrat since 1995), by far most disasters are as a result of hydrological, climatological and meteorological ('hydromet'[3]) events—361 events since 1990 are reportedly as a primary result of storms, floods, droughts, wildfires and landslides. While earthquakes have unfortunately resulted in more deaths, hydromet events have affected more than 41.6 million Caribbean people over the past 4 decades, and contributed to about US$142 billion in damages over time. It is noted that the economic and societal impacts of drought, flood and storm events in the years since 2016 have been often the highest on record for many islands.

Layered onto this challenge of natural hazards is the complexity of climate change and variability. Although small islands do not have uniform climate change risk profiles, the IPCC's 5th Assessment Report indicates that Caribbean SIDS have already experienced impacts of climate change—under the Representative Concentration Pathway 4.5 Scenario, the annual projected change for 2081–2100, relative to 1986–2005, is likely a 1.4 °C average increase in temperature, 5% decline in rainfall, and 0.5–0.6 m sea level rise (IPCC, 2014). Climate change, as well as the El Niño-Southern Oscillation (ENSO) inter-annual phenomenon, has likely affected the frequency and severity of rainfall and extreme events, including hydromet-related drought, storms, and floods, and is certain to have direct and indirect wide-reaching impacts within Caribbean small islands (Hsiang, 2010; Chapter 29 of IPCC, 2014). The IPCC reports that "relative to other areas, small islands are disproportionately affected by current hydro-meteorological extreme events, both in terms of the percentage of the population affected and losses as a percentage of GDP" (IPCC, 2014).

With this context of the nature of disasters in mind, it is not surprising that Caribbean SIDS are particularly impacted by the resulting effects of natural hazard-based disasters (Heger et al., 2008). Country size is likely to be a typical driver affecting the extent of disaster impact (Cavallo et al., 2010), and as Kahn (2005) explains, even though developed countries experienced disasters of similar frequency and severity there were less disaster-related deaths in these countries. Furthermore, disaster impact may also be lower in situations of stable political systems or improved institutional conditions (Kahn, 2005; Toya & Skidmore, 2007; Raschky, 2008; Strömberg, 2007; Felbermayr & Gröschl, 2014). Toya and Skidmore (2007) summarise that "countries with higher income, higher educational attainment, greater openness, more complete financial systems and smaller government experience fewer losses". Particularly noting the inherent vulnerability of small islands, then, it would seem imperative that there be comprehensive research on disasters, including inter-disciplinary socio-economic detail underscoring the specific challenges of disasters in the local island context. This would be essential to

[3]It should be noted that the United Nations Office for Disaster Risk Reduction (UNDRR, formerly UNISDR) differs from EM-DAT definitions by defining hydrometeorological hazards to be "of atmospheric, hydrological or oceanographic origin. Examples are tropical cyclones (also known as typhoons and hurricanes); floods, including flash floods; drought; heatwaves and cold spells; and coastal storm surges." UNDRR also classifies landslides as geological or geophysical hazards. https://www.undrr.org/terminology. Also note the section on "Selected Disaster-Related Definitions".

identify the core factors and impending impacts of disasters, offer required disaster management and climate adaptation policy direction, as well as justify the need for additional resilience-building and planning resources. However, generally, such research is still under development.

At the global level, natural scientists have contributed significantly to the understanding and prediction of hazardous events (Cavallo & Noy, 2009). Benefiting both commercial interests and general civil protection, innovative prediction models utilising technologies such as Geographical Information Systems (Carrara et al., 1999), improved radar (Creutin & Borga, 2003), and satellite imagery (Joyce et al., 2009) have vastly encouraged the monitoring and detection of, as well as response to, natural hazards. Such information has been fundamental for the building of *ex ante* disaster knowledge and policy, including preparation for disasters towards minimising the costs and extent of impacts. Noting the relationship between disaster impacts and losses, social scientists have also focused on *ex post* disaster research, generally on the macroeconomic consequences of disasters. Although one early paper by Albala-Bertrand (1993) indicated that while disasters are a problem of development, "they are not necessarily a problem for development", many recent studies suggest that disasters do play a role in observable adverse macroeconomic impacts, which may lead to negative long-term growth and development consequences (Hochrainer, 2009; von Peter et al., 2012; Hsiang & Jina, 2014), often especially within developing/low-income countries (Benson & Clay, 2004; Noy, 2009; Raddatz, 2009; Zapata & Madrigal, 2009; Strobl, 2012; Felbermayr & Gröschl, 2014). Further research in this topic has expanded to offer details on various disaster drivers and macroeconomic implications, as well as an awareness of the effects of disasters in various dimensions, including countries' labour markets, production efficiency, human capital base, gender, and fertility consequences.

Much like in the wider global lens, Caribbean region-specific literature is largely dominated by natural science research. Such studies include a better understanding of the nature and/or predictability of Atlantic storms (Walker et al., 1991; Tartaglione et al., 2003; Trenberth & Shea, 2006; Donnelly & Woodruff, 2007), observed changes in climate (Peterson et al., 2002; Karmalkar et al., 2013; Stephenson et al., 2014), as well as the likely projected climate based on downscaled regional climate models (Campbell et al., 2011; Taylor et al., 2012; Karmalkar et al., 2013). Socio-economic research has only recently begun to focus on the impact of weather, climate and disasters within the region. In a seminal article regarding the region, Hsiang (2010) uses data for 28 Caribbean basin countries and finds that cyclones and temperature increases are associated with large reductions in economic output for industries not traditionally seen as vulnerable to climate change, such as within non-agricultural sectors. Strobl (2012) also looks at the particular impacts of hurricanes on macroeconomic outcomes in the wider Central American and Caribbean region, finding an inverse relationship between average hurricane strike and economic output (also see Bertinelli et al., 2016). Moore et al. (2016) perform general equilibrium framework model simulations which suggest that not only do output losses due to hurricanes have economy-wide effects, but rural regions may suffer most.

The work of the sub-regional headquarters for the United Nations Economic Commission for Latin America and the Caribbean (UNECLAC) has offered much insight into the impact and effects of natural hazard-based disasters within this small island grouping. Their Damage and Loss Assessments (DaLAs) (see Zapata & Madrigal, 2009) have been formidable in providing governments with contextual estimates of direct damage and indirect or foregone losses. Using data from these DaLAs in a before-and-after statistical analysis, UNECLAC (2010) reviewed extreme events within the region over the period 1990–2008, finding that "declines in real GDP and employment in some sectors following disasters temporarily set back living standards in affected countries". Also including the impact of disasters on other macroeconomic elements such as external debt and the imports and exports of goods and services, it considers a serious challenge to be the increase of external debt in a disaster year compared to the year before.

2 Managing and Reducing Disaster Risk in the Global Context

Development agencies, including the United Nations Office for Disaster Risk Reduction (UNDRR, formerly UNISDR) & the World Bank, and a wide range of inter-disciplinary writing, have long suggested the need to minimise disaster impacts through an attention to risk, resilience, and adaptive capacity (Mochizuki et al., 2014) in order to maintain and sustain livelihoods, economic development and growth (Otero & Marti, 1995; Thomalla et al., 2006; McBean, 2012; Vorhies, 2012; World Bank, 2013; UNDRR, 2015c). Disasters pose a hindrance to poverty reduction and sustainable development (Pelling et al., 2002; Desai et al., 2015; Tanner et al., 2015), and the notion of disaster risk reduction (DRR) seems critical in assisting a global prosperity agenda.

A leading authority on the topic, the UNDRR defines disaster risk as a function of hazard, exposure and vulnerability, and expresses it as the probability of loss of life, injury or destroyed or damaged capital stock in a given period of time (Desai et al., 2015; also see *Selected Disaster-related Definitions (UNDRR)* in Appendix I). Examples of DRR can include "reducing exposure to hazards, lessening vulnerability of people and property, wise management of land and the environment, and improving preparedness and early warning for adverse events" (UNDRR, n.d.). The post-2015 development agenda/2030 Agenda for Sustainable Development specifically notes DRR as an important development objective due to the inextricable link between DRR and achieving sustainable development (Hillier & Nightingale, 2013; IRDR & ICSU, 2014; UNDRR, 2015a, b, c; UN, n.d.). Of the 17 internationally agreed Sustainable Development Goals (SDGs) and 169 global targets, ten goals and 25 targets are related to DRR (UNDRR, 2015a), showcasing the cross-section of aspects and sectors through which disasters affect development.

However, despite this international importance and attention, DRR advancement faces several challenges, including problems measuring the costs and benefits as well as investing in DRR (IDB, IMF, OAS and the World Bank, 2005; Hallegatte & Przyluski, 2010; CSIS, 2015; Vorhies, 2012; Shreve & Kelman, 2014). As well, there has been relatively limited funding channelled towards disaster prevention and preparedness, as noted in donors' bilateral humanitarian assistance funds (Sparks, 2012; Kellett & Caravani, 2013; GHA, 2013, 2015; Becerra et al., 2014). Figure 1 shows that over 2005–2014, the majority of humanitarian assistance from the 26 Organisation for Economic Co-operation and Development (OECD) Development Assistance Committee (DAC) countries had been channelled towards material relief and assistance (including expenditures on shelter; water, sanitation and health services; supply of medicines and other non-food relief items; and assistance to refugees and internally displaced people in developing countries other than for food or protection (GHA, 2013), and emergency food aid. An average of only 3.7% of the total bilateral humanitarian assistance funding was directed to disaster prevention and preparedness in the form of 'multi hazard response preparedness' (considering years of available disbursements, 2005–2018).

In a United Nations and World Bank (2010) report entitled 'Natural Hazards, UnNatural Disasters' (highlighting that while hazards are natural, the "*unnatural disasters* [emphasis retained] are deaths and damages that result from human acts of omission and commission"), the focus on prevention is encouraged: the "effectiveness of prevention spending is more important than its magnitude". It notes although prevention spending is less than post-disaster relief, it does not imply that there is too little of it but rather that "disasters increase spending on relief and that such expenditures remain high for several subsequent years".

2.1 *Managing and Reducing Disaster Risk in the Caribbean*

Governments of Caribbean SIDS have certainly considered managing and minimising disaster risk of vital importance, and operate both at the national and coordinated regional levels in responding to natural hazards and resulting risk. Davoli (2012) highlights that "the first response is always at the local/national level and that any [regional] response must support, rather than compete with the national response".

At the national level, there are generally similar structures of varying sizes and scopes for disaster management across Caribbean countries (Davoli, 2012): the Head of the Government is usually at the apex of the national disaster response structure, trusting information and advice from operational structures such as the national disaster management office. Davoli (2012) outlines several likely functions of national disaster management offices, including:

- Implementing government policy and programs aimed at lessening the impact of disasters;

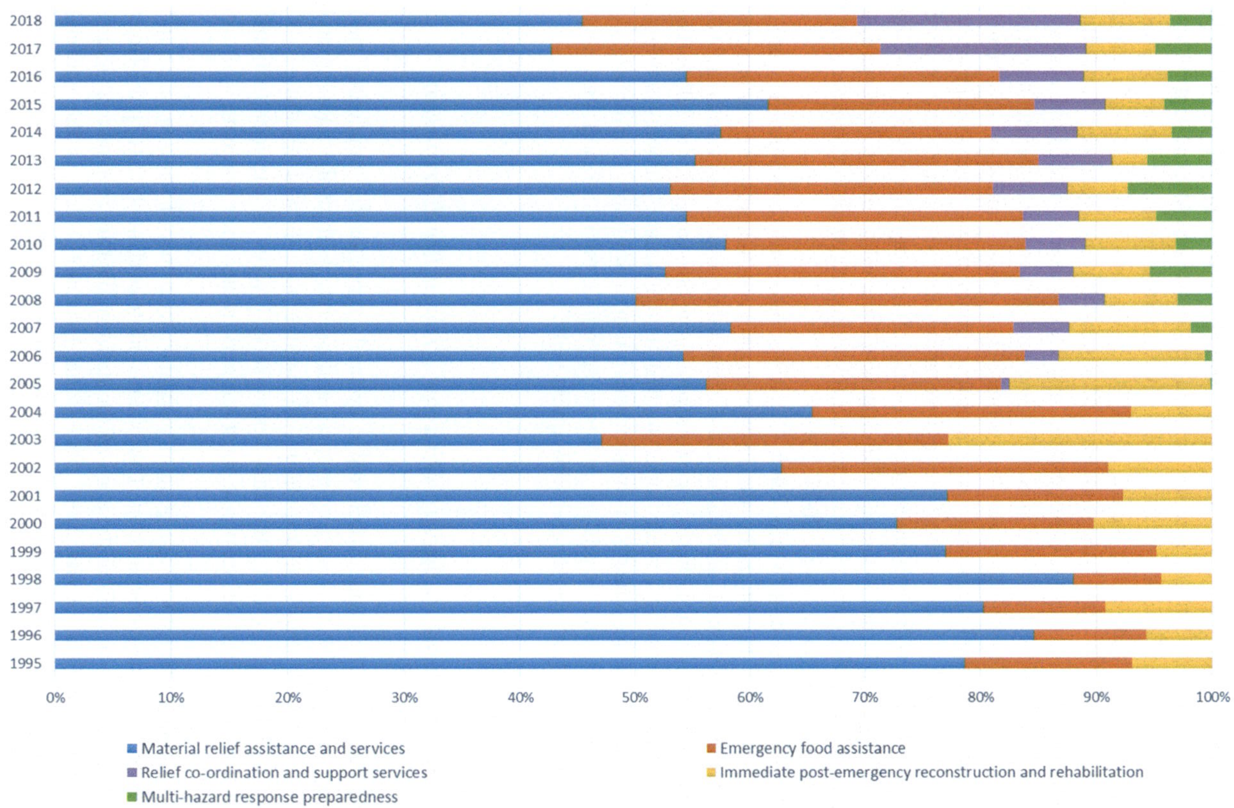

Fig. 1 OECD DAC donors' bilateral humanitarian assistance by expenditure type, 1995–2018 [Source: Authors' based on OECD Development Assistance Committee (DAC) Creditor Reporting System: https://stats.oecd.org/Index.aspx?DataSetCode=CRS1#]

- Providing training in disaster management;
- Issuing early warning of hazards to institutions and the general population;
- Calling for activation and/or deactivation of the National Emergency Response Plan; and
- Leading disaster response efforts and coordinating with other sectors and with regional and international structures.

Davoli (2012) also includes that Government Heads also receive and utilise information from policy structures such as a country's National Disaster Committee or Council. These, she mentions, are "typically a multi-agency, multi-sectoral body, which includes the private sector and non-governmental and voluntary organizations" and "serve primarily as a forum for identification of hazards and definition of policy strategies to prevent and mitigate damages and to make preparedness, response, and rehabilitation determinations", although this may vary by country. As an example, Davoli (2012) outlines some of the members of the National Disaster Committee of Barbados: the Director of Emergency Services, Director of Statistical Services, Commission of Police, Chief Medical Officer, Chief Welfare Officer, Airport Manager, Barbados Red Cross, and others.

Notwithstanding the immense efforts made at the national levels, collaboration and coordination on a variety of concerns, including disaster risk management, has also been undertaken at the regional level. Kirton (2013) reminds us that "the Treaty of Chaguaramas, which established the CARICOM [Caribbean Community] in 1973, places significant importance on collaboration and functional cooperation in DRM, establishing it as one of the key pillars of the integration movement". In reviewing the "[e]pisodes of change in DRR" over the years within the region, Collymore (2011), specifically reminds us of the Pan Caribbean Disaster Preparedness and Prevention Project (1981–1991), which came out of concern following the disasters of the late 1970s, and was the first donor-based regional effort to improve disaster preparedness in the Caribbean. Central within this decade-long effort was the shift from response and relief to preparedness and prevention, and at the national level there were initiatives to build greater consciousness, improve public awareness of hazard risk, and institutionalise disaster preparedness and emergency response coordination.

Following these efforts, there was a "move beyond preparedness" to include vulnerability assessments, information networking, and coordination amongst a wide range of partners and agencies (such as media and NGOs). As well, severe impacts of storms in the late 1980s cemented the need for a structured mechanism to respond to collective interests (Collymore, 2011). Kirton (2013) offers that political leadership within the region, "influenced by the successes of regional cooperation in other areas, came to see the strengthening of the regional disaster management process as a way of more efficiently managing the increasing threats of natural hazard-based disasters; they made DRM a "front burner" issue on the region's political agenda." Through endorsement and declaration through CARICOM, the Caribbean Disaster Emergency Response Agency (CDERA) was formed in September 1991, evolving into the current Caribbean Disaster Emergency

Management Agency (CDEMA) in 2009 (CDEMA, n.d.). CDEMA aims to assist islands to engage in Comprehensive Disaster Management, "an integrated and proactive approach to disaster management seeking to reduce the risk and loss associated with natural and technological hazards and the effects of climate change to enhance regional sustainable development" (CDEMA, n.d.).

There are presently eighteen Participating States in CDEMA: Anguilla, Antigua and Barbuda, Commonwealth of the Bahamas, Barbados, Belize, Commonwealth of Dominica, Grenada, Republic of Guyana, Haiti, Jamaica, Montserrat, St. Kitts and Nevis, Saint Lucia, St. Vincent and the Grenadines, Suriname, Republic of Trinidad and Tobago, Turks and Caicos Islands and the Virgin Islands. CDEMA lists its main functions as follows:

1. Mobilising and coordinating disaster relief;
2. Mitigating or eliminating, as far as practicable, the immediate consequences of disasters in Participating States;
3. Providing immediate and coordinated response by means of emergency disaster relief to any affected Participating State;
4. Securing, coordinating and providing to interested inter-governmental and non-governmental organisations reliable and comprehensive information on disasters affecting any Participating State;
5. Encouraging –
   (a) the adoption of disaster loss reduction and mitigation policies and practices at the national and regional level;
   (b) cooperative arrangements and mechanisms to facilitate the development of a culture of disaster loss reduction; and
6. Coordinating the establishment, enhancement and maintenance of adequate emergency disaster response capabilities among the Participating States.

As shown in Fig. 2, Davoli (2012) summarises CDEMA's Regional Response Mechanism (RRM), which is "an arrangement for the coordination of disaster response among CDEMA Participating States, regional and international agencies." The RRM is comprised of national disaster plans, regional coordination plan, regional warehouses, memoranda of understanding, standard operating procedures, and requires input from the Caribbean Disaster Relief Unit (CDRU) ("a facility created to manage the use and secure the participation of regional forces in humanitarian situation"), as well as specific input from the Eastern Caribbean Donor Group (ECDG).

The call for an improved focus on disaster management in the Caribbean has been well-suggested by these Caribbean experts. Collymore (2011) notes that despite some investment in basic institutional infrastructure over the years, the "event-driven developments in disaster risk reduction (DRR) programming and embryonic policy suggest a high measure of reactiveness", and questions whether and how the region may be more proactive in nature. He suggests a closer look at the Comprehensive Disaster Management Strategy Framework, as touted by CDEMA, in this regard. Such a framework inextricably links disaster management with sustainable

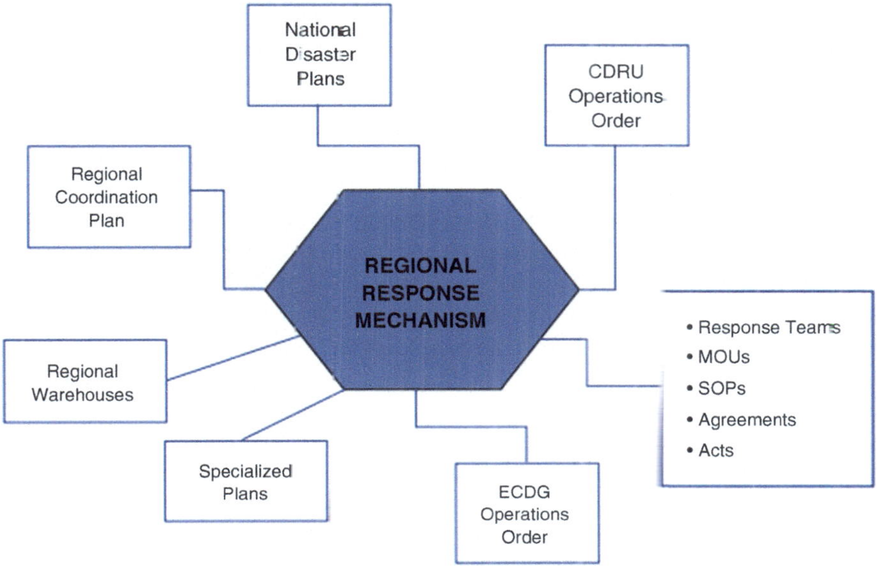

Fig. 2 Components of CDEMA's Regional Response Mechanism [*Source*: Davoli (2012)]

development through a strengthening of "regional-, national- and community-level capacity for mitigation, management and coordinated response to natural and technological hazards and the effects of climate change" (Collymore, 2011) and integration of disaster management considerations into the development planning and decision-making process (CDEMA, n.d.).

As well, after extensively reviewing the context of DRM in the Caribbean and highlighting several successes and challenges regarding its implementation, Kirton (2013) offers a few suggestions to encourage such efforts, including: a standardisation of institutional collaboration, a prioritisation of DRM at the political level, an increase in disaster response capacity (due to the absence of a structural arrangement for mobilising capacity in a timely fashion), strengthening community awareness and participation (as it was considered that "education and awareness programs are an important element in developing a culture of preparedness and safety in the region"), reduction in the duplication of efforts across Caribbean states, as well as the confirmation of DRM as a development priority.

3 Beyond Disaster Management and Recovery in the Caribbean: Developing a Wider Sense of Preparedness and Building Resilience

The above-listed recommendations and priorities are essential within an updated disaster management framework, not only within the greater Caribbean but for other small island nations. Efforts to better synthesise disaster management plans and policies towards a more comprehensive and inter-connected DRM strategy and structure would furthermore be quite positive in terms of the local socio-economic development context. Figure 3 offers a graphical depiction of the disaster-development continuum in the context of disaster management and emergency management, as cited from a WHO Training Package (WHO, 2002). The associated text of the training package mentions that "[i]t is the people who matter most, and without the people we have no disaster". It notes that "disaster prevention, mitigation and preparedness safeguard development," as shown on the left side of the continuum, while "ever-increasing resources are spent for disaster relief, at the expense of development . . . But only development can reduce vulnerabilities, and the hazards arising from the socio-economic structure" (WHO, 2002).

As the region takes stock of its inherent vulnerability, varying levels of risk exposure, trends of disaster impacts, need for future development, as well as the changing nature of the climate, it is ripe for concerted action regarding disaster risk

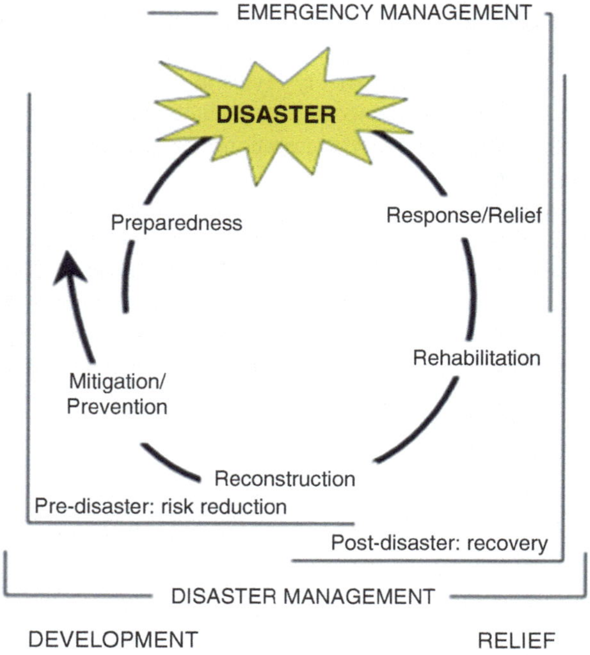

Fig. 3 Disaster-development continuum in the context of DRM and emergency management [*Source:* WHO/EHA Training Package: Disasters & Emergencies—Definitions. Panafrican Emergency Training Centre, Addis Ababa. Updated March 2002 by EHA: http://apps.who.int/disasters/repo/7656.pdf]

reduction. At this juncture, a heightened focus on building resilience through pre-disaster development-enhancing initiatives, including broadening the scope of preparedness, is critical.

It may be the perspective of many people that preparedness simply relates to the development of coordinated plans in the event of a disaster, or broader emergencies. Classic thoughts may include developing evacuation procedures, awareness of local shelters, finding sandbags and stocking up on non-perishables, and general effective communication of threats with family members, loved ones, employees and the wider public. However, the International Federation of Red Cross and Red Crescent Societies (IFRC) defines disaster preparedness as "measures taken to prepare for and reduce the effects of disasters. That is, to predict and, where possible, prevent disasters, mitigate their impact on vulnerable populations, and respond to and effectively cope with their consequences" (IFRC, 2018). As well, the UNDRR defines preparedness as "knowledge and capacities developed by governments, response and recovery organizations, communities and individuals to effectively anticipate, respond to and recover from the impacts of likely, imminent or current disasters" (see Appendix I for *Selected Disaster-related Definitions*). In this regard, preparedness is much wider than these usual (and important) actions in the relative short-term before the onset of a hazard. In the context of building resilience, or being able to "resist, absorb, accommodate, adapt to, transform and recover from the effects of a hazard in a timely and efficient manner" (see Appendix I), a wider sense of preparedness through better anticipation and prediction of hazards and potential disasters can be seen as a fundamental component of disaster risk reduction. Moreover, when considering hydromet hazards, this can as well be part of a holistic strategy to adapt to climate change and variability.

A March 2017 World Bank online news feature implores that building resilience to the growing economic, environmental, and social challenges faced by countries worldwide is of paramount importance to support the required goals of ending extreme poverty and boosting shared prosperity in a sustainable manner (World Bank, 2017). While there are certainly many ways in which this could be addressed, the article states that "improving the prediction of hydrological and meteorological (or "hydromet") hazards by getting accurate, timely predictions into the hands of decision-makers and the public, can save lives and money". More specifically, according to a 2014 World Bank "Disaster Vulnerability Reduction Project," there is an urgent need to improve weather forecasting and modelling as well as the availability, access to and delivery of the climate information base, in addition to the development of associated climate services (also see Dari (2017)).

It should be noted here that the term '*climate information*' denotes "externally provided scientific weather and climate information, which refers to processed data, products and/or evidence-based knowledge about the atmosphere-ocean system across short (hours to days) and long (seasons to decades) time scales; the term information, as opposed to data, implies that it has meaning and relevance within a given context" (Singh et al., 2018). Climate information can include historical information, present observations, as well as projections and forecasts, and is "typically produced and disseminated by scientific institutions such as national meteorological agencies, or intermediaries and boundary organizations (e.g. environmental consultancies, applied university research centres)" (Singh et al., 2018). The region's Caribbean Institute for Meteorology and Hydrology (CIMH) and Columbia University's International Research Centre for Climate and Society (IRI)[4] may be considered as boundary organisations in this context, and they assist regional and national bodies with climate information including drought, dry spell, temperature, rainfall and wet day monitoring and outlooks.[5] These outlooks combine various scientific model outputs with regional and local climate expertise to better contextualise this information before it is shared with farmers and others in the agriculture sector, water agencies, disaster managers, as well as those in the coastal management and tourism sectors. Certainly, there may be entities within the private sector sharing such information, as is found in other parts of the world (for example, Singh et al. (2018) mentions that Skymet in India provides "climate services for agriculture risk management, weather forecasting, and crop insurance and delivers short-, medium- and long-term forecasts at the district level to multiple actors").

Developing such climate information and promoting the need for "forecasting for catastrophes", through the investment in weather services and the possibility for improved temperature and weather forecasts, the World Bank notes that enhanced preparedness for hazardous events can not only reduce damage to people and livestock, but also assist health services, farmers, and the energy sector, facilitating a "move from primarily response and recovery after natural disasters to a more proactive approach of making systematic and strategic DRM decisions" (World Bank, 2017). Moreover, such actions enable "countries [to] have better information to adapt to a changing climate" (World Bank, 2017).

An increase in awareness of, and investment in, the hydromet sector can no doubt be of benefit to Caribbean SIDS. To better understand this in context, a short case study on the detrimental impacts of Tropical Storm Erika in Dominica in August 2015 is offered, highlighting certain limitations of DRM and preparedness without adequate climate information.

3.1 Dominica and Storm Disasters

Dubbed "The Nature Island" due to its unspoilt natural beauty, the Caribbean island of Dominica is located about halfway between the French islands of Guadeloupe and Martinique, with an area of 750 km^2 and a 2018 population of about 71,625 (World Bank Development Indicators). Unlike many other Caribbean islands, the island is more famous for its lush mountainous rainforests than its beaches—the island boasts the world's second-largest hot spring, Boiling Lake (located within the World Heritage Site Morne Trois Pitons National Park, and developed as a result of the

[4]See http://www.cimh.edu.bb/ and https://iri.columbia.edu/.

[5]A list of CIMH's output may be found here: https://rcc.cimh.edu.bb/climate-outlooks/.

island's formation due to geothermal-volcanic activity), and exotic flora and fauna. The island, like many others within the region and elsewhere, was traditionally dependent on agriculture, primarily bananas (CIA World Factbook, 2018). Nowadays, while such a reliance has been marred by external shocks to commodity prices and extreme weather events, agriculture is still responsible for about 40% of local employment. Local economic development has been recently boosted by an emerging tourism (specifically eco-tourism), construction and other services—the services sector contributes an estimated 71% to the national gross domestic product (CIA World Factbook, 2018).

Tropical storms and heavy rainfall events have historically affected the island. Data from NOAA's Historical Hurricane Tracks (https://coast.noaa.gov/hurricanes/) lists 96 named and unnamed tropical systems passing within 70 nautical miles of Dominica between 1851–2019, of which 30 were of hurricane strength and 11 were of Category 3 and stronger. EM-DAT statistics indicate there were at least 14 declared storm-induced disasters since 1900, collectively affecting 196,283 and killing 2140 persons, and contributing to approximately US$2.2 billion in local damages. The island sustained significant losses in 1979 from the direct hit of Category 5 Hurricane David, in 1994 from Category 4 Hurricane Luis, and in 2007 from Category 1 Hurricane Dean. However, such damages pale in comparison to the effects of more recent storms and hurricanes.

Sadly, the worst natural hazard-based disaster affecting Dominica to date has been category 5 Hurricane Maria, which made landfall on September 18th 2017. Hurricane Maria caused widespread havoc and damage—as much as 98% of the island's buildings were destroyed, thousands of trees were uprooted, perhaps 100% of the banana and tuber plantations was lost, and there was an island-wide water shortage (IFRC, 2017). The hurricane left 31 people dead and 37 persons missing (Government of the Commonwealth of Dominica, 2017). Around 80% of the population (65,000 people), were directly affected. A Post-Damage Needs Assessment (Government of the Commonwealth of Dominica, 2017) conducted by the Government of Dominica estimates while Hurricane Maria is thought to have contributed to about US$931 million in damages, there was approximately $1.37 billion in damages and losses, which is a whopping equivalent of 226% of its 2016 GDP (Government of the Commonwealth of Dominica, 2017). (It is noted that the EM-DAT database lists quite a conservative estimate of damages at US$380.5 million but at least 6000 more affected persons in total (Guha-Sapir et al., 2020)).

3.2 *Tropical Storm Erika, August 27th 2015*

> *"This took us by surprise... it is really terrible"* [6]—Lennox Linton (Dominica Opposition Leader), Aug. 27th 2015

[6]Dominica News Online (2015).

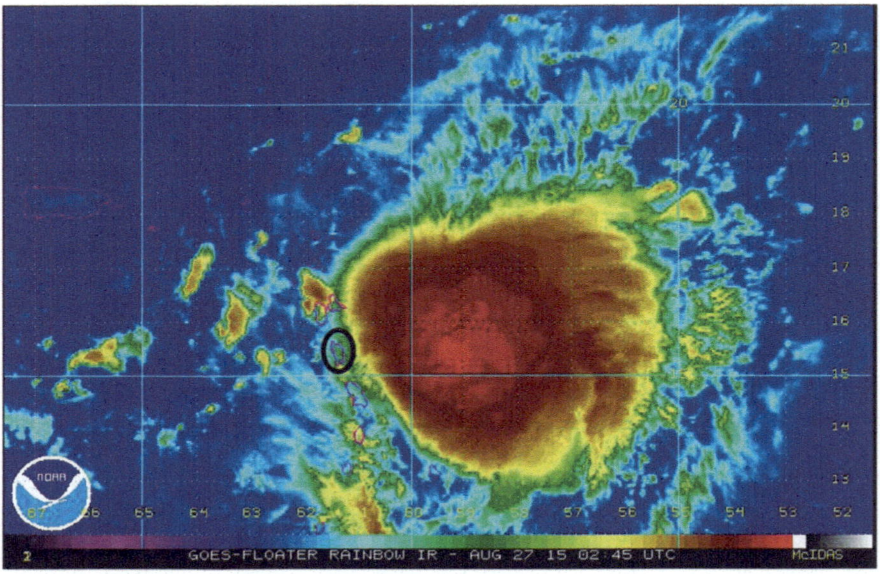

Fig. 4 Tropical storm Erika approaching Dominica, August 27th 2015 [*Source*: NOLA The Times-Picayune, posted August 25th 2015. Still of NOAA GOES-Floater satellite animation: http://www.nola.com/hurricane/index.ssf/2015/08/tropical_storm_erika_track_kee.html]

One hopes that a massive storm like Hurricane Maria is infrequent (Maria is noted among the top ten most intense Atlantic hurricanes on record). The sheer intensity of this storm would have truly pushed the limits of many a disaster plan, and made it difficult to effectively manage impending risk. However, a couple years prior to Maria, there was another relatively weaker storm which had some forewarning but which created quite a stress on the island. On August 27th 2015, Tropical Storm Erika caused extensive flooding and landslides, and resulted in the deaths of at least 30 persons and an estimated US$483 million in damages.

Even though the storm was disorganised at first, the Tropical Cyclone Report produced by the US National Hurricane Center indicates the storm's general formation around 1800 UTC on Monday 24 August 2015 around 900 nautical miles east of the Lesser Antilles, and sustained winds of about 40 knots (Pasch & Penny, 2016). The Report states that Erika's formation was reasonably well anticipated for the eastern Caribbean, and beginning 0900 UTC on Tuesday 25 August, tropical storm watches and warnings were issued for Guadeloupe and islands north of it (i.e. almost all Leeward Islands, or the northern islands of the Lesser Antilles chain, starting east of Puerto Rico and reaching southward to Dominica).

Within the sub-region, Dominica was the only island which did not receive such a forecast, perhaps due to the low probability of landfall considering the storm's best track. However, as shown in the graphic in Fig. 4, with Dominica circled in black outline, Erika was associated with a wide swath of outer rainbands, bringing heavy bursts of rain and wind. In fact, the largest rainfall amounts associated with Erika across all islands affected were observed on Dominica, with rainfall of about 32 cm mostly occurring between 0600 to 1800 UTC (2 am to 2 pm, local time) on 27 August 2015. Noting the push from the east by this northern hemisphere system, storm surges would have been stronger on the east coast, hence the damage to the island's only functioning airport at Marigot. The torrential rains produced catastrophic flooding and mud/landslides and destroyed hundreds of homes, bridges and roads (making evacuation and rescue difficult), and caused the equivalent of 90% of Dominica's GDP in damages. Of interest, other islands did not receive a similar fate—the Tropical Cyclone Report states that there was localised flooding in Guadeloupe, and the rains were "beneficial" in Puerto Rico due to previous drought conditions.

It seems that there was little local knowledge of the potential of Erika (Sun Dominica, 2016), which led to confusion over what could have been done. Given the prevalence of international and social media, neighbouring island forecasts almost 48 h before, and the understanding of secondary threats of heavy rain and winds of such storms, even for nearby systems which do not make landfall, the impacts of Erika are certainly tragic and truly unfortunate.

3.3 *Learning from Erika*

It must be mentioned that there is the requisite attention to disaster risk management in Dominica. DRM is generally governed by the Emergency Powers Act, which was established in 1951 and last revised in 1990 (GFDRR, 2017b), with DRM coordination done jointly by Dominica's Office of Disaster Management and the National Emergency Planning Organisation based on the standard DRM principles of prevention, mitigation, preparedness, response and recovery (NEPO-Dominica, 2001). According to a US$39.5 million Disaster Vulnerability Reduction Project by the World Bank in 2014 (World Bank, 2014), the island has developed a number of policies, plans and procedures relevant to disaster risk reduction, including the following: National Integration Water Resources Management Policy (Draft), 2010; Disaster Management Plan, 2009; National Emergency Management Policy, 2009; National Shelter Policy, 2009; National Environment Policy/National Environment Management Strategy, 2004; Dominica's Policy on Planning for Adaptation to Climate Change, 2002; Physical Planning Act, 2002; and Plan to Reduce the Vulnerability of School Buildings to Natural Disasters, 1998.

Despite these improvements over time, it is noted that "Dominica still faces challenges in strategically and comprehensively managing natural hazard risk, particularly in the context of a changing climatic environment that threatens to increase disaster risk, further expose existing vulnerabilities, and complicating the search for efficient long-term solutions" (World Bank, 2014).

Further,

> . . .an overall structure for analyzing and integrating disaster risk information in the development process is lacking. Development decisions in Dominica commonly do not account for disaster risk and expected climate change impacts due to a lack of available information on hazards, vulnerability, exposure, and expected climate change impacts. Secondly, information sharing among agencies is weak, largely due to limited capacity and lack of an overall mechanism to share information with low transaction costs. Finally, disaster risk management (DRM) responsibilities are dispersed among various government agencies, with limited collaboration between entities (World Bank, 2014).

The Dominican government has resolved to "build back better" post-Erika (GFDRR, 2017b), and prioritise advancing their DRM agenda through the exploration of additional options to strengthen fiscal resilience to natural hazard events, strengthening the resilience of infrastructure, and conducting additional assessments and mapping to better understand natural hazard risk. Such elements are certainly important and essential for building resilience in Dominica—it is noted that the capital, Roseau, and airport are in relatively low-lying areas. As reviewed by the Assessment Capabilities Project in October 2017, there were several general lessons learnt from the initial phases of the response to Tropical Storm Erika:

- Weather forecasting and modelling have improved tremendously over the past years, yet forecasting for small island states is a major challenge. The primary need is to improve regional and national weather forecasting, focusing on the impact of specific weather forecasts rather than generalised forecasts (Dani, 2017).
- Post Disaster Needs Assessment needs to be conducted through collaboration with international agencies. This is crucial to ensure that efforts are not duplicated and that the single assessment is accepted by all agencies (PAHO). Government experts expressed that the most important lesson TS Erika taught the island, and the rest of the region, is that the science and management of natural and weather-related events must be synchronised. Government, regional agencies and disaster management offices must collaborate and exchange information (Sun Dominica, 2016).
- Monitoring and reporting needs to be done on short, medium, and long-term recommendations made in the Disaster Needs Assessment (GFDRR, 2017a).
- Access to finance mechanisms for implementing recommendations should be identified, considering economic challenges facing the country (GFDRR, 2017a).
- A resilient approach to development needs to be adopted. This is best done through the integration of sectoral recommendations into national planning (GFDRR, 2017a).

The detrimental impacts of Maria have further created a heightened impetus to boost current plans towards building back better. In 2018, the Government of Dominica shared its intentions to hold a public consultation to develop a national resilience plan, set up an emergency relief fund, and a national disaster committee; efforts to discuss funding and other resources necessary for improved economic and social resilience are also underway (Thomson Reuters Foundation, 2018). Since then, Dominica has solidified its pathway to disrupt the business-as-usual pattern, and has committed to becoming the first climate resilient country in the world. Its recent "Climate Resilience and Recovery Plan 2020–2030" outlines various objectives and strategies towards this goal, noting partners including within the private sector, as well as possible constraints in realising this vision (Government of the Commonwealth of Dominica, 2020).

4 Towards Building Resilience in the Caribbean: The Role of Climate Information

Taking stock of the UNDRR definitions of resilience and preparedness (see Appendix I for *Selected Disaster-related Definitions*), the use of improved climate information within the aforementioned context of "forecasting for catastrophes" not only encourages an appreciation of a wider sense of disaster preparedness, but also can assist communities in being more aware of impending threats and in a better and more relevant and appropriate position to manage and reduce risk. In the case of the Caribbean, which is dependent on weather-related activities such as agriculture and tourism, and where weather-related hazards dominate past disaster events, understanding the availability and relevance of such climate information could be essential in minimising the risk of such hazards. Certainly, an improvement in hydromet service delivery has the potential to further encourage comprehensive disaster risk management elements, such as those advocated by CDEMA. In this regard, it can allow countries to better understand and implement locally-appropriate preparedness and risk communication strategies.

With this additional knowledge, for instance, more informed decisions can be made regarding local, national and regional strategies addressing monitoring and warning systems for extreme weather events. As well, fostering dialogue between national weather offices and the sectors they serve can encourage improved ownership and sharing of weather and climate information which can improve hazard risk identification and monitoring, and better information in the hands of relevant sectors can prompt a suite of prevention and mitigation strategies. Lastly, it is noted that such information can stimulate tailored and more effective disaster response, rescue and recovery.

In the context of wider preparedness efforts and building resilience, there are also a slew of related potential benefits of improving climate information (availability, delivery and use), which may include:

- *Better awareness of climate information*. Perhaps an overlooked but fundamental benefit of improving climate information availability and delivery is the inherent recognition and understanding of climate information, including what information may be available and appropriately relevant for the Caribbean islands (Dookie et al., 2019). While it is understood that "climate is the typical weather condition experienced at any location or area" (BOM, n.d.), an assessment of the local information that entail this typified climate for the region—including rainfall, wind, temperature, fog, thunder, humidity, pressure, ocean temperatures and sunshine data, in addition to the "type, frequency, duration, and intensity of weather events such as heat waves, cold spells, storms, floods, and droughts" (US-EPA, n.d.)—can assist researchers in providing forecasts and outlooks into seasonal and longer-term climate patterns. Monitoring climate information locally could also enable national agencies to better and more quickly be aware of and understand sudden changes in hazards, such as rapid intensification of storms.

- *Utility of remotely-sensed data.* The emerging role of remotely-sensed climate information for data-scarce regions highlights a possible solution to the limited availability of localised data. By being aware of the on-the-ground climate information that currently exists, this may offer some insight into which data could and should be sourced from satellite products, encouraging an informed database of climate variables and improved forecast outputs. It should be noted that further to the awareness of data, it is important to evaluate the utility of such climate information to ascertain the appropriateness and relevance of use and context, rather than assume its feasibility for a particular region or context.
- *Developing enhanced data repositories.* Following an improved awareness of climate information, both local and remotely-sensed, is the building of national and regional data repositories which should continuously be monitored and updated based on new information and technology. In this vein, information managers can track changes in the efficiency of data collection, indicate data gaps, understand changes in the accuracy of estimates, and signal the need for which additional information may be needed for policy directives. For the case of a small island, and other developing nations which have probably always relied on external data sources, this is a formidable step in moving away from being passive to active data users and perhaps providers.
- *Improving climate services.* It is noted that climate services "involve the production, translation, transfer, and use of climate knowledge and information in climate-informed decision making and climate-smart policy and planning" and "ensure that the best available climate science is effectively communicated with agriculture, water, health, and other sectors, to develop and evaluate adaptation strategies" (Climate Services Partnership, 2018). The strategic development of improved information-sharing and climate service partnerships and dialogue between local disaster and meteorological offices together with regional counterparts and agencies, boundary organisations (including the Caribbean Institute for Meteorology and Hydrology), local and international universities and research institutes (including Columbia University's International Research Institute for Climate and Society), and the US National Hurricane Center, makes "decision-relevant scientific information" accessible and timely in order to cope with current climate variability and limit the economic and social damage caused by climate-related disaster" (Climate Services Partnership, 2018). Not only does

improved climate information and climate services delivery "allow society to build resilience to future change", but also "take advantage of opportunities provided by favorable conditions" (CSP, 2018). See Mahon et al. (2018) and Mahon et al. (2019) for more details on the role of climate services within the region.

- *Rationalising 'climate science translators'*: As the awareness and development of climate information improves, the encouragement of 'climate science translators' becomes apparent and critical. While this necessarily includes the empowering of local communities with relevant and understandable information (see Parris et al., 2016), such translation refers to people and tools which can purposefully balance complex details of climate science with policy/practitioner-oriented knowledge, needs, and outputs. Streamlining the translation of climate science has the potential to enrich and better encourage the awareness and use of climate information, as well as improved targeting of climate services, and can enhance local utilisation by a wide range of users.[7]
- *Guiding local disaster risk coordination and communication*. Improving climate information availability, delivery and use helps build resilience also by being an integral part of guiding local disaster risk communication. A better sense of climate information can encourage improved coordination and communication between and amongst weather, disaster and other national and international agencies as well as the local public. The combination of relevant available climate information and the knowledge of local needs and risks can lead to enhanced locally appropriate risk reduction strategies and communication mechanisms as sub-national agencies can better understand, target and respond to impending disaster risk. Building local capacity and ownership is critical, and perhaps there is an increased role of community leaders and civil society organisations, together with institutional research and support, to encourage community participation and awareness of disaster risk.
- *Input into forecast-based early action*. The ODI shares that "FbA [forecast-based early action] programming uses forecasts to provide earlier support to at-risk communities before a disaster occurs, and the limited evidence on the costs and benefits of anticipatory action would seem to suggest that even a false early response is more than offset by the cost of a late response" (Wilkinson et al., 2018). Improving climate information and hydromet service delivery, as such, boosts preparedness and local resilience to disasters by potentially anticipating and providing for such hazard risks before onset.
- *Utility in multiple sectors*: Beyond the traditional use by national meteorological services in the present context of rainfall, temperature and storm monitoring, climate information is generally utilised to prepare and adapt for climate variability and change. As well, there are a large number of broad sectors that have been identified which may utilise climate information knowledge. Soares et al.

[7]The authors thank Dr. Markus Enenkel of Harvard Humanitarian Initiative for his discussion on this topic.

(2018) have identified and offer detailed information on a range of sectors which utilise such information including agriculture, forestry, energy, water, tourism, insurance, health, emergency services and transport sectors. Of note, index insurance may better serve some communities as it is insurance provision that "pays out benefits on the basis of a predetermined index (e.g. rainfall level) for loss of assets and investments, primarily working capital, resulting from weather and catastrophic events, without requiring the traditional services of insurance claims assessors" (IFC, 2018). Also, historical, present and projections of climate information can be utilised to better offer guidance within site selection for buildings and settlements, including post-disaster construction, land use zoning (for instance, taking into account flood maps alongside areas of potential sea level rise).

- *A critical input for inter-disciplinary scholarly research.* It is also evident that such an informed database of climate variables is essential for a plethora of related studies. Climate information could be coupled with other socio-economic indicators to better understand the vulnerability of the country to disaster risk vis-à-vis climate change and variability. As well, including the role and influence of climate information on socio-economic outcome factors such as "agricultural output, industrial output, labour productivity, energy demand, health, conflict, and economic growth, among other outcomes" (Dell et al., 2014) may offer more specific recommendations within a national sustainable development context. This not only assists researchers in understanding the inter-linkages between climate and the economy, but can play a critical part in underscoring the need for additional research, policy-making and implementation as well as likely resources required for further comprehensive review.
- *Informed science-policy dialogue.* As well, climate information can certainly inform national policy concerns. By specifically improving the science-policy dialogue and moving towards science-based policy determination, national sustainable development frameworks and holistic climate change adaptation strategies could be vastly enhanced as such a focus could highlight local vulnerabilities and policy gaps. As well, the sharing of such information between and amongst sectors and ministries could create important synergies and interactive decision-making, such as through the creation of cross-sectoral and inter-ministerial platforms which not only benefit from a plethora of backgrounds, understanding and perspectives, but which can work together to better manage and utilise scarce financial, technical and human resources.
- *Encouraging national and regional learning.* The Caribbean Institute for Metrology and Hydrology (CIMH) in Barbados has been generously pushing the boundary outward regarding the awareness and teaching of climate products and forecasts by national meteorological offices. Furthermore, their biannual Caribbean Climate Outlook Forums attract new knowledge and information-sharing not only among meteorologists, but specialists within the water, agriculture and tourism sectors, for example, as well as national and regional disaster

management experts, and regional development representatives. Such efforts should be broadened and expanded, if not mandated, where possible, perhaps to include more members of the national planning ministries so as to ensure the priority offered to the improved awareness and utilisation of climate information across the region. See Guido et al. (2016) for details on the value of the Caribbean Climate Outlook Forums.

- *Narrowing the knowledge gap in small islands.* Such benefits of developing and improving the use of climate information are not just relevant to Dominica or Caribbean islands, but lessons learnt here can certainly inform similar interventions for other small islands. As many other islands face similar vulnerability challenges, the awareness of how this information could be collated and used might be helpful for other nations to develop similar insights. However, it is noted that small nations are not a homogenous group but unique in many ways, so encouraging preparedness through improved climate information availability and delivery can assist tailored resilience-building in the wake of climate change and variability in small island states.

5 The Way Forward

The awareness and development of climate information, weather and climate forecasting, and hydromet service delivery is not a silver bullet to prevent hydromet disasters from occurring in the Caribbean. Unfortunately, certain large events, as well as repeated and multiple disaster events, will certainly take a toll on the small islands of the region. However, investing in weather forecasting and modelling, as well as the availability, access to and delivery of the climate information base, in addition to the development of associated climate services, can be a fundamental if not critical step in the direction of encouraging the adaptive and transformative nature of communities and economies within the Caribbean region to not only bounce back from but become more resilient to the impacts of natural hazard-based disasters within a changing climate.

It is understood that there is an investment needed to "forecast for catastrophes," requiring the update of software and technology and development of local technical capacity and human capability and capital to better understand, store and assess climate information towards adequately and efficiently improving forecasts and early warning of natural hazards. While such efforts towards forecast improvement are noted to be costly, the World Bank estimates that "every one dollar invested has the potential of generating at least three dollars' worth of benefits in weather and climate services – a win-win" (World Bank, 2017). Moreover, the Bank suggests that there is an increasing number of funds available to support such initiatives. In addition, to effectively utilise this information, improved coordination and communication between and amongst weather, disaster and other national and international agencies and institutions, as well as the local public, is needed.

These requirements underscore the importance of political understanding to appreciate such inter-disciplinary needs within the medium and long-term, as well as increased political motivation to ensure the uptake and success of these policies. It is possible that, given the recent impacts of storms within the region, there is a heightened sense of attention to disaster preparedness and resilience strategies, which could encourage new disaster policy directives and resource allocation planning at the national, regional and international settings, benefiting the Caribbean as well as other small island states. It is hoped that policymakers may be aware of and learn from the experiences of Dominica, among others, to better understand, monitor and effectively respond to disaster risk, especially at a time when climate information is readily available as a potentially powerful tool to better inform resilience and risk reduction strategies.

Appendix

Selected Disaster-Related Definitions (UNDRR)[8]

- **Affected**: People who are affected, either directly or indirectly, by a hazardous event. Directly affected are those who have suffered injury, illness or other health effects; who were evacuated, displaced, relocated or have suffered direct damage to their livelihoods, economic, physical, social, cultural and environmental assets. Indirectly affected are people who have suffered consequences, other than or in addition to direct effects, over time, due to disruption or changes in economy, critical infrastructure, basic services, commerce or work, or social, health and psychological consequences.
- **Build back better**: The use of the recovery, rehabilitation and reconstruction phases after a disaster to increase the resilience of nations and communities through integrating disaster risk reduction measures into the restoration of physical infrastructure and societal systems, and into the revitalization of livelihoods, economies and the environment.
- **Disaster**: A serious disruption of the functioning of a community or a society at any scale due to hazardous events interacting with conditions of exposure, vulnerability and capacity, leading to one or more of the following: human, material, economic and environmental losses and impacts.
- **Disaster impact:** the total effect, including negative effects (e.g., economic losses) and positive effects (e.g., economic gains), of a hazardous event or a disaster. The term includes economic, human and environmental impacts, and may include death, injuries, disease and other negative effects on human physical, mental and social well-being.

[8] Source: https://www.undrr.org/terminology.

- **Disaster management**: The organization, planning and application of measures preparing for, responding to and recovering from disasters.
- **Disaster risk**: The potential loss of life, injury, or destroyed or damaged assets which could occur to a system, society or a community in a specific period of time, determined probabilistically as a function of hazard, exposure, vulnerability and capacity.
- **Disaster risk information**: Comprehensive information on all dimensions of disaster risk, including hazards, exposure, vulnerability and capacity, related to persons, communities, organizations and countries and their assets.
- **Disaster risk management**: Disaster risk management is the application of disaster risk reduction policies and strategies to prevent new disaster risk, reduce existing disaster risk and manage residual risk, contributing to the strengthening of resilience and reduction of disaster losses.
- **Disaster risk reduction**: Disaster risk reduction is aimed at preventing new and reducing existing disaster risk and managing residual risk, all of which contribute to strengthening resilience and therefore to the achievement of sustainable development.
- **Exposure**: The situation of people, infrastructure, housing, production capacities and other tangible human assets located in hazard-prone areas.
- **Hazard**: A process, phenomenon or human activity that may cause loss of life, injury or other health impacts, property damage, social and economic disruption or environmental degradation.
- **Preparedness**: The knowledge and capacities developed by governments, response and recovery organizations, communities and individuals to effectively anticipate, respond to and recover from the impacts of likely, imminent or current disasters.
- **Resilience**: The ability of a system, community or society exposed to hazards to resist, absorb, accommodate, adapt to, transform and recover from the effects of a hazard in a timely and efficient manner, including through the preservation and restoration of its essential basic structures and functions through risk management.
- **Vulnerability**: The conditions determined by physical, social, economic and environmental factors or processes which increase the susceptibility of an individual, a community, assets or systems to the impacts of hazards.

References

Albala-Bertrand, J. M. (1993). *Political economy of large natural disasters: With special reference to developing countries*. OUP Catalogue.

Australian Bureau of Meteorology (BOM). (n.d.). *Climate and past weather*. Retrieved from http://www.bom.gov.au/climate/

Becerra, O., Cavallo, E., & Noy, I. (2014). Foreign aid in the aftermath of large natural disasters. *Review of Development Economics, 18*(3), 445–460.

Benson, C., & Clay, E. (2004). *Understanding the economic and financial impacts of natural disasters*. The World Bank. Retrieved from https://elibrary.worldbank.org/doi/abs/10.1596/0-8213-5685-2

Bertinelli, L., Mohan, P., & Strobl, E. (2016). Hurricane damage risk assessment in the Caribbean: An analysis using synthetic hurricane events and nightlight imagery. *Ecological Economics, 124*, 135–144.

Briguglio, L. (1995). Small island developing states and their economic vulnerabilities. *World Development, 23*(9), 1615–1632.

Briguglio, L. (2014). *A vulnerability and resilience framework for small states*. Report prepared for the Commonwealth Secretariat. Retrieved from https://www.um.edu.mt/library/oar/handle/123456789/18015

Campbell, J. D., Taylor, M. A., Stephenson, T. S., Watson, R. A., & Whyte, F. S. (2011). Future climate of the Caribbean from a regional climate model. *International Journal of Climatology, 31*(12), 1866–1878.

Caribbean Disaster Emergency Management Agency (CDEMA). (n.d.). *What is CDEMA?* Retrieved March 22, 2018, from http://www.cdema.org/index.php?option=com_content&view=article&id=89&Itemid=79

Carrara, A., Guzzetti, F., Cardinali, M., & Reichenbach, P. (1999). Use of GIS technology in the prediction and monitoring of landslide hazard. *Natural Hazards, 20*(2-3), 117–135.

Cavallo, E., Powell, A., & Becerra, O. (2010). Estimating the direct economic damages of the earthquake in Haiti. *The Economic Journal, 120*(546), F298–F312.

Cavallo, E. A., & Noy, I. (2009). *The economics of natural disasters: A survey*. Inter-American Development Bank Working Paper Series: No. IDB-WP-124. Washington, DC: Inter-American Development Bank. Retrieved from https://papers.ssrn.com/sol3/papers.cfm?abstract_id=1817217

Center for Strategic and International Studies (CSIS). (2015). *Economics of disaster prevention: Measuring the costs and benefits of disaster risk reduction*. Transcript of session by CSIS-LSU Series on Disaster Management and Emergency Response. May 23, 2011. Washington, D.C: Center for Strategic and International Studies.

CIA World Factbook. (2018). *Dominica*. Retrieved April 22, 2018, from https://www.cia.gov/library/publications/the-world-factbook/geos/do.html

Climate Services Partnership. (2018). *What are climate services?* Retrieved April 22, 2018, from http://www.climateservices.org/about-us/what-are-climate-services/

Collymore, J. (2011). Disaster management in the Caribbean: Perspectives on institutional capacity reform and development. *Environmental Hazards, 10*(1), 6–22.

Creutin, J. D., & Borga, M. (2003). Radar hydrology modifies the monitoring of flash-flood hazard. *Hydrological Processes, 17*(7), 1453–1456.

Dani, S. (2017). *A monster attacked the Caribbean: Time to rebuild thinking about the next one*. World Bank Blog, 5 October 2017. Retrieved from http://blogs.worldbank.org/latinamerica/monsterattacked-caribbean-time-rebuild-thinking-about-next-one

Davoli, M. Z. (2012). Disaster management structures in the Caribbean. In: *Mental health and psychosocial support in disaster situations in the Caribbean: Core knowledge for emergency preparedness and response*. A joint publication of the mental health program and the area on emergency preparedness and disaster relief of the pan American Health Organization. Retrieved from http://www.who.int/iris/handle/10665/173271

Dell, M., Jones, B. F., & Olken, B. A. (2014). What do we learn from the weather? The new climate-economy literature. *Journal of Economic Literature, 52*(3), 740–798.

Desai, B., Maskrey, A., Peduzzi, P., De Bono, A., & Herold, C. (2015). *Making development sustainable: The future of disaster risk management, global assessment report on disaster risk reduction*. Geneva: United Nations Office for Disaster Risk Reduction (UNDRR, formerly UNISDR).

Dominica News Online. (2015). *Focus is on search and rescue efforts following Erika, says PM, 27 August*. Retrieved from dominicanewsonline.com/news/homepage/news/accidents-tragedy/focus-is-search-and-rescue-efforts-following-erika-says-pm/

Donnelly, J. P., & Woodruff, J. D. (2007). Intense hurricane activity over the past 5,000 years controlled by El Niño and the West African monsoon. *Nature, 447*(7143), 465–468.

Dookie, D. S., Enenkel, M., & Spence, J. (2019). From science to science-based: Using state-of-the-art climate information to strengthen DRR in Small Island States. *Strengthening Disaster Resilience in Small States: Commonwealth Perspectives, 13*.

Felbermayr, G., & Gröschl, J. (2014). Naturally negative: The growth effects of natural disasters. *Journal of Development Economics, 111*, 92–106.

Global Facility for Disaster Reduction and Recovery (GFDRR). (2017a). *ACP-EU Natural Disaster Risk Reduction Program - Focus day on post disaster response and recovery frameworks*. Retrieved from https://www.gfdrr.org/sites/default/files/publication/DOMINICA%20-%20ACP-EU%20NDRR%20Focus%20Day%20presentation%20-%209%20June%202017.pdf

Global Facility for Disaster Reduction and Recovery (GFDRR). (2017b). *Dominica*. Retrieved from https://www.gfdrr.org/dominica

Global Humanitarian Assistance (GHA). (2013). *Global Humanitarian Assistance Report 2013*. Bristol: Global Humanitarian Assistance. Retrieved July 4, 2016, from http://www.globalhumanitarianassistance.org/report/gha-report-2013

Global Humanitarian Assistance (GHA). (2015). *Global Humanitarian Assistance Report 2015*. Bristol: Global Humanitarian Assistance. Retrieved July 4, 2016, from http://www.globalhumanitarianassistance.org/report/gha-report-2015

Government of the Commonwealth of Dominica. (2017). *Post-Disaster Needs Assessment (PDNA) Hurricane Maria September 18, 2017*. A Report by the Government of the Commonwealth of Dominica. Retrieved from https://resilientcaribbean.caricom.org/wp-content/uploads/2017/11/DOMINICA-EXECUTIVE-SUMMARY.pdf

Government of the Commonwealth of Dominica. (2020). *Dominica Climate Resilience and Recovery Plan 2020 – 2030*. Retrieved from http://dominica.gov.dm/images/documents/CRRP-Final-042020.pdf

Guha-Sapir, D., Below, R., & Hoyois, P. (2020). *EM-DAT: International Disaster Database* – www.emdat.be. Brussels: Université Catholique de Louvain.

Guido, Z., Rountree, V., Greene, C., Gerlak, A., & Trotman, A. (2016). Connecting climate information producers and users: Boundary organization, knowledge networks, and information brokers at Caribbean climate outlook forums. *Weather, Climate, and Society, 8*(3), 285–298.

Hallegatte, S., & Przyluski, V. (2010). *The economics of natural disasters: Concepts and methods*. The World Bank. Retrieved from http://elibrary.worldbank.org/doi/abs/10.1596/1813-9450-5507

Heger, M., Julca, A. & Paddison, O. (2008). *Analysing the impact of natural hazards in small economies: The Caribbean case* (No. 2008/25), WIDER Research Paper.

Hillier, D., & Nightingale, K. (2013). *How disasters disrupt development: Recommendations for the post-2015 development framework*. Oxfam Great Britain for Oxfam International.

Hochrainer, S. (2009). *Assessing the macroeconomic impacts of natural disasters: Are there any?* The World Bank. Retrieved from https://elibrary.worldbank.org/doi/abs/10.1596/1813-9450-4968

Hsiang, S. M. (2010). Temperatures and cyclones strongly associated with economic production in the Caribbean and Central America. *Proceedings of the National Academy of sciences, 107*(35), 15367–15372.

Hsiang, S. M., & Jina, A. S. (2014). *The causal effect of environmental catastrophe on long-run economic growth: Evidence from 6,700 cyclones* (No. w20352). National Bureau of Economic Research.

Integrated Research on Disaster Risk (IRDR) & International Council for Science (ICSU). (2014). *Issue Brief: Disaster risk reduction and sustainable development*. Prepared by experts associated with the Integrated Research on Disaster Risk (IRDR) programme for the Seventh Session of the UN General Assembly Open Working Group on Sustainable Development Goals, New York, 6–10 January 2014. Retrieved July 8, 2016, from https://sustainabledevelopment.un.org/getWSDoc.php?id=2133

Inter-American Development Bank (IDB), International Monetary Fund (IMF), Organisation of American States (OAS) & the World Bank. (2005). *The economics of disaster mitigation in the Caribbean – Quantifying the benefits and costs of mitigating natural hazard losses*. Working Paper by Inter-American Development Bank, International Monetary Fund, Organisation of American States, and the World Bank. Washington DC: Organisation of American States. Retrieved from http://www.oas.org/dsd/MinisterialMeeting/Documents/economics_disaster_mitigation_caribb_full_report.pdf

Intergovernmental Panel on Climate Change (IPCC). (2014). *Climate Change 2014: Impacts, adaptation, and vulnerability*. Contribution of Working Group II to the Fifth Assessment Report of the Intergovernmental Panel on Climate Change. In: C. B. Field, V. R. Barros, D. J. Dokken, K. J. Mach, M. D. Mastrandrea, T. E. Bilir, M. Chatterjee, K. L. Ebi, Y. O. Estrada, R. C. Genova, B. Girma, E. S. Kissel, A. N. Levy, S. MacCracken, P. R. Mastrandrea & L. L White (eds.), Cambridge, United Kingdom and New York, NY, USA: Cambridge University Press. Retrieved from http://www.ipcc.ch/report/ar5/wg2/

International Federation of Red Cross and Red Crescent Societies (IFRC). (2017). *Dominica: Hurricane Maria*. Information Bulletin, 25 September. Retrieved from https://reliefweb.int/sites/reliefweb.int/files/resources/IB25092017.pdf

International Federation of Red Cross and Red Crescent Societies (IFRC). (2018). *Disaster preparedness*. Retrieved from http://media.ifrc.org/ifrc/what-we-do/disaster-and-crisis-management/disaster-preparedness/

International Finance Corporation (IFC). (2018). *Index insurance – Frequently asked questions*. Retrieved from https://www.ifc.org/wps/wcm/connect/industry_ext_content/ifc_external_corporate_site/industries/financial+markets/retail+finance/insurance/index+insurance+-+frequently+asked+questions

Joyce, K. E., Belliss, S. E., Samsonov, S. V., McNeill, S. J., & Glassey, P. J. (2009). A review of the status of satellite remote sensing and image processing techniques for mapping natural hazards and disasters. *Progress in Physical Geography, 33*(2), 183–207.

Kahn, M. E. (2005). The death toll from natural disasters: The role of income, geography, and institutions. *Review of Economics and Statistics, 87*(2), 271–284.

Karmalkar, A. V., Taylor, M. A., Campbell, J., Stephenson, T., New, M., Centella, A., Benzanilla, A., & Charlery, J. (2013). A review of observed and projected changes in climate for the islands in the Caribbean. *Atmósfera, 26*(2), 283–309.

Kellett, J., & Caravani, A. (2013). *Financing disaster risk reduction. A 20 year story of international aid*. London: Overseas Development Institute (ODI). Retrieved from https://nab.vu/sites/default/files/nab/documents/27/09/2013%20-%2009%3A40/85741.pdf

Kirton, M., (2013). *Caribbean regional disaster response and management mechanisms: Prospects and challenges*. London: Brookings - London School of Economics Project on Internal Displacement. Retrieved from http://www.brookings.edu/research/reports/2013/07/caribbean-regional-organizations-disasters

Mahon, R., Farrell, D., Cox, S. A., Trotman, A., van Meerbeeck, C., & Barnwell, G. (2018). Climate services and Caribbean resilience: A historical perspective. *Social and Economic Studies, 67*(2/3), 239–324.

Mahon, R., Greene, C., Cox, S. A., Guido, Z., Gerlak, A. K., Petrie, J. A., Trotman, A., Liverman, D., Van Meerbeeck, C. J., Scott, W., & Farrell, D. (2019). Fit for purpose? Transforming national meteorological and hydrological services into national climate service centers. *Climate Services, 13*, 14–23.

McBean, G. A. (2012). Integrating disaster risk reduction towards sustainable development. *Current Opinion in Environmental Sustainability, 4*(1), 122–127.

Mochizuki, J., Mechler, R., Hochrainer-Stigler, S., Keating, A., & Williges, K. (2014). Revisiting the 'disaster and development' debate – Toward a broader understanding of macroeconomic risk and resilience. *Climate Risk Management, 3*, 39–54.

Moore, W., Elliott, W., & Lorde, T. (2016). Climate change, Atlantic storm activity and the regional socio-economic impacts on the Caribbean. *Environment, Development and Sustainability*, 1–20.

National Emergency Planning Organisation (NEPO-Dominica). (2001). *National Disaster Plan 2001*. Retrieved from http://odm.gov.dm/images/docs/national_disaster_plan.pdf

Noy, I. (2009). The macroeconomic consequences of disasters. *Journal of Development Economics, 88*(2), 221–231.

Otero, R. C., & Marti, R. Z.. (1995). The impacts of natural disasters on developing economies: Implications for the international development and disaster community. *Disaster Prevention for Sustainable Development: Economic and Policy Issues. Washington DC, World Bank*, 11-40. Retrieved from http://www.disaster-info.net/lideres/spanish/mexico2003/biblio/eng/doc6538/doc6538-a.pdf

Parris, A. S., Garfin, G. M., Dow, K., Meyer, R., & Close, S. L. (Eds.). (2016). *Climate in context: Science and society partnering for adaptation*. John Wiley & Sons.

Pasch, R. J., & Penny, A. B. (2016). *Tropical Storm Erika, 24–28 August 2015*. National Hurricane Center Tropical Cyclone Report. Retrieved from www.nhc.noaa.gov/data/tcr/AL052015_Erika.pdf

Pelling, M., Özerdem, A., & Barakat, S. (2002). The macro-economic impact of disasters. *Progress in Development Studies, 2*(4), 283–305.

Peterson, T. C., Taylor, M. A., Demeritte, R., Duncombe, D. L., Burton, S., Thompson, F., Porter, A., Mercedes, M., Villegas, E., Semexant Fils, R., & Klein Tank, A. (2002). Recent changes in climate extremes in the Caribbean region. *Journal of Geophysical Research: Atmospheres, 107* (D21), 1–9.

Raddatz, C. (2009). *The wrath of God: Macroeconomic costs of natural disasters*. The World Bank.

Raschky, P. A. (2008). Institutions and the losses from natural disasters. *Natural Hazards and Earth System Sciences, 8*(4), 627–634.

Shreve, C. M., & Kelman, I. (2014). Does mitigation save? Reviewing cost-benefit analyses of disaster risk reduction. *International journal of Disaster Risk Reduction, 10*, 213–235.

Singh, C., Daron, J., Bazaz, A., Ziervogel, G., Spear, D., Krishnaswamy, J., Zaroug, M., & Kituyi, E. (2018). The utility of weather and climate information for adaptation decision-making: Current uses and future prospects in Africa and India. *Climate and Development 10*(5), 389–405.

Soares, M. B., Alexander, M., & Dessai, S. (2018). Sectoral use of climate information in Europe: A synoptic overview. *Climate Services, 9*, 5–20.

Sparks, D. (2012). *Aid investments in disaster risk reduction-rhetoric to action*. Bristol: Global Humanitarian Assistance. Retrieved from http://www.globalhumanitarianassistance.org/wp-content/uploads/2012/10/Aid-investments-in-disaster-risk-reduction-rhetoric-to-action-Dan-Sparks1.pdf

Stephenson, T. S., Vincent, L. A., Allen, T., Van Meerbeeck, C. J., McLean, N., Peterson, T. C., Taylor, M. A., Aaron-Morrison, A. P., Auguste, T., Bernard, D., & Boekhoudt, J. R. (2014). Changes in extreme temperature and precipitation in the Caribbean region, 1961–2010. *International Journal of Climatology, 34*(9), 2957–2971.

Strobl, E. (2012). The economic growth impact of natural disasters in developing countries Evidence from hurricane strikes in the Central American and Caribbean regions. *Journal of Development economics, 97*(1) 130–141.

Strömberg, D. (2007). Natural disasters, economic development, and humanitarian aid. *Journal of Economic Perspectives, 21*(3), 199–222.

Sun Dominica. (2016). *Lessons learnt from Erika*. Posted 27 August. Retrieved from http://sundominica.com/articles/lessons-learnt-from-erika-3747/

Tanner, T., Lovell, E., Wilkinson, E., Chesquiere, F., Reid, R. & Rajput, S. (2015). *Why all development finance should be risk-informed.* London: Overseas Development Institute (ODI) and Global Facility for Disaster Reduction and Recovery (GFDRR). Retrieved from https://www.odi.org/sites/odi.org.uk/files/odi-assets/publications-opinion-files/9730.pdf

Tartaglione, C. A., Smith, S. R., & O'Brien, J. J. (2003). ENSO impact on hurricane landfall probabilities for the Caribbean. *Journal of Climate, 16*(17), 2925–2931.

Taylor, M. A., Stephenson, T. S., Chen, A. A., & Stephenson, K. A. (2012). Climate change and the Caribbean: Review and response. *Caribbean Studies*, 169–200.

Thomalla, F., Downing, T., Spanger-Siegfried, E., Han, G., & Rockström, J. (2006). Reducing hazard vulnerability: Towards a common approach between disaster risk reduction and climate adaptation. *Disasters, 30*(1), 39–48.

Thomson Reuters Foundation. (2018). *Hurricane-hit Dominica hurries to prepare for next storm season*. Posted January 31, 2018. https://www.reuters.com/article/us-caribbeanhurricane-dominica/hurricane-hit-dominica-hurries-to-prepare-for-next-storm-season-idUSKBN1FK30D

Thomas, A., Baptiste, A., Martyr-Koller, R., Pringle, P., & Rhiney, K. (2020). Climate change and Small Island Developing States. *Annual Review of Environment and Resources, 45*, 1–27.

Toya, H., & Skidmore, M. (2007). Economic development and the impacts of natural disasters. *Economics Letters, 94*(1), 20–25.

Trenberth, K. E., & Shea, D. J. (2006). Atlantic hurricanes and natural variability in 2005. *Geophysical Research Letters, 33*(12).

UN Economic Commission for Latin America and the Caribbean (UNECLAC). (2010). *Analysis of extreme events in the Caribbean 1990 – 2008*. LC/CAR/L.254. Retrieved from http://repositorio.cepal.org/bitstream/handle/11362/38660/1/LCCARL254_en.pdf

UN Economic Commission for Latin America and the Caribbean (UNECLAC). (2011). *Study on the vulnerability and resilience of Caribbean Small Island Developing States (SIDS)*. LC/CAR/L.354. Retrieved from http://www.cepal.org/publicaciones/xml/4/45364/LCARL.354.pdf

United Nations (UN) & World Bank. (2010). *Natural hazards, unnatural disasters – The economics of effective prevention.* Washington, DC: The International Bank for Reconstruction and Development/The World Bank. Retrieved from https://www.gfdrr.org/sites/gfdrr/files/publication/NHUD-Report_Full.pdf

United Nations Office for Disaster Risk Reduction (UNDRR). (2015a). *Disaster risk reduction and resilience in the 2030 agenda for sustainable development.* New York: United Nations Office for Disaster Risk Reduction - New York UNHQ Liaison Office. Retrieved from: https://www.undrr.org/publication/disaster-risk-reduction-and-resilience-2030-agenda-sustainable-development

United Nations Office for Disaster Risk Reduction (UNDRR). (2015b). *Economic aspects of DRR.* Background document for working session at the UN World Conference on Disaster Risk Reduction, 14–18 March, Sendai. Retrieved from http://www.wcdrr.org/wcdrr-data/uploads/871/B.%20Brief%20issue%20paper.pdf

United Nations Office for Disaster Risk Reduction (UNDRR). (2015c). *The Pocket GAR 2015. Making development sustainable: The future of disaster risk management.* Geneva: United Nations Office for Disaster Risk Reduction (UNDRR, formerly UNISDR). Retrieved from https://www.preventionweb.net/english/hyogo/gar/2015/en/gar-pdf/GAR15_Pocket_EN.pdf

United Nations Office for Disaster Risk Reduction (UNDRR) (n.d.). *What is disaster risk reduction?* United Nations Office for Disaster Risk Reduction (UNDRR, formerly UNISDR). Retrieved from https://www.undrr.org/terminology

United States Environmental Protection Agency (US-EPA). (n.d.). *Climate change indicators in the United States – Weather and climate.* Retrieved from http://www.epa.gov/climate/climatechange/science/indicators/weather-climate/index.html

United Nations (UN). (n.d.) *Disaster risk reduction. Sustainable development knowledge platform.* Retrieved July 4, 2016, from https://sustainabledevelopment.un.org/topics/disasterriskreduction

Von Peter, G., Von Dahlen, S., & Saxena, S. C. (2012). *Unmitigated disasters? New evidence on the macroeconomic cost of natural catastrophes*. Bank for International Settlements Working Paper No. 394, Basel, Bank for International Settlements.

Vorhies, F. (2012). *The economics of investing in disaster risk reduction*. Geneva: UN International Strategy for Disaster Reduction Working paper based on a review of the current literature commissioned by UNDRR, formerly UNISDR, Geneva, Secretariat to the UN International Strategy for Disaster Reduction. Retrieved from https://www.preventionweb.net/posthfa/documents/drreconomicsworkingpaperfinal.pdf

Walker, L. R., Lodge, D. J., & Waide, R. B. (1991). An introduction to hurricanes in the Caribbean. *Biotropica, 23*(4), 313–316.

Wilkinson, E., Weingärtner, L., Choularton, R., Bailey, M., Todd, M., Kniveton, D., & Cabot Venton, C. (2018). *Forecasting hazards, averting disasters: Implementing forecast-based early action at scale*. Overseas Development Institute (ODI). Retrieved from https://www.odi.org/publications/11069-forecasting-hazards-averting-disasters-implementing-forecast-based-early-action-scale

World Bank. (2013). *World Development Report 2014. Risk and opportunity. Managing risk for development*. Washington DC: International Bank for Reconstruction and Development/The World Bank.

World Bank. (2014). *Project Information Document: Third phase disaster vulnerability reduction Adaptable Program Loan (APL) for Dominica (P129992)*. Retrieved from http://documents.worldbank.org/curated/en/441531468025798014/text/PID-Appraisal-Print-P129992-03-06-2014-1394125006045.txt

World Bank. (2017). *Forecasting for catastrophes: How investment in weather services can save lives and grow economies*, 23 March. Retrieved from http://www.worldbank.org/en/news/feature/2017/03/23/forecasting-for-catastrophes-how-investment-in-weather-services-can-save-lives-and-grow-economies

World Health Organisation (WHO). (2002). *WHO/EHA training package: Disasters & emergencies – Definitions*. Addis Ababa: Panafrican Emergency Training Centre. Retrieved from http://apps.who.int/disasters/repo/7656.pdf

Zapata, R., & Madrigal, B. (2009). *Economic impact of disasters: Evidence from DALA assessments by ECLAC in Latin America and the Caribbean* (Vol. 117). United Nations Publications.

Sustainable Land Use Systems in Natural Resource Policies: The Role of Agroforestry in the Rio Conventions for Small Island Developing States

Marc Dumas-Johansen and Andreas Thulstrup

Abstract Small Island Developing States (SIDS) are amongst the most vulnerable countries in the world to both ongoing climate variability and future impacts of climate change. With growing populations and increasing demands for land and natural resources, SIDS are facing significant social, economic and environmental challenges, all of which are exacerbated by the negative impacts of climate variability and change. In order to address these challenges, a holistic and multi-sectoral approach is needed for the agriculture sectors. When promoted as part of sustainable land use regimes, agroforestry can provide social, economic and environmental services by providing improved food and nutrition, fodder, fuel and income in addition to the environmental services provided by the planting of trees. Furthermore, agroforestry can enable numerous climate change adaptation and mitigation synergies while simultaneously conserving biodiversity and restoring degraded and deforested lands. A review of the national reports to the three Rio conventions (UNFCCC, UNCCD and CBD) submitted by all SIDS was conducted in order to assess the particular climate vulnerabilities of the countries and the degree to which national governments in these countries view agroforestry as a viable strategy for absorbing the impacts of climate change.

Twenty-one SIDS refer to agroforestry in their national reports to one or several of the three Rio conventions (United Nations Convention to Combat Desertification—UNCCD, Convention on Biological Diversity—CBD and the United Nations Framework Convention on Climate Change—UNFCCC). The majority of SIDS referred to agroforestry in their National Biodiversity Strategy and Action Plans (NBSAPs) (19 SIDS), followed by National Action Plans (NAPs) (8 SIDS) and Nationally Determined Contributions (NDCs) (7 SIDS) with 11 SIDS referring to agroforestry in two or all three conventions (Comoros and Haiti).

Please note that the analysis and views expressed in this chapter are the personal views of the two authors and do not necessarily reflect the official position of the Green Climate Fund.

M. Dumas-Johansen (✉)
Agriculture and Food Security Specialist, Green Climate Fund, Incheon, South Korea

A. Thulstrup
Roskilde University, Roskilde, Denmark

S. Moncada et al. (eds.), *Small Island Developing States*, The World of Small States 9, https://doi.org/10.1007/978-3-030-82774-8_6

While many SIDS share specific vulnerabilities, their climates, ecosystems and current agroforestry practices differ substantially. The priorities of these countries range from large scale agroforestry development to the design of national agroforestry programs. The analysis shows that the benefits of agroforestry are not yet recognised, even in places where agro-ecological conditions are favourable, and that there is a need to promote stronger linkages between agroforestry and climate change adaptation and mitigation policies. The analysis of the benefits of agroforestry, and the degree to which such integrated systems are currently prioritized across the three Rio conventions can support the governments of SIDS in understanding the potential such systems have in addressing social, economic and environmental challenges faced by the agriculture sector. It also makes a good case for the development of an integrated approach to implementing all three conventions at the national level.

Keywords Agroforestry · Rio Conventions · Nationally determined contributions · National action programmes · National biodiversity strategies and action plans · Small island developing states · Natural hazards · Climate policy

1 Introduction

SIDS are considered some of the most vulnerable countries to climate change due to their size, isolation and exposure to natural hazards (Scandurra et al., 2018). Furthermore, they are known to be biodiversity hotpots, albeit often in severe need of conservation efforts (Myers et al., 2000). Land degradation has an enormous impact on SIDS which hinders their economic growth, human development and environmental sustainability (Rioux et al., 2017). Their small size combined with specific soil types, topography, climate and gaps in land use policies restrict a proper expansion of urban areas, agriculture, mining, forestry and other sectors and thus contributes to persisting conflicts between different land uses.

The Earth Summit in Rio de Janeiro, Brazil, in 1992 resulted in the formation of three United Nations Conventions: the Framework Convention on Climate Change (UNFCCC), the Convention on Biological Diversity (CBD) and the Convention on Combating Desertification (UNCCD) (Aronson & Alexander, 2013). The aims of the three conventions are to combat climate change, conserve biodiversity and reverse land degradation, respectively. Furthermore, the three conventions are interlinked in terms of their broader focus on sustainable environmental management. UNFCCC concentrates on reducing harmful human interaction with the climate and keeping greenhouse gas concentrations at a sound level, UNCCD aims at fighting desertification and land degradation (United Nations Convention to Combat Desertification (UNCCD), 2018) and the CBD seeks to highlight biological diversity as a cross cutting issue, essential to fostering sustainable development. The national reports to the three conventions present nationally determined and crafted solutions that address challenges pertaining to the three focal areas of the conventions. Hence, these national reports can serve as a very strong tool in understanding the national needs and challenges faced by each of the member countries to the conventions and how they aim to solve these challenges.

National governments of SIDS have become increasingly vocal and present on the front line in terms of addressing the broader challenges framed in the three conventions. Under the UNFCCC for example, SIDS have taken a lead role in proposing global solutions to cope with climate change impacts. For the first time in history, the Conference of the Parties to the UNFCCC 23 was under the presidency of a SIDS—Fiji—which resulted in strong efforts to continue the work of the Paris Agreement from 2015, culminating in the Fiji Momentum for Implementation (United Nations Climate Change Conference (COP23), 2017).

Integrated approaches are needed in order to tackle the numerous challenges which are faced by SIDS and presented in their national reports to the three conventions. For the agriculture sector, and given the particular constraints in terms of land use, a system that can address environmental and socioeconomic challenges simultaneously within the same unit of land is arguably an option worth considering. A pertinent example of such a system is agroforestry, a relatively new term first coined 40 years ago, which comprises a set of practices that go back more than several thousands of years (Catacutan et al., 2017). Agroforestry is practiced widely in various forms across the SIDS although little is known about the specific role it plays for the benefit of livelihoods (Drew, 2008).

We define agroforestry as the

> . . .collective name for land-use systems and technologies where woody perennials (trees, shrubs, palms, bamboos, etc.) are deliberately used on the same land-management units as agricultural crops and/or animals, in some form of spatial arrangement or temporal sequence. In agroforestry systems there are both ecological and economic interactions between the different components. Agroforestry can also be defined as a dynamic, ecologically based, natural resource management system that, through the integration of trees on farms and in the agricultural landscape, diversifies and sustains production for increased social, economic and environmental benefits for land users at all levels (Food and Agriculture Organization (FAO), 2018).

This chapter reviews evidence of the importance of agroforestry in SIDS and the degree to which it is prioritized in national reports to the three Rio conventions. This will help to support a better understanding of the current gaps and potential role agroforestry could play in the future in SIDS.

2 Agroforestry Systems in SIDS

The various types of agroforestry systems differ significantly across the SIDS. Table 1 illustrates an overview of seven SIDS and their characteristic agroforestry systems as well as the current challenges faced in implementing these systems. There is little data available on the comparison of agroforestry systems across SIDS, with the majority of studies focusing on national issues. In addition, there is a lack of data

Table 1 Selected agroforestry systems across SIDS

Country	Main system/purpose	Constraints/Challenges	References
Cabo Verde	Home gardens and multi-storey tree gardens	Loan funds a major issue	Johnson and Delgado (2003)
Comoros	Boundary and hedgerow plantings, orchards. Important species include Ylang-Ylang and Clove trees	Limited land holdings, poor access to markets, lack of diversity of seedlings in nurseries	Lufung (2016)
Haiti	Charcoal and timber based plantings		Murray and Bannister (2004)
Saint Lucia	Cocoa based intercropped with coconut, breadfruit and mango.	Agroforestry plots lacking regeneration, abandonment of plots, low rates of replanting	Walters and Hansen (2012)
Sao Tome e Principe	Cocoa and coffee intercropped with banana, taro and bordered by e.g. Coral trees		de Lima et al. (2013)
Timor Leste	Agroforestry systems centred around commercial species such as coffee, cocoa, candlenut, cashew and sandalwood species		McWilliam (2003)
Tonga	Multispecies system—usually short-term (Coconut, Ylang-Ylang, breadfruit, mango, avocado, citrus, yams etc.		Thaman et al. (2000)

The table was developed specifically for this chapter and provides an overview of selected agroforestry systems across the SIDS. The intention is not to generalise for all systems, but rather provide an overview of existing systems and their constraints and challenges

linking agroforestry, SIDS and the Rio conventions. The common denominator in Table 1 appears to be agroforestry systems characterized by mixed production and focusing on a few high value commercial species such as Ylang-Ylang or Cocoa. Table 1 does not represent all agroforestry systems across all SIDS but rather seeks to provide some examples of specific locations across various regions.

The challenges to agroforestry development as described for Cabo Verde, Comoros and Haiti in Table 1 are not new. Challenges to agroforestry development have been well documented by numerous studies (Catacutan et al., 2017). Catacutan et al. (2017) found that lack of explicit policies on agroforestry is the main hindering factor for agroforestry development. Other challenges included (1) little recognition of and attention paid to agroforestry among policy makers and government agencies; (2) shortage of market links for agroforestry products; (3) inadequate sharing of information on agroforestry; (4) little awareness and capacity within extension networks; (5) land and tree tenure insecurity; (6) lack of capital and land resources for farmers; (7) lack of planting stock; (8) few incentives and lack of financial and technical support to farmers; and (9) lack of explicit guidelines for implementing agroforestry development.

The competition for land between agroforestry and other uses is perhaps more widespread and intense in SIDS than elsewhere in the world. Given the already limited land available for agroforestry practices in SIDS, it is likely that future development will have a major impact on landscapes in general. Table 2 has been developed specifically for this chapter in order to illustrate the population growth, population density and availability of agricultural land per person from 1950 to 2050 in 38 selected SIDS. The tendencies clearly highlight that less land will be available in the future, which will have severe consequences for these countries. Antigua and Barbuda, Bahamas, Barbados, Comoros, Grenada, Maldives, Nauru, St. Kitts and Nevis, Saint Lucia and Trinidad and Tobago account for the most severe scenarios in 2050 with available agricultural land per person ranging from 0.02 hectare (ha) per person (Bahamas) to 0.08 ha per person (Comoros). This calculation is based on the recorded agricultural land in 2015, which in some cases will increase in order to allow for more agricultural production. However, the information in the table raises numerous concerns. The limited land available in 2050 for food production will still need to compete with other land uses and priorities.

If implemented successfully, agroforestry systems can cater to multiple needs simultaneously while contributing to food production. It can increase the usage of the available land to its maximum capacity without jeopardizing biodiversity conservation. It can serve to minimize land degradation and increases in GHG emissions. It can furthermore help restore and rehabilitate degraded lands which could help increase the agricultural land available in the future without jeopardizing forest land and other land uses.

Agroforestry offers numerous co-benefits that can be achieved simultaneously such as carbon sequestration and adaptation (synergies of adaptation and mitigation) (Anderson & Zerriffi, 2012; Beenhouwer et al., 2015; Jose, 2009; Kandji et al., 2006; Neufeldt et al., 2011; Schoeneberger, 2008), biodiversity conservation (Bhagwat et al., 2008; Jose, 2009; Nair, 2011) and land restoration (Hillbrand et al., 2017; Tougiani et al., 2009). Many recent studies have documented the untapped potential agroforestry has in carbon sequestration in particular (e.g. Mbow et al., 2014; Verchot et al., 2007; Zomer et al., 2016). Zomer et al. (2016) estimated that agroforestry at the global scale has the potential to sequester 45.3 PgC calculated on agricultural land with trees responsible for at least 75% of this figure. The study does not make specific references to SIDS and it is likely that due to the particular bio-geography of SIDS, the results referred to by the study might not be specific enough to fully account for the situation in SIDS.

Many agroforestry systems are also Integrated Food Energy Systems (IFES), i.e. farming systems that combine the production of food and biomass for energy generation on the same land (Bogdanski et al., 2010). Examples of such systems include the planting of *Acacia* trees alongside cassava and maize in Democratic Republic of Congo (Bogdanski et al., 2010) and the planting of *Gliricidia sepium* trees with pigeon pea and maize in southern Malawi (Chirwa et al., 2003). In addition to the products and services provided by agroforestry, these types of

Table 2 Relationship between population growth, density and agricultural land per person from 1950 to 2050

	Country size (km^2)	Agricultural land 2015 (ha)	Population, density and agricultural land per person (ha)											
			1950			*2015*			*2030*			*2050*		
Antigua and Barbuda	440	9000	46000	104.5	0.20	92000	209.1	0.10	105000	238.6	0.09	114000	259.1	0.08
Bahamas	11,010	14000	79000	7.2	0.18	388000	35.2	0.04	446000	40.5	0.03	489000	44.4	0.03
Bahrain	771	8600	116000	150.5	0.07	1377000	1786.0	0.01	1642000	2129.7	0.01	1822000	2363.2	0.00
Barbados	430	14000	211000	490.7	0.07	284000	660.5	0.05	290000	674.4	0.05	282000	655.8	0.05
Belize	22810	160000	69000	3.0	2.32	359000	15.7	0.45	472000	20.7	0.34	588000	25.8	0.27
Cabo Verde	4030	79000	178000	44.2	0.44	521000	129.3	0.15	614000	152.4	0.13	707000	175.4	0.11
Comoros	1861	133000	156000	83.8	0.85	788000	423.4	0.17	1081000	580.9	0.12	1502000	807.1	0.09
Cuba	104020	6240300	5920000	56.9	1.05	11390000	109.5	0.55	11237000	108.0	0.56	10339000	99.4	0.60
Dominica	750	25000	51000	68.0	0.49	73000	97.3	0.34	76000	101.3	0.33	74000	98.7	0.34
Dominican Republic	48310	2352000	2365000	49.0	0.99	10528000	217.9	0.22	12087000	250.2	0.19	13238000	274.0	0.18
Fiji	18270	425000	289000	15.8	1.47	892000	48.8	0.48	940000	51.5	0.45	924000	50.6	0.46
Guinea-Bissau	28120	1630000	535000	19.0	3.05	1844000	65.6	0.88	2541000	90.4	0.64	3564000	126.7	0.46
Grenada	340	8000	77000	226.5	0.10	107000	314.7	0.07	112000	329.4	0.07	110000	323.5	0.07
Haiti	27560	1840000	3221000	116.9	0.57	10711000	388.6	0.17	12578000	456.4	0.15	14189000	514.8	0.13
Jamaica	10830	444000	1403000	129.5	0.32	2793000	257.9	0.16	2867000	264.7	0.15	2710000	250.2	0.16
Kiribati	810	34000	33000	40.7	1.03	112000	138.3	0.30	142000	175.3	0.24	178000	219.8	0.19
Maldives	300	7900	74000	246.7	0.11	364000	1213.3	0.02	437000	1456.7	0.02	494000	1646.7	0.02
Marshall Islands	180	11500	13000	72.2	0.88	53000	294.4	0.22	56000	311.1	0.21	67000	372.2	0.17

Federated states of Micronesia	700	22000	32000	45.7	0.69	104000	148.6	0.21	118000	168.6	0.19	129000	184.3	0.17
Mauritius	2030	85000	493000	242.9	0.17	1273000	627.1	0.07	1310000	645.3	0.06	1249000	615.3	0.07
Nauru	20	400	3000	150	0.13	10000	500	0.04	11000	550.0	0.04	11000	550.0	0.04
Palau	460	5000	7000	15.2	0.71	21000	45.7	0.24	25000	54.3	0.20	28000	60.9	0.18
PNG	452860	1190000	1708000	3.8	0.70	7619000	16.8	0.16	10057000	22.2	0.12	13240000	29.2	0.09
Samoa	2830	35000	82000	29.0	0.43	193000	68.2	0.18	210000	74.2	0.17	241000	85.2	0.15
Sao Tome e Principe	960	48700	60000	62.5	0.81	190000	197.9	0.26	256000	266.7	0.19	353000	367.7	0.14
Singapore	709	700	1022000	1441.5	0.00	5604000	7904.1	0.00	6418000	9052.2	0.00	6681000	9423.1	0.00
St. Kitts and Nevis	260	6000	46000	176.9	0.13	56000	215.4	0.11	63000	242.3	0.10	68000	261.5	0.09
Saint Lucia	610	10600	83000	136.1	0.13	185000	303.3	0.06	202000	331.1	0.05	207000	339.3	0.05
St. Vincent and the Grenadines	390	10000	67000	171.8	0.15	109000	279.5	0.09	112000	287.2	0.09	109000	279.5	0.09
Seychelles	460	1600	36000	78.3	0.04	96000	208.7	0.02	101000	219.6	0.02	100000	217.4	0.02
Solomon Islands	27990	108000	90000	3.2	1.20	584000	20.9	0.18	757000	27.0	0.14	992000	35.4	0.11
Suriname	156000	88200	215000	1.4	0.41	543000	3.5	0.16	599000	3.8	0.15	624000	4.0	0.14
Timor-Leste	14870	380000	433000	29.1	0.88	1185000	79.7	0.32	1577000	106.1	0.24	2162000	145.4	0.18
Tonga	720	33000	47000	65.3	0.70	106000	147.2	0.31	212000	294.4	0.16	140000	194.4	0.24

(continued)

Table 2 (continued)

	Country size (km^2)	Agricultural land 2015 (ha)	Population, density and agricultural land per person (ha)											
			1950			*2015*			*2030*			*2050*		
Trinidad and Tobago	5130	54000	646000	125.9	0.08	1360000	265.1	0.04	1372000	267.4	0.04	1291000	251.7	0.04
Tuvalu	30	1800	5000	166.7	0.36	10000	333.3	0.18	11000	366.7	0.16	11000	366.7	0.16
Vanuatu	12190	187000	48000	3.9	3.90	265000	21.7	0.71	354000	29.0	0.53	476000	39.0	0.39

The table was developed specifically for this chapter. The three columns per year reflect population size, population density (number of persons per km^2) and agricultural land per person (ha). Data from UNDESA (2015) and World Bank (2017)

systems are already well-established in many societies and small-scale farming communities across the world. Hence, agroforestry has been shown to offer countries the opportunity to help achieve NDC targets due to the widespread use and familiarity of small-scale farmers and practitioners with agroforestry techniques (Duguma et al., 2017).

Given the general scarcity of land and the increasing demand for agricultural land in particular, SIDS are facing significant natural resource management challenges with social, environmental and economic implications. Hence, a land use system that meets multiple needs simultaneously, such as agroforestry, could potentially address these challenges. Several studies have already documented the role of agroforestry, agriculture, urban forestry and wood energy in the NDCs (e.g. Richards et al., 2015; Bervoets et al., 2016, 2017; Food and Agriculture Organization (FAO), 2016; Duguma et al., 2017; Amugune et al., 2017; Dumas-Johansen et al., 2018). However, few studies have analysed the role of agroforestry in SIDS as an integrated solution to climate, biodiversity and land degradation issues. Furthermore, there are very few studies which have studied the national reports submitted by SIDS to the three conventions (e.g. Scandurra et al., 2018).

3 Methods

The national reports of the 38 SIDS[1] to the UNFCCC, UNCCD and CBD were used for the analysis. The following three national reports where used: Nationally Determined Contributions (NDCs) or the Intended Nationally Determined Contributions (INDCs); National Action Programmes (NAPs) and National Biodiversity Strategies and Action Plans (NBSAPs), respectively. The reports were downloaded from the official convention websites.[2,3,4] For the NAPs and NBSAPs the latest submitted reports were used. The original versions of all reports were used when available, i.e. English, French or Spanish and translations were avoided to be sure to capture original meanings and scope. Reports were screened regarding any type of references made to agroforestry and associated terms, e.g. alley cropping and silvicultural techniques. A frequency count was conducted for each country similar to the approach used in recent studies, e.g. Richards et al. (2015) and Dumas-Johansen et al. (2018). Finally, a database for each country was constructed to allow for an in-depth analysis of the agroforestry references within each national report and in order to compare them across all three conventions.

---

[1] https://sustainabledevelopment.un.org/topics/sids/list.

[2] UNFCCC: http://www4.unfccc.int/ndcregistry/Pages/Home.aspx; http://www4.unfccc.int/submissions/indc/Submission%20Pages/submissions.aspx.

[3] UNCCD: https://knowledge.unccd.int/search?f%5B0%5D=type%3Aaction_programmes.

[4] CBD: https://www.cbd.int/nbsap/about/latest/default.shtml.

The three conventions and their national reports have been initiated under different time frames, with the NDCs being the newest initiative (from 2015 onwards) while the UNCCD and CBD date back to the 1990s and onwards. In some cases, initiatives mentioned older reports may have already been implemented. It goes beyond the scope of this analysis to study the impacts of past references to agroforestry. This study aims only to study the references made to agroforestry in order to understand the priority SIDS place on agroforestry.

Only SIDS with references to agroforestry across the three Rio conventions were included in the various analyses and illustrated in tables and figures.

4 Results

Agroforestry already plays an important role in the NDCs. Dumas-Johansen et al. (2018) found that 56 countries highlighted agroforestry in their NDCs relating mainly to adaptation to climate change. With an increasing population, at least until 2050, and high population densities very little available land is left for farming and forestry activities at present and in the future (see Table 1 for data from all SIDS). It is therefore unavoidable that agricultural expansion will take place, resulting in deforestation and forest degradation. This underlines the urgency of promoting integrated approaches such as the establishment or strengthening of agroforestry systems.

The following sections present the results of the analysis of agroforestry aspects in the SIDS reports submitted under each of the three conventions.

4.1 Agroforestry and NDC Targets

Seven SIDS out of a total of 38 refer to agroforestry in their Nationally Determined Contributions. These seven SIDS are spread across Africa (3), Asia-Pacific (2) and the Caribbean (2). The references to agroforestry differ substantially between the seven countries, ranging from very specific and concrete targets to more vaguely defined targets and references. Table 3 illustrates the targets across the seven SIDS. Comoros, Haiti and Sao Tome e Principe contain the most concrete targets aiming at converting 200 ha annually and 60,000 ha by 2030 for Comoros and Haiti, respectively. Sao Tome and Principe highlight the goal of developing a national agroforestry program by 2025. Other targets include the development of specific agroforestry systems (e.g. Timor Leste) and use of integrated approaches (Tonga and Timor Leste). These specific references to agroforestry and integrated approaches could present a very promising approach to supporting SIDS in addressing climate change, conserving biodiversity and decreasing land degradation in the context of a decreasing cultivable area in the future (see Table 2).

Table 3 References to agroforestry in the NDCs

Countries	NDC agroforestry targets/references
Cabo Verde	Cabo Verde's forestry sector includes the Economic Transformation Strategy (TEE), which proposes, among others the development of agroforestry
Comoros	The area covered by agroforestry and aboriculture should increase with 200 ha per year from 2018 until 2030. Technology transfer needed for natural resources including agroforestry
Haiti	Restore, evaluate and extend existing agroforestry systems (minimum an additional 60,000 ha between 2020 and 2030)
Saint Lucia	Mentioned as part of measuring uncertainties
Sao Tome e Principe	Under adaptation goals: Develop a national program for sustainable management of the forest and agro forestry ecosystems by 2025
Timor Leste	Develop integrated agroforestry and watershed management including climate change dimensions
Tonga	Promoting integrated agroforestry in areas earmarked for agriculture;

Several challenges remain for countries to achieve these targets, besides those already listed in literature (c.f. Table 1). Duguma et al. (2017) highlight the following issues to be a constraint for meeting the NDC targets with respect to agroforestry: financial, policy, technology, land and tree tenure and carbon rights, climate change effects and variability and limited source of quality germsplasm. Concrete targets as put forward by Comoros and Haiti offer the countries the opportunity to monitor progress over time, and attract financial investments to help meet such targets. Countries will be more likely to attract funding for the implementation of the NDCs if targets and goals are as concrete and transparent as possible (Averchenkova & Bassi, 2016).

The Global Environment Facility (GEF) and the Green Climate Fund (GCF) are both designated as primary financial mechanisms for the UNFCCC and for the implementation of the NDCs. The GEF is also a designated financial mechanism for the other two Rio conventions (UNCCD and CBD). The GEF and GCF present strong initiatives to pilot agroforestry programmes. A blended financial construction with private sector playing a key role is needed for implementation of the NDCs (Duguma et al., 2017; Espagne, 2015).

4.2 *Agroforestry and NAP References*

Seven SIDS make reference to agroforestry in their National Action Programmes to the UNCCD. Table 4 illustrates the various agroforestry targets and references. The SIDS are spread across the same regions as for the NDC references encompassing the Pacific (3), Caribbean (3) and Africa (1). The references are less specific and concrete as compared to the NDC targets. Only Samoa highlights an increase in the area devoted to agroforestry as a priority but does not specify how much. Palau is actively seeking to promote and test agroforestry approaches. Haiti and Comoros are

Table 4 UNCCD references to agroforestry and year of submission

Countries	UNCCD agroforestry targets/references
Cabo Verde	Agroforestry mentioned in connection with the need for technical references for agroforestry and other land use systems
Comoros (2013)	References made to the NAPA which mentions agroforestry
Cuba (2003)	Agroforestry included in several planned NAP projects
Dominica (2004)	Agroforestry described as part of the existing forest resources
Fiji (2007)	References made to a past but long lasting agroforestry programme
Haiti (2015)	Agro-pastoralism and agroforestry in general are mentioned numerous times in connection with the various agro ecological zones. Agroforestry is highlighted as part of the planned NAP activities with the need to promote agroforestry further in particular in relation to land restoration
Palau (2005)	Agroforestry described as part of existing forests covering 2700 acres and several activities planned envisaging to promote and test agroforestry
Samoa (2015)	A priority is to increase area of agroforestry

the only two countries that refer to agroforestry in both their NDCs and their NAPs. It is worthwhile mentioning that the NAP reports date back to 2003 in one case (Cuba), thus more than 10 years earlier than when the NDCs were being formulated from 2015 onwards. However, Haiti's NAP report is from 2015 and coincides with the NDC process. Agroforestry and agro-pastoralism are mentioned extensively in the NAP report of Haiti, referring in particular to land restoration, which coincides well with the NDC targets of increasing the area of agroforestry to at least 60,000 ha by 2030. UNCCD has over the last few years launched the Land Degradation Neutrality (LDN) targets, as a mechanism to support countries preserve biodiversity and increase resilience to climate change (United Nations Convention to Combat Desertification (UNCCD) and Food and Agriculture Organization (FAO), 2020). UNCCD has worked with SIDS to identify their national LDN targets which are a key cornerstone in continuing to manage land degradation and furthermore identify valid investment opportunities.

The UNCCD recently initiated a very innovative funding mechanism to help raise finance for member countries' priorities. The Land Degradation Neutrality Fund is unique in that it is being managed by the private sector, thus blending public and private finances aiming at rehabilitating degraded lands. This includes a focus on agroforestry practices. As highlighted earlier, the GEF is also a financial mechanism for the UNCCD in implementing the NAPs.

4.3 Agroforestry and NBSAP References

Of the 38 SIDS within the UN, a total of 19 countries mentioned agroforestry in their NBSAPs. The regional distribution of these SIDS included countries in the

Caribbean (5), Pacific (11) and AIMS[5] (3). Caribbean countries that mention agroforestry in their NBSAP include the Bahamas, Cuba, Dominica, Haiti and Jamaica. While they do not mention agroforestry specifically Barbados, Guyana and Saint Lucia make reference to agro-ecosystems being threatened. Pacific countries that mention agroforestry include Fiji, the Federated States of Micronesia, Nauru, Palau, Papua New Guinea, Samoa, Solomon Islands Timor-Leste, Tonga, Tuvalu and Vanuatu. Of the SIDS in the AIMS region Comoros, Guinea-Bissau and Sao Tomé and Principe mention agroforestry specifically.

References ranges from minor mention of the importance of agroforestry to the inclusion of objectives, targets and indicators pertaining directly to agroforestry. For example, one report referred to the preparation of an advanced training program in agroforestry, ethno-botany and pharmacopoeia and institutionalization of the process of eco-certification of agro-forestry products (Sao Tomé and Principe) while Samoa's NBSAP for the years 2015–2020 mentions measurable indicators for agroforestry such as the number of farmers practising agroforestry plots, the number of households engaged in agroforestry, the area under agroforestry and woodlots and the number of agroforestry projects and programmes implemented. Haiti's NBSAP mentions catchment afforestation and the revitalization of farming systems aiming to increase forest cover and arresting soil degradation through production of seedlings, tree planting and agroforestry techniques (Table 5).

Niue, a non-UN SIDS, also mentions agroforestry in its NBSAP report, referring to the encouragement and promotion of agroforestry systems, awareness raising on agroforestry and the establishment of guidelines for the development of agroforestry systems.

In total 21 SIDS referred to agroforestry in their national reports to the three Rio conventions. Table 6 illustrates the breakdown of references within the 21 SIDS across the three Rio conventions. 19 SIDS referred to agroforestry in their NBSAPs, eight SIDS referred to agroforestry in their NAPs and 7 SIDS to agroforestry in their NDCs. Comoros and Haiti were the only SIDS that referred to agroforestry in all three conventions. Nine SIDS referred to two conventions (Cuba, Cabo Verde, Domenica, Fiji, Palau, Samoa, Sao Tome e Principe, Timor Leste and Tonga) referred to two conventions and the remaining eight countries to one convention.

5 Discussion

Out of the 21 SIDS referring to agroforestry, only Comoros and Haiti referred to agroforestry in all three conventions. Both countries are considered some of the poorest nations in the world, and are certainly among the poorest countries within the 21 SIDS. The question is whether this would suggest that poorer nations see the added value of agroforestry to a larger extent than more developed nations.

[5] Atlantic, Indian Ocean, Mediterranean and South China Sea.

Table 5 Reference to agroforestry in the NBSAPs

Country	NBSAP agroforestry targets/references
Bahamas	Report refers to the Department of Agriculture objectives to develop and evaluate low-input sustainable agriculture and agroforestry systems
Cuba	Mentions the conservation of wild species, management of silvo-pastoral and agroforestry systems and sustainable land management
Dominica	Refers to synergistic actions that will buttress the NBSAP, including agro-forestry, food security and soil stabilization, as well as other conservation efforts implemented as a direct result of the NBSAP including agro-forestry and replanting of trees on a small scale and on private lands
Haiti	Refers to ensuring the best available tree germplasm being made available to peasant farmers for reforestation and agroforestry purposes. Also refers to catchment afforestation and revitalization of farming systems aiming to increase forest cover and arrest soil degradation through production of seedlings, tree planting and agroforestry techniques
Jamaica	Mentions past efforts to implement an agroforestry programme through which seedlings (fruit and timber trees) were provided to farmers
Fiji	Mentions the establishment of a working Agricultural Herbarium which would provide reference materials for future research work and for students studying multispecies agricultural and agroforestry supplements. Also mentions the development of brochures and manuals on agroforestry techniques
Micronesia	Refers to Micronesia's rich and diverse agro-forests and related traditional agricultural systems as possible models for sustainable agricultural development
Nauru	Mentions increasing agricultural production and protecting native biodiversity through sustainable agricultural practices such as agroforestry. Also refers to the promotion of environmentally sound agricultural practices such as agroforestry and the development of a national Forestry and Agroforestry Development Plan
Palau	Mentions specific traditional agroforestry systems such as taro patch farming areas (mesei) and coconut forest areas, and family farms on which agroforestry is practiced
Papua New Guinea	Refers to the introduction of a system of direct incentives to promote the conservation and sustainable use of biodiversity such as agroforestry
Samoa	Mentions specific objectives and measurable indicators for agroforestry under several target areas such as the number of farmers with agroforestry plots and the area under agroforestry
Solomon Islands	Mentions increasing areas devoted to agroforestry with the introduction of species such as eucalyptus, teak and mahogany
Timor-Leste	Mentions the promotion of agroforestry activities and techniques, e.g. in coffee plantations, to increase tree cover, provide income, improve food security and control erosion. Also mentions the use of agroforestry techniques such as integrating woody perennials in shifting cultivation areas to enhance carbon storage capacity
Tonga	Mentions promoting the use of traditional and non-traditional agroforestry systems of mixed species planting as a buffer for protected and other sensitive areas including habitats for threatened species and water catchments. Also mentions conducting awareness activities targeting local farmers to promote tree planting and agroforestry systems. Refers to a

(continued)

Table 5 (continued)

Country	NBSAP agroforestry targets/references
	strategy of promoting mixed cropping and agroforestry farming practices and providing training to local farmers on appropriate agroforestry techniques. Refers to a specific agrobiodiversity target of establishing mixed planting and agroforestry farms and privately managed agroforestry or mixed planting farms
Tuvalu	Refers to fostering and promoting traditional agroforestry and raising awareness and understanding of the importance of agro forestry and its contribution to biodiversity
Vanuatu	Refers to a project that is demonstrating sustainable agroforestry in association with local management of forest resources
Comoros	Refers to the reinforcement of agroforestry systems and the development of a programme for promoting agro-silvipastoral systems
Guinea-Bissau	Refers to the effective operation of six nurseries at village level for forestry and agroforestry
Sao Tomé and Principe	Refers to the preparation of an advanced training program in agroforestry and the institutionalization of the process of eco-certification of agro-forestry products

Table 6 Overview of references to agroforestry across the three Rio conventions for the 21 SIDS mentioning agroforestry in one or more conventions

	NBSAP (CBD)	NAP (UNCCD)	NDC (UNFCCC)
Bahamas	1		
Cuba	1	1	
Cabo Verde		1	1
Comoros	1	1	1
Domenica	1	1	
Fiji	1	1	
Guinea Bissau	1		
Haiti	1	1	1
Jamaica	1		
Micronesia	1		
Nauru	1		
Palau	1	1	
PNG	1		
Saint Lucia			1
Samoa	1	1	
Sao Tome e Principe	1		1
Solomon Islands	1		
Timor Leste	1		1
Tonga	1		1
Tuvalu	1		
Vanuatu	1		
Total	**19**	**8**	**7**

Dumas-Johansen et al. (2018) found that a majority of NDCs submitted referring to agroforestry were from Least Developed Countries (LDCs), in particular those found in Africa. Agroforestry has often been associated with poverty alleviation and a number of successful interventions in this regard have been documented (e.g. Garrity, 2004; Leakey et al., 2005). These findings challenge standard and linear theories of economic development, such as the Environmental Kuznets Curve, which suggest that economic development initially leads to deterioration in the environment. Further research is needed to monitor these trends on the ground, in countries such as the Comoros and Haiti, in order to verify whether the trends documented on paper are indeed unfolding in reality. In this regard, it should be noted that there are likely gaps between national communications and monitoring and evaluation on the ground in SIDS. Communications often tend to overestimate, and may sometimes refer only to projections and future actions, the performance vis-à-vis requirements derived from the Conventions. While beyond the scope of this article, there is a need to carry out further research comparing 'declared' actions and the actual situation on the ground.

The reasoning behind why some SIDS mention agroforestry and others not is not fully transparent. It is likely that the definition of agroforestry itself is at the root of this gap in understanding (Torquebiau, 2000) since the definition is being debated regularly (Leakey, 1996). Another challenge could be linked to the role of agroforestry within the three conventions. Minang et al. (2014) found that the unclear role of agroforestry within the UNFCCC is both an advantage due to its ambiguity, allowing agroforestry to be a flexible mechanism, while the disadvantage is that there are few immediate benefits.

The lack of references to agroforestry, especially for the NDCs as the newest reporting scheme, could also be due to lack of guidance and templates. Food and Agriculture Organization (FAO) (2016) and Hedger and Nakhooda (2015) found that the lack of clarity on guidelines and templates could explain the significant variability of NDC goals across countries. Atteridge et al. (2020) found from examining the NDCs of seven SIDS that the key concerns for SIDS mainly related to governance and economic management, land use planning were not included in the NDCs and thus a concrete opportunity was missed for enhancing the development agenda.

A majority of the SIDS referred to agroforestry in their NBSAPs, which could imply that promoting or maintaining agroforestry systems is regarded as an important factor in biodiversity conservation. The important role of agroforestry in biodiversity conservation in general is well documented (Bhagwat et al., 2008; McNeely & Schroth, 2006; Schroth et al., 2004). However, little research has focused on agroforestry and biodiversity in SIDS.

Agroforestry can cater to the challenges raised by the three Rio conventions and offers countries a toolbox with numerous possibilities to help them craft proper solutions. However, agroforestry is not a panacea. There is a need to strike a balance between economic and ecological needs in the management of agroforestry systems (Tscharntke et al., 2011) as agroforestry can result in both positive and negative outcomes that affect crops, water, livestock and users (German et al., 2006). Some trade-offs include the conflict between carbon sequestration and aspects pertaining to other sustainable development goals (Foster & Neufeldt, 2014) or the benefits of using trees for shade versus the achievement of maximum yield in coffee and cocoa based systems (Vaast & Somarriba, 2014). Another trade-off pertains more generally to the loss of income generated from agricultural production while attaining benefits from having trees on the farm (Rahman et al., 2016). Further research and pilot studies are needed to fully understand the role agroforestry plays in SIDS and how and under which conditions it can best be promoted, used and implemented. It would be crucial to ensure consistency between the three national reports in order to maximize the use of the limited public funding available and to channel that into larger scale programmes that are more effective. The authors argue that by building a global or regional partnership focusing on agroforestry, SIDS will have a venue through which to share knowledge and best practices. In 2020 countries to the UNFCCC are to submit their second NDC. This offers a unique opportunity for all SIDS to further align their agroforestry commitments.

The authors argue that a global partnership could be fostered to help countries prepare for this next round of submissions, including by providing access to financial support. A first step would be to integrate agroforestry into existing natural resource management policies. There are several existing platforms on which it would be ideal to build on. The first step would be to conduct an in- depth review of such platforms to determine the best practices on which to build and existing gaps exist that are not currently addressed and funded. This analysis would also reveal the role of the SIDS in general as they engage across the three Rio conventions and how agroforestry could be better targeted as a cross-cutting solution to bring to the attention of higher political and negotiation levels.

One promising programme is the recently launched Food and Agriculture Organization (FAO) programme targeting food security in SIDS, entitled the "Global Action Programme on Food Security and Nutrition in Small Island Developing States" (GAP). This programme aims to build an enabling environment for food security, nutrition and the creation of sustainable food systems (Food and Agriculture Organization (FAO), 2017). The programme offers an opportunity to highlight agroforestry as one of many important approaches to help achieve food security and nutrition. But it does not solve the problem that SIDS face alone. Strong national programmes should be developed to target both policies and measures as well as to put financial structures in place to support and promote agroforestry in a more structured manner at the national level.

We argue that a combination of strategic programmes at the global, regional and national level is needed. The FAO programme is one of many needed platforms that can help build a strong evidence base and lessons learned on agroforestry across the numerous SIDS and their specific national contexts. Other important platforms to include in these structures would be strong regional bodies, such as the Pacific Community, that can finance, fundraise for and implement several innovative natural resource projects and programmes. Given that many of the members of the pacific community are SIDS, there is a strong evidence base to build on.

The overall architecture should begin with "The Small Island Developing States Partnership Framework", a global partnership for development of SIDS.[6] The partnership would be in the best position to coordinate and oversee ongoing projects and programmes and facilitate further partnerships and collaboration from SIDS to SIDS. The FAO programme as well as other existing programmes should be better tied to the SIDS partnership. The growing demand for research from these programmes as well as upcoming research would add tremendous value to the limited knowledge and research on agroforestry and SIDS at the global level. This would allow for improvement of policies and strategies, particularly with regards to the three conventions.

We argue that a special investment fund should be established to accompany the SIDS partnership in order for it to be fully functional. An innovative investment fund could be created to attract private investments, public funds and a combination of the two. This would truly support SIDS by not only investing in building resilience and mitigating climate change through agroforestry measures but also strengthening the voices of the SIDS across the three conventions.

6 Conclusion

SIDS are considered some of the most vulnerable countries in terms of climate change. With limited land areas and agricultural land, population pressure and overuse of natural resources, challenges are increasing. Furthermore, SIDS are known to be biodiversity hotspots, but because of the limited land, the pressure for additional agricultural land has taken its toll on biodiversity. Land degradation is severe and expected to continue. There is need for a land use system targeting all three issues simultaneously. Agroforestry offers a unique opportunity to address all these in parallel. The three Rio conventions, UNCCD, CBD and UNFCCC deal with national responses to exactly these issues. To understand the current role agroforestry plays in this regard, an analysis of the national reports to the three Rio conventions was carried out. A total of 21 SIDS referred to agroforestry in their national reports to one or several of the three Rio conventions (UNCCD, CBD and UNFCCC). The majority of SIDS referred to agroforestry in their NBSAPs (19 SIDS), followed by NAPs (8 SIDS) and NDCs (7 SIDS) with 11 SIDS referring to two or all three conventions. Comoros and Haiti referred to agroforestry in all three conventions.

The growing number of programmes and platforms that include SIDS need to be better structured and coordinated. The SIDS partnership is a global partnership that offers a strong evidence base on which to build and help structure future coordination and support measures. We argue that a specific investment fund should be created to help support future agroforestry interventions in SIDS and accompany the

[6] https://sustainabledevelopment.un.org/sids/partnershipframework.

SIDS partnership and its many sub-platforms. This would help bridge funding gaps across the three conventions and avoid competition for funding.

Detailed List of Agroforestry References in the Three Rio Conventions

Country	Conventions
	NBSAP—CBD
Bahamas (1999)	Report refers to the Department of Agriculture objectives to develop and evaluate low-input sustainable agriculture and agroforestry systems
Cuba	Mentions the conservation of wild species, management of silvo-pastoral and agroforestry systems and sustainable land management
Dominica	Refers to synergistic actions that will buttress the NBSAP, including agro-forestry, food security and soil stabilization, as well as other conservation efforts implemented as a direct result of the NBSAP including agro-forestry and replanting of trees on a small scale and on private lands
Haiti	Refers to ensuring the best available tree germplasm being made available to peasant farmers for reforestation and agroforestry purposes. Also refers to catchment afforestation and revitalization of farming systems aiming to increase forest cover and arrest soil degradation through production of seedlings, tree planting and agroforestry techniques
Jamaica	Mentions past efforts to implement an agroforestry programme through which seedlings (fruit and timber trees) were provided to farmers
Fiji	Mentions the establishment of a working Agricultural Herbarium which would provide reference materials for future research work and for students studying multispecies agricultural and agroforestry supplements. Also mentions the development of brochures and manuals on agroforestry techniques
Micronesia	Refers to Micronesia's rich and diverse agro-forests and related traditional agricultural systems as possible models for sustainable agricultural development
Nauru	Mentions increasing agricultural production and protecting native biodiversity through sustainable agricultural practices such as agroforestry. Also refers to the promotion of environmentally sound agricultural practices such as agroforestry and the development of a national Forestry and Agroforestry Development Plan
Palau	Mentions specific traditional agroforestry systems such as taro patch farming areas (mesei) and coconut forest areas, and family farms on which agroforestry is practiced
Papua New Guinea	Refers to the introduction of a system of direct incentives to promote the conservation and sustainable use of biodiversity such as agroforestry
Samoa	Mentions specific objectives and measurable indicators for agroforestry under several target areas such as the number of farmers with agroforestry plots and the area under agroforestry
Solomon Islands	Mentions increasing areas devoted to agroforestry with the introduction of species such as eucalyptus, teak and mahogany

(continued)

Country	Conventions
	NBSAP—CBD
Timor-Leste	Mentions the promotion of agroforestry activities and techniques, e.g. in coffee plantations, to increase tree cover, provide income, improve food security and control erosion. Also mentions the use of agroforestry techniques such as integrating woody perennials in shifting cultivation areas to enhance carbon storage capacity
Tonga	Mentions promoting the use of traditional and non-traditional agroforestry systems of mixed species planting as a buffer for protected and other sensitive areas including habitats for threatened species and water catchments. Also mentions conducting awareness activities targeting local farmers to promote tree planting and agroforestry systems. Refers to a strategy of promoting mixed cropping and agroforestry farming practices and providing training to local farmers on appropriate agroforestry techniques. Refers to a specific agrobiodiversity target of establishing mixed planting and agroforestry farms and privately managed agroforestry or mixed planting farms
Tuvalu	Refers to fostering and promoting traditional agroforestry and raising awareness and understanding of the importance of agro forestry and its contribution to biodiversity
Vanuatu	Refers to a project that is demonstrating sustainable agroforestry in association with local management of forest resources
Comoros	Refers to the reinforcement of agroforestry systems and the development of a programme for promoting agro-silvipastoral systems
Guinea-Bissau	Refers to the effective operation of six nurseries at village level for forestry and agroforestry
Sao Tomé and Principe	Refers to the preparation of an advanced training program in agroforestry and the institutionalization of the process of eco-certification of agroforestry products
Countries	*NAP—UNCCD*
Cabo Verde	Agroforestry mentioned in connection with the need for technical references for agroforestry and other land use systems
Comoros (2013)	References made to the NAPA which mentions agroforestry
Cuba (2003)	Agroforestry included in several planned NAP projects
Dominica (2004)	Agroforestry described as part of the existing forest resources
Fiji (2007)	References made to a past but long lasting agroforestry programme
Haiti (2015)	Agro-pastoralism and agroforestry in general are mentioned numerous times in connection with the various agro ecological zones. Agroforestry is highlighted as part of the planned NAP activities with the need to promote agroforestry further in particular in relation to land restoration
Palau (2005)	Agroforestry described as part of existing forests covering 2700 acres and several activities planned envisaging to promote and test agroforestry
Samoa (2015)	A priority is to increase area of agroforestry
Countries	*NDC—UNFCCC*
Cabo Verde	Cabo Verde's forestry sector includes the Economic Transformation Strategy (TEE), which proposes, among others the development of agroforestry
Comoros	The area covered by agroforestry and aboriculture should increase with 200 ha per year from 2018 until 2030. Technology transfer needed for natural resources including agroforestry

(continued)

Country	Conventions
	NBSAP—CBD
Haiti	Restore, evaluate and extend existing agroforestry systems (minimum an additional 60.000 ha between 2020 and 2030)
Saint Lucia	Mentioned as part of measuring uncertainties
Sao Tome e Principe	Under adaptation goals: Develop a national program for sustainable management of the forest and agro forestry ecosystems by 2025
Timor Leste	Develop integrated agroforestry and watershed management including climate change dimensions
Tonga	Promoting integrated agroforestry in areas earmarked for agriculture

References

Amugune, I., Cerutti, P., Baral, H., Leonard, S., & Martius, C. (2017). *Small flame but no fire: Wood fuel in the (Intended) Nationally Determined Contributions of countries in Sub-Saharan Africa* (p. 45) (Working Paper 232). Bogor, Indonesia:CIFOR.

Anderson, E., & Zerriffi, H. (2012). Seeing the trees for the carbon: agroforestry for development and carbon mitigation. *Climatic Change, 115*(3-4), 742–757.

Aronson, J., & Alexander, S. (2013). Ecosystem restoration is now a global priority: Time to roll up our sleeves. *Restoration Ecology, 21*(3), 293–296.

Atteridge, A., Verkuijl, C., & Dzebo, A. (2020). Nationally determined contributions (NDCs) as instruments for promoting national development agendas? An analysis of small island developing states (SIDS). *Climate Policy, 20*(4), 485–498. https://doi.org/10.1080/14693062.2019.1605331

Averchenkova, A., & Bassi, S. (2016). Beyond the targets: Assessing the political credibility of pledges for the Paris Agreement. Policy Brief, p. 60.

Beenhouwer, M. D., Geeraert, L., Mertens, J., Van Geel, M., Aerts, R., Vanderhaegen, K., & Honnay, O. (2015). Biodiversity and carbon storage co-benefits of coffee Agroforestry across a gradient of increasing management intensity in the SW Ethiopian highlands. *Agriculture, Ecosystems and Environment, 222*, 193–199.

Bervoets, J., Boerstler, F., Borelli, S., Dumas-Johansen, M., Thulstrup, A., & Zhuzang, X. (2017). Achieving food and wood security in the context of climate change: The role of urban forests and agroforestry in the NDCs in Sub-Saharan Africa. *Nature and Faune, 31*, 22–25.

Bervoets, J., Boerstler, F., Dumas-Johansen, M., Thulstrup, A., & Zhuzang, X. (2016). The climate change energy nexus: how the wood energy sector is addressed in the INDCs. *Unasylva, 67* 53–60.

Bhagwat, S. A., Willis, K. J., Birks, H. J. B., & Whittaker, R. J. (2008). Agroforestry: A refuge for tropical biodiversity? *Trends in Ecology and Evolution, 23*(5), 261–267.

Bogdanski, A., Dubois, O., Jamieson, C., & Krell, R. (2010). *Making integrated food-energy systems work for people and climate: An overview*. Food and Agriculture Organization of the United Nations (FAO).

Catacutan, D., van Noordwijk, M., Nguyen, T., Öborn, I., & Mercado, A. (2017). *Agroforestry: Contribution to food security and climate-change adaptation and mitigation in Southeast Asia*. World Agroforestry Centre (ICRAF) Southeast Asia Regional Program & Jakarta, Indonesia, ASEAN-Swiss Partnership on Social Forestry and Climate Change

Chirwa, P. W., Black, C. R., Ong, C. K., & Maghembe, J. A. (2003) Tree and crop productivity in gliricidia/maize/pigeonpea cropping systems in southern Malawi. *Agroforestry Systems, 59*(3), 265–277.

de Lima, R. F., Dallimer, M., Atkinson, P. W., & Barlow, J. (2013). Biodiversity and land-use change: Understanding the complex responses of an endemic-rich bird assemblage. *Diversity and Distributions, 19*(4), 411–422.

Drew, M. W. (2008). *Socioeconomic analysis of agroforestry and livelihoods on a small island developing state: a case study of Pohnpei, Federated states of Micronesia* (p. 128). PhD Desertation, University of Florida

Duguma, L. A., Nzyoka, J., Minang, P. A., & Bernard, F. (2017). *How agroforestry propels achievement of nationally determined contributions* (p. 8) (ICRAF Policy Brief no. 34). Nairobi, Kenya: World Agroforestry Centre

Dumas-Johansen, M., Simelton, E. & Borelli, S. (2018). *The role of agroforestry in the Paris Agreement. An analysis of the Nationally Determined Contributions (NDCs)*, Personal communication, 19 January 2018.

Espagne, E. (2015). *Climate finance at COP21 and after: Lessons learnt* (p. 12) (CEPII Policy Brief)

Food and Agriculture Organization (FAO). (2016). *The agriculture sectors in the intended nationally determined contributions: Summary* (p. 16). Food and Agriculture Organization.

Food and Agriculture Organization (FAO). (2017). *Global action programme on Food Security and nutrition in small Island developing states: Supporting the implementation of the Samoa pathway* (p. 8). Food and Agriculture Organization.

Food and Agriculture Organization (FAO). (2018). *Agroforestry*. Food and Agriculture Organization. [online] Available from: https://www.fao.org/forestry/agroforestry/80338/en/. Accessed 22 Sept 2018

Foster, K., & Neufeldt, H. (2014). Biocarbon projects in agroforestry: Lessons from the past for future development. *Current Opinion in Environmental Sustainability, 6*, 148–154.

Garrity, D. P. (2004). Agroforestry and the achievement of the millennium development goals. *Agroforestry Systems, 61*, 5–17.

German, L. A., Kidane, B., & Shemdoe, R. (2006). *Social and environmental trade-offs in tree species selection. A methodology for identifying niche incompatibilities in agroforestry* (AHI Working Papers).

Hedger, M., & Nakhooda, S. (2015). *Finance and Intended Nationally Determined Contributions (INDCs): enabling implementation* (ODI working paper), pp. 12.

Hillbrand, A., Borelli, S., Conigliaro, M., & Olivier, A. (2017). *Agroforestry landscape restoration. Exploring the potential of Agroforestry to enhance the sustainability and resilience of degraded landscapes* (p. 28). Food and Agriculture Organization.

Johnson, J. E., & Delgado, O. J. (2003). 'Farmer perspectives on agroforestry opportunities and constraints in Cape Verde' *Small-scale Forest Economics. Management and Policy, 2*(3), 343–355.

Jose, S. (2009). Agroforestry for ecosystem services and environmental benefits: An overview. *Agroforestry Systems, 76*, 1–10.

Kandji, S. T., Verchot, L. V., Mackensen, J., Boye, A., van Noordwijk, M., Tomich, T. P., Ong, C., Albrecht, A., & Palm, C. (2006). Opportunities for linking climate change adaptation and mitigation through Agroforestry systems. In D. Garrity, A. Okono, M. Grayson, & S. Parrott (Eds.), *World Agroforestry into the future* (pp. 113–122). World Agroforestry Centre (ICRAF).

Leakey, R. R. B. (1996). Definition of Agroforestry revisited. *Agroforestry Today, 8*(1), 5–7.

Leakey, R. R. B., Tchoundjeu, Z., Schreckenberg, K., Shckleton, S. E., & Shackleton, C. M. (2005). Agroforestry tree products (AFTPs): Targeting poverty reduction and enhanced livelihoods. *International Journal of Agricultural Sustainability, 3*(1), 1–23.

Lufung, B. R. (2016). *Local ecological knowledge of tree-crop-livestock interactions on smallholder farms around Adda-Daoueni in Anjouan, Comoros,* Dissertation submitted in partial fulfillment of the requirements for the degree of Master of Science (MSc) in Sustainable Tropical Forestry (SUTROFOR), Bangor, UK, Bangor University, p. 94.

Mbow, C., Smith, P., Skole, D., Duguma, L., & Bustamante, M. (2014). Achieving mitigation and adaptation to climate change through sustainable Agroforestry practices in Africa. *Current Opinion in Environmental Sustainability, 6*, 8–14.

McNeely, J., & Schroth, G. (2006). Agroforestry and biodiversity conservation, traditional practices, present dynamics, and lessons for the future. *Biodiversity and Conservation, 15*, 549–554.

McWilliam, A. (2003). New beginnings in East timorese forest management. *Journal of Southeast Asia Studies, 34*(2), 307–327.

Minang, P. A., Duguma, L. A., Bernard, F., Mertz, O., & van Noordwijk, M. (2014). Prospects for agroforestry in REDD+ landscapes in Africa. *Current opinion in Environmental Sustainability, 6*, 78–82.

Murray, G. F., & Bannister, M. E. (2004). Peasants, agroforesters and anthropologists: A 20-year venture in income-generating trees and hedgerows in Haiti. *Agroforestry Systems, 61*, 383–397.

Myers, N., Mittermeier, R. A., Mittermeier, C. G., da Fonseca, G. A. B., & Kent, J. (2000). Biodiversity hotspots for conservation priorities. *Nature, 403*, 853–858.

Nair, P. K. R. (2011). Agroforestry systems and environmental quality: Introduction. *Journal of Environmental Quality, 40*, 784–790.

Neufeldt, H., van de Sand, I., Dietz, J., Hoang, M. H., Yatich, T., Lasco, R., & van Noordwijk, M. (2011). Climate change, climate variability and adaptation options. In M. van Noordwijk, M. H. Hoang, H. Neufeldt, I. Öborn, & T. Yatich (Eds.), *How trees and people can co-adapt to climate change: Reducing vulnerability through multifunctional Agroforestry landscapes* (pp. 15–36). World Agroforestry Centre (ICRAF).

Rahman, S. A., Sunderland, T., Kshatriya, M., Roshetko, J. M., Pagella, T. & Healey, J. R. (2016). Towards productive landscapes: Trade-offs in tree-cover and income across a matrix of smallholder agricultural land-use systems. *Land Use Policy, 58*, 152–164.

Richards, M., Bruun, T.B., Campbell, B., Gregersen, L. E., Huyer, S., & Kuntze, V. (2015). *How countries plan to address agricultural adaptation and mitigation: An analysis of Intended Nationally Determined Contributions.* Climate Change, Agriculture and Food Security (CCAFS) Info Note, p. 8.

Rioux, J., Roopnarine, R., Biancalani, R., & Petri, M. (2017). *Land degradation assessment in Small Island Developing States (SIDS). Land and Water Division Working Paper 17* (p. 134). Food and Agriculture Organization.

Scandurra, G., Romano, A. A., Ronghi, M., & Carfora, A. (2018). On the vulnerability of Small Island Developing States: A dynamic analysis. *Ecological Indicators, 84*, 382–392.

Schoeneberger, M. M. (2008). Agroforestry: Working trees for sequestering carbon on agricultural lands. *Agroforestry Systems, 75*(1), 27–37.

Schroth, G., da Fonseca, G. A. B., Harvey, C. A., Gascon, C., Vasconcelos, H. L., & Izac, A. M. N. (Eds.). (2004). *Agroforestry and biodiversity conservation in tropical landscapes* (p. 537). Island Press.

Thaman, R. R., Elevitch, C. R., & Wilkinson, K. M. (2000). *Multipurpose trees for agroforestry in the Pacific Islands, Agroforestry guides for Pacific Islands no. 2* (p. 50). Permanent Agriculture Resources.

Torquebiau, E. M. (2000). A renewed perspective on agroforestry concepts and classification. *Comptes Rendus de l Académie des Sciences - Series III - Sciences de la Vie, 323*(11), 1009–1017.

Tougiani, A., Guero, C., & Rinaudo, T. (2009). Community mobilisation for improved livelihoods through tree crop management in Niger. *GeoJournal, 74*, 377.

Tscharntke, T., Clough, Y., Bhagwat, S. A., Buchori, D., Faust, H., Hertel, D., Hölscher, D., Juhrbandt, J., Kessler, M., Perfecto, I., Scherber, C., Schroth, G., Veldkamp, E., & Wanger, T. C. (2011). Multifunctional shade-tree management in tropical agroforestry landscapes – a review. *Journal of Applied Ecology, 48*, 619–629.

UNDESA. (2015). *World Population prospects. The 2015 revision. Key findings and advance tables* (p. 66). Department of Economic and social affairs. Population division.

United Nations Climate Change Conference (COP23) (2017) *Draft decision 1/CP.23*, Fiji Momentum for Implementation, p. 9. [online] Available from: https://cop23.com.fj/wp-content/uploads/2018/01/Fiji-Momentum-for-Implementation.pdf

United Nations Convention to Combat Desertification (UNCCD) (2018) *The Land Degradation Neutrality Fund*, Bonn, Germany, United Nations Convention to Combat Desertification. [online] Available from: https://www2.unccd.int/actions/impact-investment-fund-land-degradation-neutrality

United Nations Convention to Combat Desertification (UNCCD) & Food and Agriculture Organization (FAO). (2020). *Land degradation neutrality in small island developing states*. United Nations Convention to Combat Desertification.

Vaast, P., & Somarriba, E. (2014). Trade-offs between crop intensification and ecosystem services: the role of Agroforestry in cocoa cultivation. *Agroforestry Systems, 88*, 947–956.

Verchot, L. V., van Noordwijk, M., Kandji, S., Tomich, T., Ong, C., & Albrecht, A. (2007). Climate change: Linking adaptation and mitigation through Agroforestry. *Mitigation and Adaptation Strategies for Global Change, 12*, 901. https://doi.org/10.1007/s11027-007-9105-6

Walters, B. B., & Hansen, L. (2012). Farmed landscapes, trees and forest conservation in Saint Lucia (West Indies). *Environmental Conservation, 40*(3), 211–221.

World Bank. (2017). *World Bank Open Data*. World Bank. [online] Available from: https://data.worldbank.org/indicator/AG.LND.TOTL.K2

Zomer, J. R., Neufeldt H., Xu, J., Ahrends, A., Bossio, D., & Trabucco, A. (2016) 'Global tree cover and biomass Carbon on agricultural land. The contribution of Agroforestry to global and national carbon budgets' *Nature*, Scientific Reports 6, Article number: 29987

Economic Impacts of Climate Change on Agriculture: Insights from the Small Island Economy of Mauritius

Riad Sultan

Abstract As a small island developing state, the Republic of Mauritius is highly vulnerable to the impacts of climate change in the agricultural sector. Agriculture activities constitute 43.3 per cent of total land surface in the mainland of Mauritius and employs 7.2 per cent of the labour force. Being self-sufficient in a few food crops, a decline in agricultural output caused by climate change poses a threat, not only to the livelihoods of farmers, but also to food security. The present study attempts to estimate the economic impacts of climate change on agriculture for the Island of Mauritius. It applies the Ricardian approach on a small territorial scale, motivated by the microclimate system of the island. Using cross-sectional farm data from a sample of 392 farmers, the Ricardian estimates show that the agriculture sector responds negatively to changes in mean summer temperature and precipitation. The economic impacts of a rise in mean temperature by 1 °C is USD26 6 per acre per year with an elasticity of −0.13 evaluated at mean temperature. The elasticity of net farm revenue evaluated at mean precipitation is 0.03. The findings show the vulnerability of a small island to climate change and suggest a need to invest in adaptation measures for farmers in Mauritius.

Keywords Climate change · Economic impacts · Agriculture · Cross-section survey · Ricardian approach · Mauritius · Revenues · Precipitations · Temperatures

1 Introduction

The Republic of Mauritius is a small island developing state of about 2040 km^2 in area which comprises of Mauritius, Rodrigues, Agalega, Tromelin, Cargados, Carajos and the Chagos Archipelago. Agriculture activities constituted 43.3 per cent of total land surface in the mainland and contributed 3.2 per cent to GDP in

R. Sultan (✉)
Department of Economics and Statistics, University of Mauritius, Réduit, Mauritius
e-mail: r.sultan@uom.ac.mu

S. Moncada et al. (eds.), *Small Island Developing States*, The World of Small States 9, https://doi.org/10.1007/978-3-030-82774-8_7

2018 (Statistics Mauritius, 2018). While the share of the agricultural sector has declined over the years, it currently employs 7.2 per cent of the labour force. Yet, Mauritius is self-sufficient in a few food crops while it relies on imports for the remaining needs. A decline in agricultural output poses a serious threat to the livelihoods of farmers. At the same time, it can impact negatively on food security.

Mauritius faces the inherent environmental vulnerabilities, associated with the characteristics of a small island developing state. The island has a small land area with sensitive ecosystems. It is also susceptible to natural disasters and suffers from geographical isolation. As a result of its economic success which has been termed by many as the *Mauritian Miracle* (Subramanian, 2001; Stiglitz, 2011; Svirydzenka & Petri, 2014), there is considerable pressure on its limited natural resources. Increasing demand for land, coastal degradation, rising pollution and inadequate waste disposal systems, are major constraints which influence the development of Mauritius (Republic of Mauritius, 2016). Further, Mauritius has a relatively high degree of reliance on international trade with a trade deficit which has widened sharply during the past decades due to rising food costs. Consequently, it is highly exposed to exogenous economic shocks and global economic environment.

There is a consensus that the world is observing changes in climatic conditions caused by high concentration of greenhouse gases (GHGs) (IPCC, 2001a, 2001b, 2007a, 2007b, 2014, 2018). In fact, accumulated GHGs were the highest in the history from 2000 to 2010 according to the IPCC fifth assessment report (IPCC, 2014). The IPCC further reaffirms that climate change is expected to cause widespread negative impacts on human and natural systems (IPCC, 2014, 2018). Global warming is expected to manifest itself in a variety of ways including changes in cloud cover and related mean annual rainfall, depending on the geographical location. It can lead to a variety of negative impacts such as reduced productivity of natural resources, damaged human-built environments, increase in health hazards, among others (Kolstad & Toman, 2005; Wheeler & von Braun, 2013). There has been extensive debate on the likely impacts of climate change on different sectors of the economy including agriculture, forestry, water, energy, coasts, ecosystem and health (Boko et al., 2007; Mendelsohn & Massetti, 2017; IPCC, 2014). The agricultural sector is among the most vulnerable to climate change. Changes in the surface evaporation, soil moisture and surface atmospheric humidity are likely to have significant impacts on agricultural yield in particular (Mendelsohn & Massetti, 2017; Wheeler & von Braun, 2013; Lamboll et al., 2017). Cui (2020) shows that climate change explains 10–35 per cent of observed US corn and soybean expansion over the past 30 years.

Small island states, such as Mauritius, have greater concerns with climate change which has far reaching effects on their environment and economic prospects (Republic of Mauritius, 2016). Climate change induced effects on Small Island Development States (SIDS) include changes in the patterns of precipitation which impact on water resources, loss of land along coastline, rising food insecurity and degradation of biodiversity, among others (Leal Filho et al., 2020). Agricultural activities, which play an important role in the economy of many SIDS, are highly vulnerable to current climate conditions. At the same time, SIDS have weak financial position with restricted human capacity (Veitayaki, 2010). Consequently, their adaptive capacities are very limited. The impacts of climate change on agriculture can impede development efforts geared towards improving the livelihoods of poor farmers as well as the population.

Mauritius is classic example where the effects of global warming can cause severe negative consequences on the livelihood of farmers (Republic of Mauritius, 2016). Changes in climate conditions are clearly visible over the island. For instance, the temperature for the period 1950–2015 has increased on average by 0.022 °C per year while annual precipitation shows a decreasing long term trend by 16.2 per cent between 1950 and 2015.

In order to design and invest on adaptation strategies for farmers, it is important to estimate the likely economic impacts of climate change on the agricultural sector. The estimates can be useful for international donors to better understand the climate needs and provide an optimal financial assistance to SIDS as established under the Paris Agreement of 2015 (Scandurra et al., 2020).

Attempts to quantify the changes in climate conditions on agriculture have concentrated on crop modelling (Bozzola et al., 2018). However, agro-economists point out that when faced with the effects of climate change, farmers exert efforts to implement some adaptation measures, which eventually attenuate the impacts on farming (Mendelsohn & Nordhaus, 1994; Wheeler & von Braun, 2013). The method to quantify climate change impacts must therefore account for these adaptation measures. Crop modelling, which fails to consider these measures, would over-estimate climate change impacts.

The Ricardian method has been proposed as an alternative to study the long-term impacts of climate change on agriculture (Mendelsohn & Nordhaus, 1994). The method examines the extent to which observed cross-sectional variations in land values (or net farm revenues) can be explained by long-term climatic conditions while controlling for other relevant factors. A major strength of the Ricardian method is its ability to measure the long-run impacts of climate change taking into account the behaviour and capacity of each farmer to adapt. While adaptation practices cannot be accurately measured to adjust crop models, they significantly influence agricultural production (Cui, 2020). The Ricardian method has been applied worldwide from the US by Mendelsohn and Nordhaus (1994), to Europe by Maddison (2000), Lippert et al. (2009) and Bozzola et al. (2018), as well as in Africa by Gbetibouo and Hassan (2005), Seo et al. (2009), Kabubo-Mariara and Karanja (2007), Nhemachena et al. (2010) and Fonta et al. (2018), among others. Several studies in Asian countries are also observed such as Cui et al. (2004) and Mendelsohn (2014) in China, Mishra et al. (2016) in India and Hossain et al. (2019) in Bangladesh. Studies in South America include Sanghi and Mendelsohn (2008) for Brazil and by Seo and Mendelsohn (2008) for South American farms.

The present study contributes to the literature by applying the Ricardian method to a small island state, Mauritius. This is in line with De Salvo et al. (2013) who show that the Ricardian model can also be applied on a small territorial scale. Using climate variables and farm data from a sample of 392 farmers, the study estimates the marginal impact of a change in temperature by 1 °C and a change in precipitation by

1 mm on net farm revenue, characterising the welfare of farmers. The elasticities of net farm revenue per acre with respect to changes in temperature and precipitation are also estimated. The main purpose of the study is to show the sensitivity of climate change on agricultural activities and to provide insights on the likely threat posed by climate change on farmers.

2 Climate and Agroecosystem

The agro-ecosystem is a complex interaction between atmosphere and climate, nutrients and soils, and biological factors. (Adams et al., 2001; Lamboll et al., 2017). Crops respond directly to changes in temperature, moisture, precipitation and carbon dioxide. Temperature affects the rate of photosynthesis and hence, the rate at which plants absorb (and respire) carbon dioxide from (and to) the atmosphere. The optimum temperature varies for different crops and crop varieties. For example, the optimal range for many crops classified as C3 is 15 °C–20 °C. These include soybean, rice, barley, potatoes sugar beet, and wheat. Other crops in the category of C4 plants such as maize, sugar-cane, millet and sorghum have higher optimum temperature in the range of 25 °C–30 °C. Rising long-term temperature leads to higher respiration rates, thereby reducing crop yields (Adams et al., 2001). Temperature changes also accelerate the decomposition rate of soil nutrients. At some critical stages of crop development, a few hours of temperature above the threshold may damage crops irreversibly (Challinor, 2005).

Precipitation directly influences soil moisture. When climate change leads to a fall in precipitation, rising temperature will decrease soil moisture which directly affects plant growth. The combined effect of rainfall and temperature changes on soil moisture will vary spatially which can alter the geographic suitability of crops, leading to changes in the types and extent of crops in some areas.

3 Modelling Climate Impacts on Agriculture

There are four main methods to measure the relationship between climate change and agriculture: controlled laboratory experiments, agronomic crops models, panel weather studies and cross-sectional model (or the Ricardian model) (Mendelsohn & Massetti, 2017) The first three methods attempt to model yields and climate variables (Mendelsohn & Dinar, 2003; Gbetibouo & Hassan, 2005; Mendelsohn & Massetti, 2017), though panel weather studies may use net farm revenue but are data intensive. Models based on yields do not measure welfare directly (Mendelsohn & Massetti, 2017). Given the high level data required, the three methods are limited to few sites in the world. It is not appropriate to generalise the findings to SIDS.

The Ricardian method uses cross-sectional data on agricultural activities to analyse the sensitivity of agriculture to climate change (Mendelsohn & Nordhaus,

1994; Lui et al., 2004; Seo & Mendelsohn, 2008; Lippert et al., 2009; Jawid 2020). The method is based on the premise that a scientific analysis of spatially distributed agricultural output across various types of climate conditions may reveal the climate sensitivity of farms (Mendelsohn & Massetti 2017). In this respect, it estimates the direct impact of changes of climate condition on crop yields as well as the indirect substitution among different inputs. The method focuses on land productivity, rather than solely on crop yields. Land productivity, in turn, would be reflected in the value of land at a particular site, assuming a competitive land market which has already adjusted to any shock. Consequently, land values would equal to the sum of discounted future benefits or net revenue which are derived from the land uses. Any factor that influences the productivity of land, including climate condition, will be reflected in changes in land values or net revenues. (Mendelsohn & Nordhaus, 1994). In many countries, data on land values are limited and if they exist, they may not emanate from a long run land market. Thus, many studies have used net revenue from farms as substitutes to land values.

The Ricardian model assumes that each farmer, being a rational economic agent, maximises the net revenue of agricultural activities which are conducted on a plot of land. The farmer takes into account the expenditure arising from production activities together with the revenues generated from traded products and optimises the net benefit subject to the endogenous and exogenous conditions of the farm. The exogenous conditions include climate, soil and other constraints falling outside their control (Seo et al., 2009). Each farmer is assumed to opt for a combination of agricultural activities and inputs that give the highest return. In this respect, the relationship between agricultural revenue and climate conditions (for example long term temperature and precipitation) can provide an estimate of climate change impacts on agriculture, taking into account the adaptation behaviour of farmers (Kurukulasuriya & Ajwad, 2007; Bozzola et al. 2018).

Adaptation strategies are thus part of an optimizing behaviour. This characteristic makes Ricardian approach more suitable over the experimental approach. When farmers respond to climate change through their adaptation strategies, they reduce the climate sensitivity of agricultural production (Mendelsohn & Massetti 2017; Kurukulasuriya & Mendelsohn, 2008). Ignoring these responses is likely to overestimate the impacts of climate change in the agricultural sector. Economists' treatment of climate change assessment emphasises the adaptive capacity and adaptation strategies. Farmers, according to the economists' point of view, will adopt new methods only if the benefits exceed the costs.

The IPCC (2014) defines adaptation as the actions of adjusting practices, processes and capital in response to the actuality or threat of climate change as well as changes in the decision environment, such as social and institutional structures, and altered technical options. It is a dynamic process which occurs in the context of other endogenous processes including population growth, migration, technological change, economic growth, and structural transformation (Hertel & Rosch, 2010). Environmental factors influence the types of adaptation. Depending on the effects of a warmer climate and/or changes in rainfall, farm level adaptations include planting and harvest dates, crop rotations, selection of crops and crop varieties for cultivation,

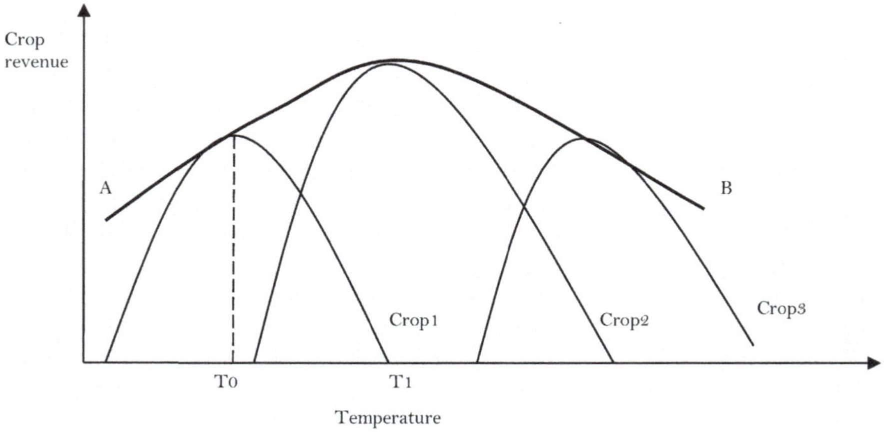

Fig. 1 Effective impact of climate change on agriculture. Source: Mendelsohn and Nordhaus (1994) and Mendelsohn and Dinar (1999)

improving irrigation facility, using different fertilizers, and adopting tillage practices (Gbetibouo & Hassan, 2005; Kurukulasuriya & Mendelsohn 2008). Farm households adopt a variety of risk minimization techniques in order to diversify against climate risks. For instance, farmers may cultivate a mix of crops that hedge climate risks. These adaptations are the natural consequence of producers' goals of maximizing returns to their land resources. However, these strategies can result in expected profits (Hertel & Rosch, 2010).

Figure 1 has been used by Mendelsohn and Nordhaus (1994) and Mendelsohn and Dinar (1999) to explain the importance of taking adaptation strategies by farmers in climate change assessment models. Accordingly, Crop1 represents a crop activity with its economic revenue on the vertical axis and temperature on the horizontal axis. With rising temperature, farmers attempt to diversify and use different methods so as to adjust to climate change and hence, shift to another crop activity, denoted by curve Crop2. Further adaptation leads to Crop3. Crop simulation will provide estimate specifically for Crop1, Crop2 and Crop3, while the effective climate change response is shown by the envelope curve AB.

Mendelsohn and Nordhaus (1994) used the Ricardian approach to evaluate the economic impacts of climate in the United States, and observed that for all seasons excluding autumn, an increase in temperature reduces farmland values while a rise in precipitation increases farmland values. When estimates from the Ricardian model were compared to agronomic studies, Mendelsohn and Nordhaus (1994) show they were significantly lower. The estimates captured the adaptation strategies which agronomic studies have proved unsuccessful. The Ricardian model consequently estimates the relationship as shown by the curve AB in Fig. 1.

Reinsborough (2003) develops a comparative static Ricardian model to examine the variation in agricultural land values and net farm revenues for Canada and finds that the covariates of climate variables were statistically significant along with those associated with latitude and population density. Reinsborough (2003) also finds that global warming is likely to be beneficial for the Canadian agriculture.

The impact of rising mean temperature and precipitation on agriculture depends on the geographical location and natural systems (Seo et al., 2009). Gbetibouo and Hassan (2005) and Benhin (2008) used net farm revenues within the Ricardian model of climate change for South African farms and concluded that net farm revenues were more responsive to changes in temperature rather than precipitation and an increase in temperature would be beneficial in summer but harmful in winter whereas a reduction in rainfall would affect net farm revenues negatively. Kabubo-Mariara and Karanja (2007) also concluded that, in Kenya increased winter temperatures would increase net crop revenue, while high summer temperatures would decrease it. However, when the Ricardian technique was applied in Ethiopia by Deressa and Hassan (2009), the study found that marginally increasing temperature during both summer and winter would significantly reduce crop net revenue per hectare whereas marginally increasing precipitation during spring would significantly increase net crop revenue per hectare. Kurukulasuriya and Mendelsohn (2008) applied the method at a larger scale to African countries and concluded that climate change could have strong negative impacts on currently dry and hot locations.

Liu et al.'s (2004) study in China conclude that under the most probable climate change scenarios, both higher temperatures and more precipitation would have an overall positive impact on China's agriculture with the impacts varying seasonally and regionally. Similar conclusions are obtained by Hossain et al. (2019) in Bangladesh and Jawid (2020) in Afghanistan. However, Mendelsohn (2014) and Mishra et al. (2016) find that rising mean temperature has a negative impact on the net farm revenues in India.

The Ricardian method has its limitations. Crops evaluated are not subject to controlled experiments across farms (Hassan, 2010). Thus, it is important to include other important variables such as soil quality and irrigation in the model (Fonta et al., 2018). Future changes that influence the agricultural sector, such as technical change, policies and institutions, are ignored since the approach make prediction using present farming practices (Gbetibouo & Hassan, 2005; Mendelsohn & Massetti 2017). Kurukulasuriya and Mendelsohn (2008) note that the Ricardian approach considers the cost of different options but it fails to measure transition costs. Hence, estimates from the model should be interpreted with caution. Another weakness of Ricardian studies is that the method does not include the effect of changing prices as it assumes that input and output prices remain constant (Mendelsohn, 2014). If the output of a certain crop decreased (increased) leading to a rise (fall) in the price of that crop, the effect on net revenue would therefore be less than predicted by the Ricardian method. A price increase would partially offset the lost revenue. The Ricardian method consequently overestimates both benefits and damages (Mendelsohn & Nordhaus, 1994). The Ricardian method does not also measure the effect of carbon dioxide concentrations which could benefit crops through carbon fertilization (Mendelsohn, 2014). The impacts may be overestimated since controlled laboratory experiments consistently report that higher carbon dioxide levels increase yields in most crops as shown by Kimball (1983).

4 Climate Change in Mauritius

Mauritius lies within the tropics and temperature distribution is determined mainly by altitude. Since precipitation depends largely on temperature, it implies that the latter may also be influenced more by elevation than by latitude. There are two seasons in Mauritius: the warm wet summer and the cooler dry winter. The former starts from November to April while the latter begins from June to September. October and May are usually the transition period.

Seasonally, temperature varies by only a few degrees, lowest on average in July, August and September at around 22 °C, and highest on average in January, February and March (Meteorological Services, 2009). The warmest months are January and February with average day maximum temperature reaching 29.2 °C and the coolest months are July and August when average night minimum temperatures drop down to 16.4 °C. The wet season peaks between January and March, when 200–250 mm of rain fall per month. This wet season rainfall is controlled by the seasonal migration of the tropical rain belt or the Inter-Tropical Convergence Zone (ITCZ), which oscillates between the northern and southern tropics over the course of the year, reaching its southern-most position over the Islands of Mauritius in January and February. Variations in the intensity and timing of the movements of the ITCZ from 1 year to the next cause inter-annual variation in the wet season rainfall. The most well documented cause of this variability is the El Niño Southern Oscillation (ENSO) which causes warmer and drier than average conditions in the wet season of Eastern Africa in its warm phase (El Niño) and relatively cold and wet conditions in its cold phase (La Niña).

The average temperature for the period 1950–1959 is by 1.69° C higher than the average for the period 2006–2015. The observed rate of temperature change is on average 0.022 °C per year. According to the Third National Communication on Climate Change, projections made on the basis of Representative Concentration Pathway (RCP) 4.5 and RCP 8.5 (the *business as usual* scenario and the *worst case* scenario, respectively) indicate an increase in temperature of up to 2 °C in Mauritius over the period 2051–2070 (Republic of Mauritius, 2016) (Fig. 2).

Over the same period, the central plateau has witnessed a decrease in annual precipitation from a maximum of 4000 mm/year to 3800 mm/year with drying being more pronounced to the north and west of the island. When a long-run trend is plotted using the monthly precipitation, as show in Fig. 3, the change from 1950 to 2015 stands at 16.2 per cent. The precipitation seasonal cycle for Mauritius shows an increase in monthly precipitation during the period from May to October.

Some major observed climate impacts are: a lengthening of the intermediate dry season; a change in the transition period between winter and summer; a shift in the start of the summer rains; an increasing number of consecutive dry days and

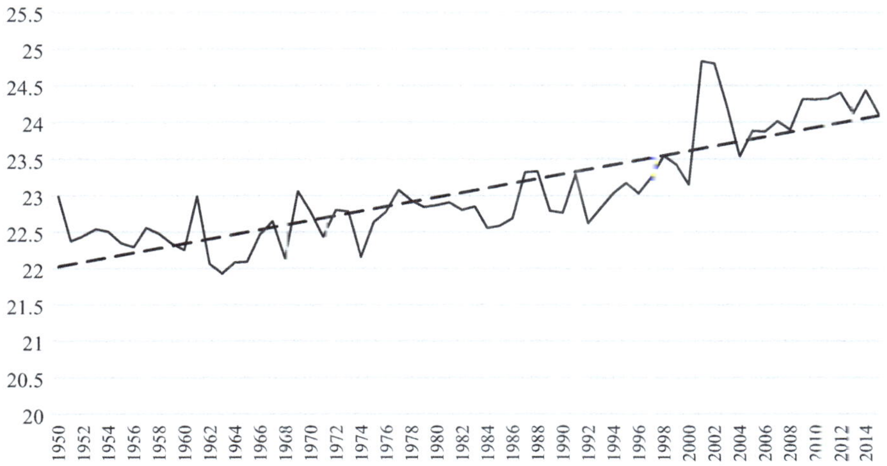

Fig. 2 Average temperature 1950–2015. Source: World Bank Climate Portal

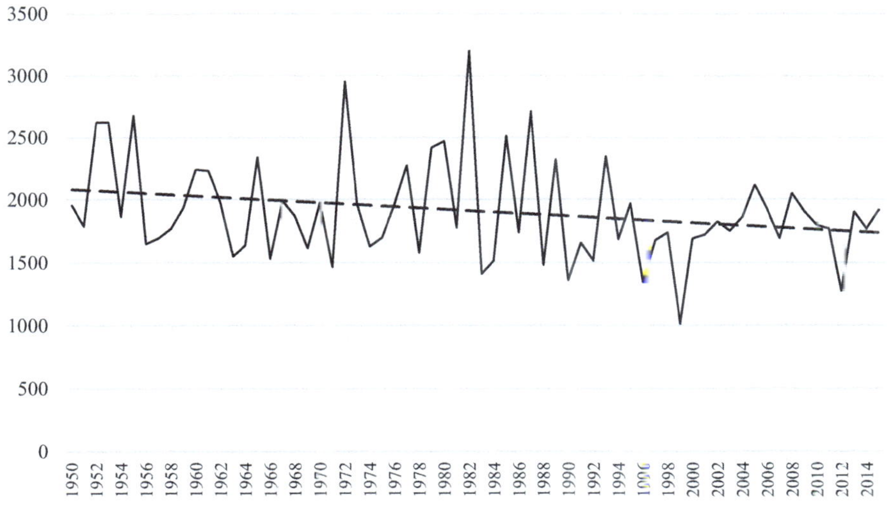

Fig. 3 Annual Precipitation 1950–2015. Source: World Bank Climate Portal

deceasing number of rainy days; a rise in episodes of flash floods; increased frequency of extreme weather events and heavy rains; bigger tropical cyclones; and sea level rise (Meteorological Services, 2009; Republic of Mauritius, 2016).

5 Materials and Methods

5.1 Conceptual Framework

Following Mendelsohn and Nordhaus (1994), the Ricardian approach involves specifying a net farm revenue function per acre (V) of the form:

$$V = \sum P_i Q_i(X, F, Z, G) - \sum P_X X \qquad (1)$$

where V is net farm revenue per acre, P_i is the market price of crop i, Q_i is output of crop i, X is a vector of purchased inputs (other than land), F is a vector of climate variables such as temperature and precipitation, G is a set of socio-economic variables, Z is a set vector of control variables such as soil types and irrigation, and P_x is a vector of input prices. A major advantage of using net farm revenue is that it accounts for the direct impacts of climate on yields of different crops as well as the indirect substitution of different inputs, the introduction of different activities and other possible adaptations by farmers to different climatic shocks (Mendelsohn & Dinar, 1999). Under this framework, a farmer is assumed to maximise net farm revenue per acre by choosing inputs (X) subject to F, G and Z. The Ricardian approach implies a reduced form model that examines how exogenous variables such as F, G and Z affect net farm revenues. Eq. (1) can be specified as follows:

$$V = \beta_0 + \beta_1 F + \beta_2 F^2 + \beta_3 G + \beta_4 Z + u \qquad (2)$$

where u is the error term and β_i represents the regression coefficients to be estimated including the intercept term. In order to capture the nonlinear relationship between net farm revenues and climate variables, the estimation specifies the vector of climate variables as a quadratic formulation. The expected marginal impact of a single climate variable (denoted by f_k) on net farm revenues, evaluated at the mean is:

$$E\left[{}^{\partial V}\!/_{\partial f_k}\right] = b_{1,k} + 2b_{2,k}E[f_k] \qquad (3)$$

where $b_{1,\,k}$ and $b_{2,\,k}$ are the regression coefficients for f_k and f_k^2 respectively. When the quadratic term is positive, the net revenue function is U-shaped and when the quadratic term is negative, the function is hill-shaped. Based on agronomic research and previous cross-sectional analyses, farm revenue is expected to have a hill-shaped relationship with temperature (Deressa & Hassan, 2009; Kurukulasuriya & Mendelsohn, 2008). For each crop, there is a known temperature at which that crop grows best across the seasons. The relationship of seasonal climate variables, however, is more complex and may include a mixture of positive and negative coefficients across seasons.

5.2 Data Collection Strategy

To estimate the Ricardian model, the required key data are temperature and precipitation in different regions with varied altitudes. Given that climate change involves long-term trends, the study uses monthly climate normal. The mean monthly temperature and precipitation (which also reflect the winter and summer season) should vary across regions. The use of the Ricardian approach in Mauritius is motivated by the fact the island exhibits micro-climatic conditions (Cheeneebash & Rughooputh, 1997). There are at least 27 microclimates which have been identified to represent the climate system of the island (Padya, 1984; Cheeneebash & Rughooputh, 1997). According to De Salvo et al. (2013), when the geographical scale is small, it is important to select the proper functional form and control variables.

The Mauritius Meteorological Services publishes a monthly bulletin on weather conditions for each month. Temperature is collected at around 28 sites over the island of Mauritius. Figure 4 shows the distribution of locations where temperature data are collected on a daily basis.

Table 1 shows the mean annual average, mean summer average and mean winter average in 22 regions (the remaining have missing data during some month and have therefore been excluded in the analysis). Table 1 indicates that there is a significant degree of variation in mean temperature across regions and over the seasonal cycle. This variation in temperature is important for the Ricardian econometric model

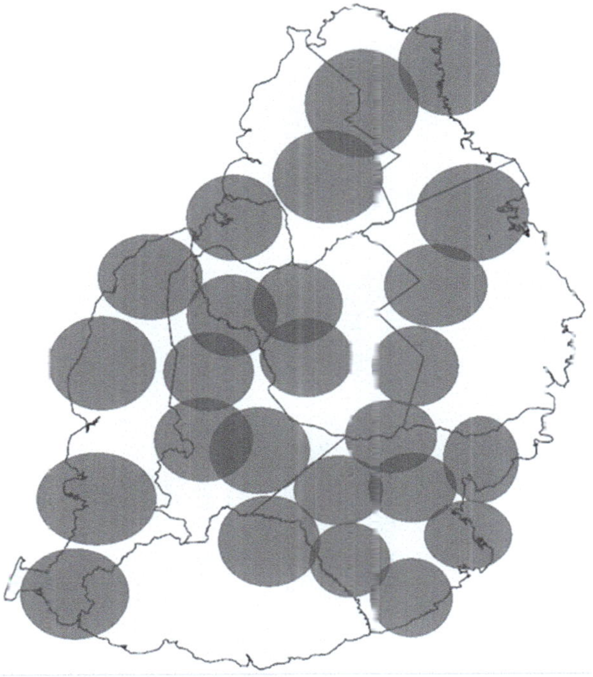

Fig. 4 Temperature data collection sites in the Island of Mauritius. Source: author (QGIS)

Table 1 Average temperature 1980–2010

Regions	Mean annual average	Mean summer average	Mean winter average	Regions	Mean annual average	Mean summer average	Mean winter average
Bois Cheri	20.5	22.4	17.8	Mont desert Alma	21.7	23.5	19.3
Britannia	22.0	23.8	19.6	Plaisance	23.6	25.6	21.4
Case Noyale	25.4	27.4	22.6	Plamplemousse	23.9	25.7	21.4
Constance	24.2	26.3	21.6	Quantre Bornes	22.7	24.6	20.1
Digue Seche	24.4	26.2	22.3	Riche en Eau	23.4	25.2	21.2
Fort William	25.4	27.2	23.3	Rose Belle	21.2	23.2	18.6
FUEL	23.1	24.9	20.7	Sans souci	21.8	23.9	19.4
La Barraque	23.4	25.3	21.0	Trou aux Cerf	20.8	22.6	18.3
Labourdonnais	23.8	25.5	21.3	Union Park	21.2	23.0	18.6
Le Morne	26.5	28.2	23.8	Vacoas	21.8	23.7	18.9
Le Val	21.3	23.2	18.6				

Source: Monthly Bulletin of Meteo

Table 2 Precipitation 1980–2010

Regions	Mean Annual	Mean Summer	Mean Winter	Regions	Mean Annual	Mean Summer	Mean Winter
Balaclava/ solitude	79.9	713.7	153.8	Hermitage	205.4	1699.8	513.7
Beau Rivage	118.0	987	255.3	Highlands	159.6	1396	348.6
Beau Sejour	120.8	1006	287.7	La Barraque	156.6	1260.1	391.1
Bel Ombre	112.5	910.1	275.2	La Marie	191.7	1580.2	513.1
Bell Vue	113.8	941.1	274.8	Moka	134.7	1169.2	324.6
Bell Vue Maurel	114.1	948.7	254.5	Mont Choisy	87.4	747.3	192.8
Belle Rose	164.6	1340.5	386.9	Notre Dame	116.8	1028.8	238
Bois Cheri	325.9	2383.8	1041.9	Nouvelle Decouvert	267.3	2002.1	915.8
Bras D'eau	104.6	863.6	231.2	Piton du Milieu	263.5	1950.7	814.1
Britannia	213.0	1658.9	587	Plaisance	143.6	1166.5	342.9
Cap Malheureux	96.0	806.5	216.9	Plamplemousse	112.4	939.7	262.4
Case Noyale	94.8	878	163.6	Providence	232.7	1912.4	529.4
Constance	130.4	1095.6	281.9	Quatre Bornes (la Louise)	121.4	1113.6	224.2
Corps de Garde	84.0	774.2	145.3	Riviere des Anguilles	139.4	1113.3	357.2
Digue Seche	110.3	893.6	264.9	Rose-Belle	181.[illegible]	1491.8	429.9
Fond du Sac	91.3	793.6	201.6	Trou aux Cerf	218.3	1508.2	841.9
Fort William	59.1	568.8	77.8	Vacoas	166.5	1413.1	402.2
FUEL	163.5	1350.4	381.5	Valetta	233.5	1918.4	595.3
Grand Port	127.1	1048.5	283.5	Valton	122.3	1069.9	256.8
Grand retraite	144.9	1191.8	329.4	XVI Mile	230.0	1471.3	943.2

Source: Monthly Bulletin of Meteo

Data on precipitation is collected over more than 300 sites. Table 2 shows 43 sites and associated mean precipitation for 1980–2010 which have been selected for the study.

A survey questionnaire was prepared to collect data on farm revenues and socioeconomic characteristics of farmers. The questionnaire was adapted from the one designed jointly by the School of Forestry and Environmental Studies of Yale University and the Center of Environmental Economics and Policy in Africa (CEEPA), University of Pretoria. Around 30 interviewers were trained for the study. Each interviewer was given a street within a list of 80 regions which was selected at random within the proximity of the sites where temperature and precipitation data are collected by the Meteorological Services. Interviewers were informed to administer the questionnaire to a farmer whose activities are conducted in that region and located on that street. The interviewer was instructed to then move at least 100–500 m from the first location to select the second farmer, until 20–25 farmers are thus selected.

5.3 Econometric Modelling

Based on the conceptual framework and the data collected for the purpose of the study, the econometric equation is as follows:

$$\begin{aligned} NREV_{ij} = {} & b_0 + b_1 ALT_j + b_2 LAB_{ij} + b_3 HSIZE_{ij} + b_4 AGE_i + b_5 TSUM_j \\ & + b_6 TSUM_j^2 + b_7 TWIN_j + b_8 TWIN_j^2 + b_9 PSUM_j + b_{10} PSUM_j^2 \\ & + b_{11} PWIN_j + b_{12} PWIN_j^2 + u_{ij} \end{aligned} \quad (4)$$

The definition of the covariates are as follows:

$NREV_{ij}$ = net revenue per acre of farm i in region j; ALT_j = altitude of region j; LAB_{ij} = labour employed in farm i; $HSIZE_{ij}$ = household size in farm i; AGE_i = age of farmer in farm i; $TSUM_j$ = mean summer temperature in region j; $TWIN_j$ = mean winter temperature in region j; $PSUM_j$ = mean summer precipitation in region j; $PWIN_j$ = mean winter precipitation in region j; u_{ij} = error term.

6 Findings

The survey questionnaire was administered to 450 farmers in relation to the sites as identified in Sect. 5. The survey took place from November 2012 to April 2013. A total of 392 responses were deemed appropriate for the analysis since the remaining had missing information. From the survey questions, the net annual revenue for each farmer was calculated by the taking into account the amount of agricultural output harvested and the relevant prices. The costs of fertilisers and other inputs used by farmers were eventually subtracted. Wage costs were not considered in the net revenue since the contribution of labour is specifically captured by household size and direct labour employed in the farm. Some selective descriptive statistics from the survey are shown in Table 3.

The econometric results are provided in Table 4. The climate variables, namely temperature and precipitation, are added in linear form in Model 1. There is a positive relationship between altitude and net revenue and the coefficient is statistically significant ($p < 0.001$). The result confirms that region in high altitude is

Table 3 Description statistics from survey

	Mean	Standard Deviation	Minimum	Maximum
Net farm revenue per acre (Rs000)	232.69	25.50	153	342
Labour	2.19	1.83	0	10
Household size	2.74	1.20	0	7
Age	50.9	11.5	23	82

Source: survey

Table 4 Ricardian regression estimates—linear and non-linear cross section response function

Covariates	Model 1	Model 2
CONS	269.65 (32.54)***	−15.88 (576.49)
ALT	0.072 (0.013)***	0.09 (0.01)***
LAB	−7.19 (0.86)***	−7.10 (0.80)***
HSIZE	3.86 (2.17)*	4.01 (2.17)*
AGE	0.14 (0.18)	0.14 (0.18)
TSUM	−7.42 (2.26)***	−141.48 (48.76)***
TSUM2		2.53 (0.93)***
TWIN	6.65 (1.83)***	194.0 (33.89)***
TWIN2		−4.22 (0.74)***
PSUM	−0.043 (0.03)	−0.21 (0.13)*
PSUM2		0.0004 (0.0002)**
PWIN	0.06 (0.05)	0.07 (0.15)
PWIN2		−0.00009 (0.0004)
Number of observation	392	392
F-statistics	F(8,383) = 21.33***	F(13,75)33.9***
Adj R-square	0.31	0.34

Source: author
Standard errors are robust regression results
*, **, *** mean significant at 10%, 5% and 1% level

expected to be more productive. Higher the number of labour employed, the lower is the net revenue of the farm ($p < 0.001$). Household size, which reflects farm labour from the family, increases net revenue ($p < 0.1$). When temperature is added in linear form, the econometric results show that a rise in summer temperature reduces net farm revenue per acre while an increase in winter temperature raises net farm revenue per acre. This result is consistent with the study by Gbetibouo and Hassan (2005) and many others. The coefficients of *TSUM* and *TWIN* are highly statistically significant.

The coefficients of the precipitation variables are not statistically significant. It appears that changes in rainfall do not have major effects on farm revenue. A simple observation is that since Mauritius is a small island, the water storage system allows water to be distributed to regions facing a scarcity of water through the irrigation network. Nevertheless, Model 2 provides a better picture when the quadratic terms of the climate variables are added in the estimation.

Table 5 Marginal effects from a rise in temperature 1 °C (Net change in welfare per acre)

	Increase in summer temperature	Increase in winter temperature	Overall effect	% Change in welfare (1 °C rise)
Linear/non-linear model	−7400***	6600***	−800***	0.3

Source: author

Model 2 is a better statistical fit of the data in terms of the F and adjusted R-square statistic. The quadratic relationship between net farm revenue and climate variables is more suitable for further analysis. The winter temperature covariate (*TWIN*) exhibits a hill-shaped relationship whereas the summer temperature covariate (*TSUM*) has U-shaped relationship with net revenue. The sign of the coefficients of the temperature covariates is consistent with the study by Gbetibouo and Hassan (2005) in South Africa, Kabubo-Mariara and Karanja (2007) in Kenya, Sanghi and Mendelsohn (2008) in Brazil, and Deressa et al. (2009) in Ethiopia. However, while there is an inverted U relationship between mean rainfall and net farm revenues, the coefficient is not statistically significant. Net farm revenues are particularly responsive to changes in summer precipitation.

The regression result provides a cross-sectional response function which can be used to estimate the effect of an infinitesimal change in temperature and rainfall. Using Eq. (3), Table 5 calculates the marginal effects of an increase in mean temperature by 1 °C and a rise in mean precipitation by 1 mm. A rise in the mean summer temperature decreases net farm revenue by Rs7400 per acre while a rise in temperature in winter enhances farm revenue by Rs6600 per acre. The marginal effect of a rise in temperature evaluated at mean level is consistent for both the linear and non-linear model. Overall effect shows a fall in crop revenue of around Rs800 (0.03%). Using the 2012–2013 exchange rate for MUR-USD,[1] the effect of a temperature rise on net revenue stands at USD26.6 per acre or USD66 per hectare.[2] This figure can be compared with USD28.5 by Kurukulasuriya and Mendelsohn (2008) for Africa, USD57.8 by Mano and Nhemachena (2007) for Zimbabwe and USD175.5 by Seo and Mendelsohn (2008) in Latin America. In Nigeria, Fonta et al. (2011) estimate the loss in net revenue to be USD38 per hectare per annum.

The marginal effect of a 1 mm rise in precipitation is shown in Table 6. A rise in precipitation in summer decreases net revenue by Rs43 and Rs12 respectively for the linear and non-linear model. In winter, a rise in precipitation increases net revenue by Rs61 and Rs53 respectively for the two models. However, as mentioned earlier, the coefficients of winter precipitation are not statistically significant. In using the non-linear response function, both the summer and winter effects are taken into account since the coefficient of a particular co-variate in the regression estimate responds to all covariates included in the estimation.

[1] USD1 = MUR30.1 (World Bank Indicator, https://data.worldbank.org/indicator/PA.NUS.FCRF?locations=MU, accessed 2 June 2018).

[2] Using the formula 1 hectare = 2.47105 acres.

Table 6 Marginal effects from a rise in precipitation by 1 mm

	Increase in summer precipitation	Increase in winter precipitation	Overall effect	% change in welfare
Linear model (Rs)	−43	61	20	0.008
Non-linear model (Rs)	−12	53	41	0.017

Source: author

Table 7 Elasticity of a rise temperature and precipitation

	Mean summer temperature	Mean winter temperature	Mean summer precipitation	Mean winter precipitation
Elasticity	−0.84***	0.63*	−0.02*	0.02

Source: author

The non-linear response function is used to calculate the marginal effect of rise in precipitation. The estimate gives a figure of Rs41 (USD1.4) per acre or USD3.5 per hectare annually.

The elasticity of temperature and precipitation is shown in Table 7. Elasticity is calculated as the percentage change in net farm revenue as a result of one percentage increase in climate variables.

The findings show that net farm revenue is more sensitive to temperature than precipitation. An increase in mean temperature in winter is likely to increase agricultural productivity. However, a rise in mean summer temperature will lead to fall in net revenue. A rise in the mean summer precipitation reduces net revenue of farmers. According to the econometric estimates, farmers are likely to be worst off in warm humid summer which extends from November to April, and the gain in winter do not offset the loss in summer. The most probable explanation is that since the island is already witnessing a temperature which is close to the tolerance level of the plants, any further increase is likely to affect agricultural productivity. The findings suggest that there is a need for intervention particularly in summer season.

The relatively lower elasticity of agricultural output with respect to precipitation warrants an explanation. In winter, since there is less rainfall compared to summer, it was expected that agriculture output would be seriously affected when there is a fall in long run precipitation level The econometric analysis shows that agricultural output is not responsive to changes in precipitation in winter. In fact, Mauritius' relatively cool dry winter runs from June to September. A change is winter rainfall is expected to impact significantly on agriculture output. The results appear inconsistent when referring to the Northern Plains of the island where 50% of the land is devoted to agriculture and which is generally dry with open evaporation exceeding rainfall. A possible explanation is that a small island like Mauritius is better able to organise its irrigation network such that in winter, when water is scarce, farmers can still be supplied with water from other regions. In fact, the Irrigation Authority, a parastatal organisation, was established under the Irrigation Authority Act of 1978 and functions under the aegis of the Ministry of Agro-Industry and Food Security. Its main functions are to study the development of irrigation and to make proposals to the Central Water Authority for the preparation of schemes for the irrigation of specific areas. It also aimed to implement and manage irrigation projects in every irrigation area and to undertake research into the optimum use of water made available by the Central Water Authority for irrigation. In the Northern Plains, there are currently several projects to improve the irrigation facilities to small farmers.

7 Conclusion and Implications for Policy

The study quantifies the potential economic impacts of climate change on the welfare of farmers in the agricultural sector. The results show that the change in net farm revenue (marginal welfare effect) following a rise in mean temperature by 1 °C is −0.3 per cent ($66 per hectare) while a fall in mean precipitation by 1 mm leads to a fall in net farm revenue by 0.008 per cent. The agriculture sector is negatively affected by changes in mean summer temperature and precipitation with an elasticity of −0.13. The elasticity of changes in mean temperature and precipitation evaluated at the mean net farm revenue per acre is −0.13 and 0.03 respectively.

Agriculture remains an economic activity for thousands of farmers and their families in Mauritius. Climate change, therefore, takes on economic significance, with the expected changes in agrosystems and processes. Adaptation strategies among the farming community are therefore essential. Adaptation measures in Mauritius are constrained by a lack of information, financing mechanisms and proper incentives to change farmers' conventional practices. Adaptation capabilities of farmers are limited. While there is a consensus that adaptation strategies must be mainstreamed in all facets of development strategies in the agricultural sector, the accompanying measures such as training and skill development of farmers are very often ignored. The relevant institutions in the agricultural sector face a big challenge to educate, train and develop the necessary capabilities of thousands of farmers in Mauritius.

Even if the island is small, the microclimate system and the geographical topography imply that climate change is expected to have varied impacts on farmers across regions. Those regions which are located in lower altitudes are likely to be severely affected by rising mean temperature. Consequently, those regions which lack proper irrigation network will also be highly affected by falling mean precipitation. Thus, a spatial vulnerability assessment of the agricultural sector may assist in defining a targeted approach to farmers in different regions.

Consequences of climate change in the agricultural sector may also impact to the wider Mauritian population given that the sector ensures a certain level of food security. This is not helped by the fact the island is situated far from the continents and food producing countries, and is self-sufficient in a few food crops only. Thus, the current study shows that climate change may have negative macroeconomic implications through import bills of food product given its adverse impacts on current production of foodcrops. Policy-makers take the issue of global warming seriously in defining a long term Mauritian economic strategy to reduce the vulnerability of the island.

An important caveat of this study is that the findings are dependent on the quality of the data collected and the sample size of the survey. The results can be compared to studies using different methods such as experimental agro-economic models. The study shows that the Ricardian approach may be used for a small island state if the climate conditions vary across regions significantly such as Mauritius.

Acknowledgments The survey was finance by a research grant from the Mauritius Research Council (MRC GA URGS 99) under the Africa Adaptation Programme. The study has also benefited from the discussion of participants on climate change training workshops (2011–2014) organised by the Centre for Environmental Economics and Policy in Africa (CEEPA), University of Pretoria with the support of the Swedish International Development Cooperation Agency (SIDA) and International Development Research Centre (IDRC-Canada). A special thank goes to Professor. R. Hassan from CEEPA for the Ricardian questionnaire and his insights on the Ricardian model.

References

Adams, R. M., Chen, C. C., McCarl, B. A., & Schimmelpfennig, D. A. (2001). *Climate variability and climate change: Implications for agriculture. The long-term economics of climate change* (pp. 95–113). Elsevier Science B.V.

Benhin, J. K. A. (2008). South African crop farming and climate change: An economic assessment of impacts. *Global Environmental Change, 18*(4), 666–678.

Boko, M., Niang, I., Nyong, A., Vogel, C., Githeko, A., Medany, M., Osman-Elasha, B., Tabo, R., & Yanda, P. (2007). Africa. In M. L. Parry, O. F. Canziani, J. P. Palutikof, P. J. van der Linden, & C. E. Hanson (Eds.), *Climate change 2007: Impacts, adaptation and vulnerability. Contribution of working group II to the fourth assessment report of the intergovernmental panel on climate change (IPCC)* (pp. 433–467). Cambridge University Press.

Bozzola, M., Masseti, E., Mendelsohn, R., & Capitanio, F. (2018). A Ricardian analysis of the impact of climate change on Italian agriculture. *European Review of Agricultural Economics, 45* (1), 57–79.

Challinor, J. (2005). Aspects of climate change prediction relevant to crop productivity. *Philosophical Transactions: Biological Sciences, 360*(1463), 1999–2009.

Cheeneebash, J., & Rughooputh, S. (1997). Analysis of the relationship between rainfall over Mauritius and physical site characteristics. *Revue Agricole et Sucrière de l'Ile Maurice, 75*(1), 36–42.

Cui, X. (2020). Climate change and adaptation in agriculture: Evidence from US cropping patterns. *Journal of Environmental Economics and Management, 101*(102306), 1–24.

De Salvo, M., Raffaelli, R., & Moser, R. (2013). The impact of climate change on permanent crops in an alpine region: A Ricardian analysis. *Agricultural Systems, 118*, 23–32.

Deressa, T. T., & Hassan, R. M. (2009). Economic impact of climate change on crop production in Ethiopia: Evidence from cross-section measures. *Journal of African Economies, 18*(4), 529–554.

Fonta, W. M., Ichoku, E. H., & Urama, E. N. (2011). Climate change and plantation agriculture: A Ricardian analysis of farmlands in Nigeria. *Journal of Economics and Sustainable Development, 2*(4), 63–75.

Fonta, W. M., Kedir, A. M., Bossa, A. Y., Greenough, K. M., Sylla, B. M., & Ayuk, E. T. (2018). A Ricardian valuation of the impact of climate change on Nigerian cocoa production: Insight for adaptation policy. *International Journal of Climate Change Strategies and Management, 10*(5), 689–710.

Gbetibouo, G. A., & Hassan, R. M. (2005). Measuring the economic impact of climate change on major South African field crops: A Ricardian approach. *Global and Planetary Change, 47*(2), 143–112.

Hassan, R. (2010). Implications of climate change for agricultural sector performance in Africa: Policy challenges and research agenda. *Journal of African Economies, 19*(2), ii77–ii105.

Hertel, T. W., & Rosch, S. D. (2010). *Climate change, agriculture and poverty* (World Bank Research Working Paper No. 5468). World Bank.

Hossain, M. S., Arshad, M., Qian, L., Zhao, M., Mehmood, Y., & Kächele, H. (2019). Economic impact of climate change on crop farming in Bangladesh: An application of Ricardian method. *Ecological Economics, 184*(106354), 1–11.

Intergovernmental Panel on Climate Change (IPCC). (2001a). *Climate change 2001: The scientific basis. Contribution of working group I to the third assessment report of the intergovernmental panel on climate change*. Cambridge University Press.

Intergovernmental Panel on Climate Change (IPCC). (2001b). *Climate Change 2001: Synthesis Report. A Contribution of Working Groups I, II, III to the third assessment report of the intergovernmental panel on climate change*. Cambridge University Press.

Intergovernmental Panel on Climate Change (IPCC). (2007a). *Climate change 2007: Synthesis report. An assessment of the intergovernmental panel on climate change*. Cambridge University Press.

Intergovernmental Panel on Climate Change (IPCC). (2007b). *The Physical Science Basis*. Fourth Assessment Report. Cambridge University Press.

Intergovernmental Panel on Climate Change (IPCC). (2014). *Climate Change 2014: Synthesis Report*. In Core Writing Team, R. K. Pachauri, & L. A. Meyer (Eds.), *Contribution of Working Groups I, II and III to the Fifth Assessment Report of the Intergovernmental Panel on Climate Change*. IPCC.

Intergovernmental Panel on Climate Change (IPCC). (2018). In Core Writing Team: V. Masson-Delmotte, P. Zhai, H.-O. Pörtner, D. Roberts, J. Skea, P.R. Shukla, A. Pirani, W. Moufouma-Okia, C. Péan, R. Pidcock, S. Connors, J.B.R. Matthews, Y. Chen, X. Zhou, M.I. Gomis, E. Lonnoy, T. Maycock, M. Tignor, & T. Waterfield (Eds.). *An IPCC Special Report on the impacts of global warming of 1.5°C above pre-industrial levels and related global greenhouse gas emission pathways, in the context of strengthening the global response to the threat of climate change, sustainable development, and efforts to eradicate poverty*. IPCC.

Jawid, A. (2020). A Ricardian analysis of the economic impact of climate change on agriculture: Evidence from the farms in the central highlands of Afghanistan. *Journal of Asian Economics, 67*(c), 1–16.

Kabubo-Mariara, J., & Karanja, F. (2007). *The economic impact of climate change on Kenyan crop agriculture: A Ricardian approach*. (World Bank Policy Research Working Paper No. 4334). World Bank.

Kimball, B. A. (1983). Carbon dioxide and agricultural yield: An assemblage and analysis of 430 prior observations. *Agronomy Journal, 75*, 779–786.

Kolstad, C. D., & Toman, M. (2005). The economics of climate policy. In K. G. Maler & J. R. Vincent (Eds.), *Handbook of environmental economics* (Vol. 3, pp. 1562–1593). Elsevier.

Kurukulasuriya, P., & Ajwad, M. I. (2007). Application of the Ricardian technique to estimate the impact of climate change on smallholder farming in Sri Lanka. *Climate Change, 81*(1), 39–59.

Kurukulasuriya, P., & Mendelsohn, R. O. (2008). A Ricardian analysis of the impact of climate change on African cropland. *African Journal of Agriculture and Resource Economics, 2*(1), 1–23.

Lamboll, R., Stathers, T., & Morton, J. (2017). Chapter 13: Climate change and agricultural Sytems. In S. Snapp & B. Pound (Eds.), *Agricultural systems - agroecology and rural innovation for development* (2nd ed., pp. 441–490). Academic.

Leal Filho, W., Ha'apio, M. O., Lütz, J. M., & Li, C. (2020). Climate change adaptation as a development challenge to small island states: A case study from the Solomon Islands. *Environmental Science and Policy, 107*, 179–187.

Lippert, C., Krimly, T., & Aurbacher, J. (2009). A Ricardian analysis of the impact of climate change on agriculture in Germany. *Climatic Change, 97*(593), 593–610.

Lui, H., Li, X., Fischer, G., & Sun, L. (2004). Study on the impacts of climate change on China's agriculture. *Climate Change, 65*(1–2), 125–148.

Maddison, D. (2000). A hedonic analysis of agricultural land prices in England and Wales. *European Review of Agricultural Economics, 27*(4), 519–532.

Mano, R., & Nhemachena, C. (2007). *Assessment of the economic impacts of climate change on agriculture in Zimbabwe: A Ricardian approach*. (World Bank Policy Research Working Paper No. 4292). World Bank.

Mendelsohn, R. O. (2014). The impact of climate change on agriculture in Asia. *Journal of Integrative Agriculture, 13*(4), 660–665.

Mendelsohn, R. O., & Dinar, A. (1999). Climate change, agriculture and developing countries: Does adaptation matter? *The World Bank Research Observer, 14*(2), 277–293.

Mendelsohn, R. O., & Dinar, A. (2003). Climate, water, and agriculture. *Land Economics, 79*(3), 328–341.

Mendelsohn, R. O., & Massetti, E. (2017). The use of cross-sectional analysis to measure climate impacts on agriculture: Theory and evidence. *Review of Environmental Economics and Policy, 11*(2), 280–298.

Mendelsohn, R. O., & Nordhaus, W. D. (1994). The impact of global warming on agriculture: A Ricardian analysis. *The American Economic Review, 84*(4), 753–771.

Meteorological Services. (2009). *Climate change impacts on Mauritius*. Government of Mauritius.

Mishra, D., Sahu, N. C., & Sahoo, D. (2016). Impact of climate change on agricultural production of Odisha (India): A Ricardian analysis. *Regional Environmental Change, 16*(2), 575–584

Nhemachena, C., Hassan, R., & Kurukulasuriya, P. (2010). Measuring the economic impact of climate change on African agricultural production systems. *Climate Change Economics, 1*(1), 33–35.

Padya, B. M. (1984). *The climate of Mauritius*. Meteorological Services, Government of Mauritius.

Reinsborough, M. J. (2003). A Ricardian model of climate change in Canada. *The Canadian Journal of Economics/Revue canadienne d'Economique, 36*(1), 21–40.

Republic of Mauritius. (2016). *Third National Communication: Report to the United Nations Framework Convention on Climate Change*. Government of Mauritius.

Sanghi, A., & Mendelsohn, R. O. (2008). The impacts of global warming on farmers in Brazil and India. *Global Environmental Change, 18*, 655–665.

Scandurra, G., Thomas, A., Passaro, R., Bencini, J., & Carfora, A. (2020). Does climate finance reduce vulnerability in Small Island developing states? An empirical investigation. *Journal of Cleaner Production, 256*(120330), 1–12.

Seo, N., & Mendelsohn, R. O. (2008). A Ricardian analysis of the impact of climate change on south American farms. *Chilean Journal of Agricultural Research, 68*(1), 69–79.

Seo, N. S., Mendelsohn, R. O., Dinar, A., Hassan, R., & Kurukulasuriya, P. (2009). A Ricardian analysis of the distribution of climate change impacts on agriculture across agro-ecological zones in Africa. *Environmental and Resource Economics, 43*(3), 313–332.

Statistics Mauritius. (2018). *Digest of Agricultural Statistics*. Government of Mauritius.

Stiglitz, J. E. (2011). The Mauritius miracle. *Project Syndicate A World of Ideas*, 7 March. [online] Retrieved February 3, 2012, from http://www.project-syndicate.org/commentary/the-mauritius-miracle

Subramanian, A. (2001). Mauritius: A case study. *Finance and Development*, 38(4). [online] Retrieved December 13, 2012, from http://www.imf.org/external/pubs/ft/fandd/2001/12/subraman.htm

Svirydzenka, K. & Petri, M. (2014). *Mauritius the drivers of growth—Can the past be extended?* (International Monetary Fund (IMF) Working Paper No. 14/134). International Monetary Fund.

Veitayaki, J. (2010). Chapter 17: Climate Change Adaptation Issues in Small Island Developing States. *Climate Change Adaptation and Disaster Risk Reduction: Issues and Challenges - Community, Environment and Disaster Riak Management, 4*, 369–391.

Wheeler, T., & von Braun, J. (2013). Climate change impacts on global food security. *Science, 341* (6145), 508–513. https://doi.org/10.1126/science.1239402

The Early Development of the Small Island Developing States' Climate Governance: A Disproportionate Impact on UN Climate Negotiations

Athaulla A. Rasheed

Abstract The policy significance of climate change was realised by the international community in the 1980s. As some of the countries most affected by climate change, Small Island Developing States (SIDS) have influenced the United Nations (UN) climate negotiations from the very beginning. This chapter analyses the climate foreign policy purposes of SIDS and their early impacts on the UN climate governance system. It is argued that, despite their weak material capabilities to shape international affairs, SIDS have made a notable and disproportionate impact on the UN climate negotiations to address their special case. Using constructivist approach to foreign policy analysis, this chapter explains how the ideas about common but differentiated responsibilities promoted in international climate negotiations have shaped SIDS' climate agenda during pre-and post-UNFCCC negotiations in driving climate governance for them. It is argued that an understanding of this disproportionate impact helps to better understand present and future trends in SIDS climate politics.

Keywords Small island developing states · Governance · Climate negotiations · Constructivism · Foreign policy · UNFCCC · Climate ideas · AOSIS

1 Introduction

Climate solutions remain at the core of the foreign policy agendas of Small Island Developing States (SIDS). Although SIDS are the least responsible for climate change, they are the places that are most gravely affected by it (Parry et al., 2007; Voccia, 2012, pp. 102, 104; Thomas et al., 2020). In times of global crises including the recent COVID-19 pandemic, SIDS have been collectively raising their voice against disproportionate impacts they incur because of their unique vulnerabilities to

A. A. Rasheed (✉)
Australian National University, Canberra, ACT, Australia
e-mail: athaulla.rasheed@anu.edu.au

S. Moncada et al. (eds.), *Small Island Developing States*, The World of Small States 9, https://doi.org/10.1007/978-3-030-82774-8_8

climate change (Mead, 2020; Thomas, 2020). Domestic policy programmes or mere internationally-proclaimed external policy agendas are insufficient on their own to address national or local problems linked to climate change. Climate change is a problem encountered by all states and therefore presents more complex issues than many of the traditional security issues and crises the global community has encountered. Correlation must be established between individual (state-level) efforts and international (system-level) efforts to fully comprehend the many different problems and projected solutions to climate change. Moreover, international collective action has become important to meaningfully address the issue of climate change (Rowlands, 2001; Barkdull & Harris, 2009; p. 22; Rasheed, 2020). In a complex international climate system, achieving results has largely remained a very slow and unresolved process (Barnett & Campbell, 2010; Murphy et al., 2009).

In a traditional sense, the capacity of SIDS to influence international policy in an anarchical and power-oriented system of states is insignificant, if not redundant (Paterson, 1996; Rowlands, 2001; Barkdull & Harris, 2009, p. 22; Underdal, 2017). However, over the years SIDS have successfully pursued their own interests in United Nations (UN) climate negotiations. SIDS have infiltrated the structural boundaries of the broader anarchical international system, showing their significance in the development and shaping of a climate governance system (Grecequet et al., 2017; Hoad, 2015). In that sense, the UN climate system has become the primary international platform for SIDS to govern their climate issues. Local efforts in SIDS are unlikely to succeed without external support or aid programmes due to internal weakness and susceptibility to international and external shocks (Barnett & Campbell, 2010), thus a stronger international action plan is necessary. The UN provides such a collective action platform for SIDS to pursue international cooperation to curb climate impacts (UNFCCC, 2005; Sadat, 2009). Indisputably, UN climate negotiations have been subjugated by the strategic diplomacy of these island states (Grecequet et al., 2017; Sadat, 2009). On reflection, the special consideration given to SIDS in the 1992 UN Framework Convention on Climate Change (UNFCCC), including its subsequent 1997 Kyoto Protocol and the recently enacted 2015 Paris Agreement, is testimony to SIDS having a disproportionate impact on UN climate governance programmes (UNFCCC, 2005, 2015a; Ourbak & Magnan, 2017).

This chapter presents a conceptual framework to explain how SIDS have managed to keep their climate agenda real and sustainable in a complex international environment where ideas and interests about climate solutions remain largely divided (Rasheed, 2019, 2020). It claims that continued dialogue through foreign policy engagements in the UN climate system has been the key to their strategic climate diplomacy. Therefore, it is important to understand the impact of SIDS on collective action programmes as a process-oriented structural development involving the interplay of climate foreign policy and UN negotiations. This chapter considers foreign policy as a phenomenon that is not bounded by domestic affairs alone (Flockhart, 2016; Smith et al., 2012). The chapter employs a constructivist approach to foreign policy analysis to illustrate that the UN climate governance system for SIDS has been constructed by the interplay of states' interests and collective understanding among UN climate negotiators. It is argued that 'constructivist ideas' can shape both domestic and international structures and the interests of climate negotiators to engender international collective action (see Flockhart, 2016; Rasheed, 2019). Defining foreign policy as a process-oriented practice of interplay between states in regional or international settings, this chapter claims that ideas about climate problems and anticipated solutions are generated and shared among states to shape their climate policy directions and policy outcomes in the international system (Rasheed, 2019). In that sense, the intersubjective understanding among actors defines the nature of the structure they interact in. The resulting 'intersubjective structure' becomes a means to address collective action problems and reach collective decisions (Blyth, 2002; Wendt, 1999). Mutual understanding reached by shared ideas can merge the interests of heterogeneous actors and help define policy actions that meet mutual goals in climate politics.

This chapter highlights the important role that SIDS have played in the early development of UN climate governance system and their disproportionate impact on UN climate negotiations during the 1980s, 1990s and 2000s. The UN climate governance system is defined in a descriptive manner (see Goodman, 1965). It includes UN member states, the climate negotiation processes and agreements or treaties reached among member states. These negotiation processes include the UN General Assembly (UNGA) and its subsidiary climate conferences held under guidelines of international climate change conventions. This chapter argues that understanding this disproportionate impact helps to better understand present and future trends in SIDS climate politics. A constructivist observation of their climate foreign policy and UN climate negotiations illustrates the success of merging interests to engender collective action. Section 2 presents an overview of SIDS as vulnerable but resilient actors by highlighting practical elements affecting the nature of SIDS in the UN climate regime. Section 3 develops the hypothesis of an analytical model to explain the causal effect of 'constructivist ideas' in shaping the UN climate governance system for SIDS. Based on the premise that shared ideas have the power to shape interests and influence systemic changes, Sections 4 and 5 illustrate how SIDS generate and share ideas to create coalitions and generate conventions to shape and manage an intersubjective structure of UN-based climate governance for SIDS. Section 6 concludes with an analytical overview of the chapter.

2 SIDS as Vulnerable But Resilient Actors

The unique situation of SIDS was first endorsed by the UN in its *Agenda 21* during the UN Conference on Environment and Development in Rio de Janeiro in 1992. SIDS comprise a group of 38 UN member states from different geographical regions including the Caribbean, Pacific, Atlantic, Indian Ocean, Mediterranean and South China Sea, and 20 non-UN members and associate members of regional commissions (OECD, 2015; UN, 2018; Voccia, 2012). Among other developing countries, SIDS experience climate change not only as an environmental issue but also as a development challenge that threatens their survival and sustainability. SIDS are among the locations that are most gravely affected by climate change. Research shows that SIDS' carbon dioxide (CO2) emissions account for less than 1% of global emissions contributing to climate change (Betzold et al., 2012; Voccia, 2012, p. 102). According to UN statistics,

> between 1990 and 2006, CO2 emissions of SIDS increased annually at an average rate of 2.3%, ranging from as high as 25 tons of CO2 per capita in Trinidad and Tobago, to as low as 0.16 tons in Timor-Leste in 2006 (Voccia, 2012, p. 102).

Despite their limited contribution to emissions, SIDS are disproportionately affected by increases in sea levels and frequent storm surges due to their low-lying island nature. Increasing intensity and frequency of catastrophic events such as rising sea levels and tropical storms associated with global warming and the increase of greenhouse gases (GHGs) have costly consequences for SIDS (Parry et al., 2007; Voccia, 2012). Climate change hence is realised as an existential threat to SIDS (Ourbak & Magnan, 2017; Thomas et al., 2020).

Certain scientific projections estimate that rising sea levels are the most dangerous and alarming consequence of climate change for SIDS as a rise of merely 1 m could completely inundate most SIDS that are physically small and low-lying. While states such as Kiribati, Tonga, Marshall Islands, Tuvalu and the Maldives could suffer land losses from rising sea levels, other SIDS such as Samoa and Fiji still face severe damage to their infrastructure and settlements located near the coasts from increasing storm surges (Voccia, 2012, p. 104). Although there has been significant uncertainty about the modelled projections of the Intergovernmental Panel on Climate Change (IPCC), the practical experiences of SIDS confirm the reality of the dangers they face due to rising sea levels (Barnett & Campbell, 2010, pp. 9, 11–12).

While being geographically distant and culturally diverse from each other, SIDS share similar environmental concerns and development challenges linked to climate change (UNFCCC, 2005; Betzold et al., 2012; Mol, 2014). As Voccia (2012, p. 101) describes:

> Scarce resources, remoteness from large markets, susceptibility to natural disasters, lack of economies of scale, small populations, dependence on international trade, high transportation and communication expenses and disproportionately expensive public administration and infrastructure severely constrain the development opportunities of SIDS.

With growing scientific evidence, particularly the findings of the IPCC about the negative effects of climate change including natural disasters, storm surges and rising sea levels (Parry et al., 2007), SIDS have raised deep concerns about their disproportionate vulnerabilities and their need for special consideration at the international level (UNFCCC, 2005; Voccia, 2012, p. 103). With scarce resources and lack of capacity to control the causes of climate change, SIDS are in a disadvantaged situation to combat climate change. Barnett and Campbell (2010, p. 9) asserted:

> The degree to which people are at risk from damages caused by changes in climate depends on: the extent to which they are dependent on ecosystems for their livelihoods (fishers are more dependent than soldiers, for example); the extent to which the ecosystems they depend

> on are sensitive to climate change (glaciers are assumed to be more sensitive than deserts, for example); and their capacity to adapt to these changes (people with insurance cover are better able to recover from an extreme event than those without insurance cover, for example). Capacity to adapt is a function of many factors, including: access to economic resources, technologies, information and skills; the degree of equity in a society; risk perception; and the quality of governance.

More importantly, physical vulnerability decreases their own likelihood of adapting to and covering the costs of damage caused by climate change. In addition to this disadvantage, as places with the least material powers, SIDS can play only a minimal role in shaping international climate politics for their advantage. Practical and theoretical questions remain as to the capacity of SIDS to influence international settings to fulfil their purposes. As climate change creates higher level complexities and uncertainties for the international system, ideas about solutions and mitigation efforts continue to be widely divided between developed and developing countries (Rowlands, 2001; Barkdull & Harris, 2009, p. 22).

The anarchical nature, physical and material capabilities, and path dependent character of industrialisation have remained factors that affect states' capacity and willingness to implement long-term mitigation measures (Underdal, 2017). While developed and industrialised countries are inclined towards measures less harmful to their industrialisation process, developing countries claim their equitable right to achieve sustainable development. While Bahamas, Bahrain, Brunei Darussalam, Cyprus, Israel, Kuwait, Malta, Oman, Qatar, Saudi Arabia, Singapore and United Arab Emirates favour positive incentives to encourage their commitment to reduce GHG emissions, members of the Organization of Petroleum Exporting Countries (OPEC) call to minimise the impact of developed countries' mitigation measures on developing countries' economies. Brazil, China, India, Indonesia, South Africa, South Korea and Mexico on the other hand stress that initial actions need to come from the developed countries. Argentina, Chile, Malaysia, Thailand, Morocco and Nigeria advocate for developed countries to lead efforts to combat climate change and support the efforts of developing countries. While the United States (USA) seeks demonstrations of 'significant actions' by countries such as China (Murphy et al., 2009, p. 13, 22), the European Union (EU) is largely seen as a leader in international climate change negotiations (Murphy et al., 2009, p. 24). However, as the countries most affected by climate change and lacking capacity to address the issue by their own means, the Least Developed Countries (LDCs) and SIDS call for developed countries to ensure that the negative impacts caused by the build-up of GHGs do not adversely affect the development and survival of any country, especially those that are most vulnerable (Murphy et al., 2009, p. 14).

The asymmetries between the climate interests in the foreign policies of developed and developing countries create challenges in reaching mutually agreeable solutions (Harris, 2009, p. 8). In this system of anarchy, SIDS can only call upon the international community to adopt mitigation measures and support their efforts to minimise the climate impacts. Significant effort has been put into directing global climate debates to show greater concern to the life-threatening effects of climate change for SIDS (Bishop, 2012; Grecequet et al., 2017; Hoad, 2015). Such efforts have been evident from the very initial stages of SIDS engagement in UN climate negotiations to the subsequent development of climate institutions including the UNFCCC, Kyoto Protocol and the Paris Agreement (AOSIS, 2015; Ourbak & Magnan, 2017). SIDS have been successful so far in pushing an agenda based on ideas of the differentiated responsibilities of states. The notion of the 'rich' supporting the efforts of the 'poor' in dealing with environmental issues has been well established in international practices since the 1970s (Bodansky, 2001; Stone, 2004, p. 79). Since the 1980s, SIDS have raised concerns about their lack of financial resources, technology and effective policy and climate governance frameworks, and called to set mitigation targets and timetables for developed countries while supporting the adaptation efforts of SIDS through financial assistance and technology transfer (Bodansky, 2001; UNFCCC, 2005). Therefore, such international institutional developments are important for SIDS as they create support mechanisms through collective action and resource distribution.

After 1988, when the UN declared climate change as a 'common concern of mankind' (Bodansky, 2001, p. 28), major developments were made in the UN climate framework leading to major international conferences including the UNGA special sessions, 1990 Second World Climate Conference in Geneva, 1992 Earth Summit in Rio and the 1994 Barbados conference. The first major milestone was the UNFCCC that established a comprehensive set of provisions with policy targets and timelines for states to implement mitigation and adaptation measures (Bodansky, 2001). Building on these institutional foundations, the Kyoto Protocol and pre-2015 Paris negotiations became key international platforms to address present and future problems. The UN climate negotiations sought a collective action approach to compliance with mitigation and adaptation efforts, mirroring ideas about equitable treatment between developed and developing countries to overcome the challenges of climate change (Bodansky, 2001; Paterson, 2001;Wiegandt, 2001; UNFCCC, 2005). The ability of SIDS to make a difference in the battle against the complex situation created by climate change should not be underestimated, but clearer insights into theoretical gaps and practical explanations are needed to observe their systemic impact.

3 A Constructivist Hypothesis for Climate Change Politics

Studies on the climate politics of SIDS have expanded over the past 20 years. Initially, studies focussed on scientific factors associated with climate change impacts became key empirical foundations for subsequent political and policy studies on climate solutions for these states (Barnett & Campbell, 2010; Paterson, 1996). The growing body of literature on small island states and climate change is mostly concentrated on success stories and practical challenges at domestic and international policy levels. While realising that SIDS are most vulnerable to the adverse effects of climate change, studies about their role in climate efforts seldom highlight their disproportionate systemic impact on international climate politics.

This is predominantly due to the influence of dominant (traditional) theoretical conceptions on these studies, including neorealism and neoliberalism, effectively eliminating the observance of the systemic impacts of materially insignificant small (island) states in the international systems design (see Keohane, 1969, 1984, 1989; Keohane & Nye, 1977; Waltz, 1979, Walt, 1987; Hellmann & Wolf, 1993; Paterson, 1996; Rowlands, 2001; Hey, 2003, p. 188; Browning, 2006). However, as with the general development of small-state studies since the 1960s, the development of literature on small island states (or SIDS) and climate change has also been affected by the development of international relations (IR) and international political economy (IPE) schools (Rowlands, 2001, pp. 43–64; Widmaier, 2005, 2007; Bishop, 2012; Flockhart, 2016). These schools have moved beyond traditional conceptualisations of international politics. Some studies are accounting for the reality of SIDS' role in the international politics of climate change, contrary to what neorealist and neoliberal theories could explain about these states. According to such studies, today, SIDS have moved to the frontline of international climate negotiations (Hey, 2003; Simpson, 2006; Harris, 2009; Chong & Maass, 2010; Lee & Smith, 2010; UN-OHRLLS, 2011; Bishop, 2012; Panke 2012; Hoad, 2015; Ourbak & Magnan, 2017). Foreign policy responses of SIDS to international action against climate change tend to challenge traditional policy and theoretical approaches in IR. Their foreign policy responses are not necessarily shaped by the powerful states or the system they put in place. Rather, their role in international climate politics is often recognised by these studies as a causal variable in international climate negotiations (see Ourbak & Magnan, 2017).

This chapter argues for a constructivist approach to the climate politics of SIDS in contrast to the traditional IR approaches including neorealist and neoliberal conceptions of international security and cooperation (Rasheed, 2019; Keohane, 1969, 1984, 1989; Waltz, 1979, pp. 84–85, 195; Walt, 1987; Hellmann & Wolf, 1993; Browning, 2006, pp. 670–673). This constructivist approach is supported on the basis that traditional liberal institutional settings and a realist or neorealist anarchical system of states are limited in their ability to holistically explain the nature of states' cooperation on the issue of climate change, especially the role of SIDS being instrumental to the design and governance of international climate regime (Ourbak & Magnan, 2017).

In the neorealist or neoliberal institutional viewpoint, international action is divided and disdained by powerful states (Keohane & Nye, 1977; Walt, 1987; Waltz, 1979). Under the neorealist theory, the design of the international system is directly linked to material factors. Therefore, a lack of material property including military and economic capabilities can limit a state's capacity to influence other states and international norms. The reality in this state of anarchy is that the materially powerful states and the international system they control can shape the foreign policy functions of the weaker states within the system (Waltz, 1979, Walt, 1987; Danilovic, 2002; Telbami, 2002; Browning, 2006; Harris, 2009,). Small states and SIDS are generally categorised as materially weak states and hence will have little ability to cause systemic change in IR. For example, discussing Walt's (1987, pp. 21–31) conception of small states' foreign policy, Elman (1995, p.177) stresses that in this system based on relative power, small states are more likely to 'bandwagon with an aggressive great power than balance against it', especially based on 'geographic proximity' and the availability of an alternative ally. The level and type of foreign policy functions of weaker or smaller states are dependent on the 'benevolence of larger powers, the relationships between them, or the nature of the balance of power (whether conservative or competitive)' (Browning, 2006, p. 670). As rational actors, they will choose to maximise their interests with the help of powerful states to disproportionately gain from the international system (Browning, 2006; Gvalia et al., 2013). This is mainly because they do not have materially significant bargaining or cohesive powers to influence the interests of the resourceful states that effectively control their system design (Keohane, 1984, 1989; Hey, 2003, pp. 2–9). In this system of states, SIDS—with their vulnerability associated with 'smallness'—may not be able to shape the interests of the powerful states to design an international system to their benefit. However, as noted above, the role of SIDS in the development of an international climate regime shows otherwise. Hence, a neorealist account of international system change may not explain this phenomenon.

Similarly, neoliberalism also does not present a complete theoretical base to explain the unorthodox impact of SIDS on international climate politics. Neoliberalism analyses the role of (liberal) international institutions and norms in determining absolute gain rather than the relative gain of states (Keohane & Nye, 1977; Rowlands, 2001, p. 54). Under this approach, international agreements and conventions such as world trade agreements are drawn up among states to promote the individual but different interests of states. International institutions, including the UNFCCC, Kyoto Protocol and the Paris Agreement, along with organisational foundations including the International Negotiating Committee for UNFCCC (NIC), Conference of Parties (COP) and international negotiation platforms for the Paris Agreement, form a collection of liberal institutions, setting guidelines and obligations for states to address climate change. In this setting, materially weak states or small islands can favour liberal international institutions as they create opportunities and constraints for them to meet individual and collective interests of states that have agreed to those institutions (Browning, 2006; Hellmann & Wolf, 1993; Keohane, 1969, 1984, 1989).

In this neoliberal system of states, institutions are formed by states to address issues such as international trade or environmental issues. Institutions provide information for states to make rational decision and hence cooperate to maximise benefits. Therefore, in this system, smaller states may have the opportunity seek their foreign policy objectives despite the anarchical system of states. However, as with the neorealist international system, the better-resourced states can shape those institutions to better serve their self-interests (Rowlands, 2001; Hey, 2003, p. 188). International institutions can at best create incentives to cooperate, but states' decisions to cooperate depend on their various self-interests. This means that the practical implications of international institutions can often vary with powerful states (Hey, 2003, p. 188; Browning, 2006). For example, this explains the challenges experienced in the 1990s climate negotiations when the USA or its 'Umbrella Group' caused the weakness in the UN-based conventions and treaties by making reservations to international mitigation actions (Coghlan, 2002). This means that SIDS are unlikely to make an impact on those climate-related institutions. However, as discussed further in this chapter, SIDS did make important contributions to the 1990s negotiations, plus SIDS shaped the design of the UN-based climate regime to support their efforts to address the problems they faced against climate change (Ourbak & Magnan, 2017). Therefore, the neoliberal theory of international cooperation does not explain the exact nature of states' cooperation on the issue of climate change (Rasheed, 2019).

Climate change is characterised by complexities and uncertainties because it affects all states (Rowlands, 2001, pp. 56–57). Many states with divergent interests are unlikely to come to mutual terms. For example, no less than 193 states are involved in building institutions and setting norms to deal with climate change (Bodansky, 2001, p. 28; Rowlands, 2001, pp. 54–60; Betzol, 2015). Furthermore, because climate change is a complex issue profoundly linked to globalisation and the inter-connectedness of world economies, it poses further challenges in defining and categorising the accountability of each states and their capacity to adopt cooperative approaches (Rowlands, 2001; Underdal, 2017, pp. 172–174). Historically, the North-South divide has caused challenges as to the effective implementation of international climate agreements. While developed states including the USA remain hesitant to make policy change due to economic reasons, developing states are interested to see the developed states take the lead in making policy changes. This challenges the feasibility of international agreements (see Coghlan, 2002). Yet, despite such challenges, SIDS seem to have found their way through international climate negotiations. States have cooperated to find solutions to climate change and agreed to consider SIDS as a special case (Ourbak & Magnan, 2017). Furthermore, rather than prescriptive liberal institutions and fixed 'balance of power' structures, continuous negotiation between states appears to have had an important impact on international climate policy outcomes, such that the international climate agenda has become a continuous process of consensus building among these states. Climate consensus-building has not only been shaped by powerful states such as the USA but has also been driven by the smaller ones. In this system of states, international negotiations shaping climate institutions have become an important variable in the process of consensus building. However, this social variable is seldom explained by neo-liberal insights or realist analysis to explain the disproportionate role of SIDS in the development of an international climate regime (Rowlands, 2001, pp. 54–60).

3.1 *Social Construction of International Affairs: The Constructivist Turn*

It can be argued that this phenomenon of disproportionate impact of SIDS is linked to the function of intersubjective understanding that social constructivists in IR and IPE scholarship have used to explain complexities and asymmetries in the international system since the post-Cold War era studies (Blyth, 2002; Flockhart, 2016; Houghton, 2007; Wendt, 1999). This contemporary body of IR and IPE literature on global climate change stresses critical approaches including social constructivism to understand the balance between theory and foreign policy practices (Houghton, 2007, pp. 24–45; Hameiri & Jones, 2015; Flockhart, 2016, p. 79). The argument is that systemic or structural change and stability, understood in the post-Cold War international context, is no longer an automatic effect of hegemonic power and predetermined rules and norms, and hence there is merit for alternative approaches (Rowlands, 2001, pp. 43–63; Barkdull & Harris, 2009, p. 33). Social constructivist, or constructivists, see the international system as a social construction of states rather than an outcome of material structures (Wendt, 1999, p. 22). It is a system composed of social relations and interactions between states where their foreign policy interests and politics with respect to each other are shaped by ideas generated and shared among them. This ideational phenomenon involves a process of intersubjective understanding and construction of meaning among states (Adler, 2012; Blyth, 2002; Flockhart, 2016; Wendt, 1999; Wicaksana, 2009; Widmaier, 2005). As Alexander Wendt (1999, p. 22) explains:

> The character of international life is determined by the beliefs and expectations that states have about each other, and these are constituted largely by social rather than material structures. This does not mean that material power and interests are unimportant, but rather that their meaning and effects depend on the social structure of the system . . .

This means that the foreign policy process is a social construct (Doty, 1993; Flockhart, 2016). This theory eliminates the state-centric and structure-centric powers determining the international system; instead of privileging either actors or structures, constructivism focuses on ideas and beliefs as the dominant causal variables of change and stability in international system. This allows constructivists to engage in foreign policy analysis, privileging neither structure-centric nor agency-centric approaches (Flockhart, 2016; Houghton, 2007).

This means that the international climate system is 'what states make of it', as Wendt (1992, p. 395) writes. In this ideational occurrence, the process of foreign policy making and international system design will coexist but can also function independent of each other in the presence of ideas. Ideas generated and shared enable mutual constitution of foreign policy and the international system (Barkdull & Harris, 2009; Flockhart, 2016, pp. 86–87). The mutual constitution of structure and actors' interests is achieved by routine social interactions or foreign policy practices exchanged between states, influenced by ideas and beliefs that they have about each other (Flockhart, 2016, p. 86). In such an intersubjective structure, ideas shared among states about climate change can influence their perceptions and expectations about one's self with respect to others, and shape their climate foreign policy interests. This can make the international climate system a product of intersubjective understanding (Busse, 1999; Smith, 2001; Wicaksana, 2009) involving a continuous process of state interactions. In this way foreign policy becomes a sequence of decisions towards building consensus through negotiations among states in the international system (see Carlsnaes, 2008). However, the outcome of the negotiations is not shaped by material factors, but ideas generated and shared between them.

This intersubjective structure understands foreign policy making and international system design as a two-way but sequential, interactive process (Flockhart, 2016, pp. 87–89). Therefore, the capacity of states to influence system change depends on the ideas commonly held and shared. As ideas can also change through time and space with changes in external conditions, foreign policy practices also become a continuous process; thus the intersubjective structure is an ongoing, living system of socially connected states. Flockhart (2016, p. 88) explains that actors will choose practices within the existing structure when the situation assures certainty and continuity of the system to meet their desires. This can entail that the climate foreign policy of SIDS and the UN climate system they interact with are neither predetermined nor shaped by material effects of the international system, but by ideas they hold about climate change.

This ideational process of intersubjective understanding can lessen the effect of collective action problems in dealing with climate change by eliminating the material constructs that cause asymmetries in the international system. This constructivist hypothesis of climate foreign policy and international climate system design can explain the unorthodox impact of an individual state or a group of states on the international system. Structural developments can be understood as a sequence of different but intertwining ideational impacts. Initially, in times of crisis including disasters caused by of climate change, despite the pre-existing structural environment (Browning, 2006; Danilovic, 2002; Telbami, 2002; Waltz, 1979), ideas held by concerned actors about the problem and desired solutions will have a causal effect on the choices they make about the issue. As ideas affect perceptions and expectations about one's self with respect to others, they in turn determine policy goals and resulting intersubjective structures (Busse, 1999; Wendt, 1999, p. 20; Smith, 2001; Blyth, 2002, pp. 35–37; Wicaksana, 2009). In the foreign policy context, ideas are shared by more than one concerned actor, leading to an intersubjective consensus about the crisis and expected interstate or international solutions. This process causes a mutual understanding among actors and reduces uncertainties in reaching desired solutions (Blyth, 2002, pp. 35–36). In this way SIDS, as the states most affected by climate change, can share ideas about climate impact to seek solutions their special circumstances related to effects of climate change.

Mutual understanding further enables coordinated decision making among actors. Coordinated decision making can support collective action whereby interests are merged to seek common goals. On issues of climate change, it becomes less likely for states to reach common goals as their interests are widely divided (Mayer & Arndt, 2009, pp. 74–85). The heterogeneous behaviour of states can limit the capacity of climate negotiators to reach common solutions, let alone the difficulty encountered by smaller states with limited material contributions to offer (Rowlands, 2001; Wiegandt, 2001, p. 127). However, ideas acting as an interpretive framework can help merge those divergent interests through the negotiation process. This group awareness can form a group of homogenous actors who become a coalitional force to pursue a common agenda (seeBlyth, 2002, pp. 37–39; Constantini et al., 2007;

Bishop, 2012). As more actors agree on common goals, it also reduces complexities about the type of solutions that can be adopted by the international system. For SIDS, a coalition can help minimise uncertainties for the UN climate negotiators to understand and determine the institutional solutions that best suits their desires.

Shared ideas remain at the core of the new institutions put in place to address problems, where they also determine change and continuity of institutions (Blyth, 2002, p. 39). When foreign policy becomes a social construction of states, the institutions put in place to address foreign policy issues can also become an ideational construction (Flockhart, 2016, p. 84). Wendt (1992, p. 399) explains:

> Institutions are fundamentally cognitive entities that do not exist apart from actors' ideas about how the world works. This does not mean that institutions are not real or objective, that they are "nothing but" beliefs. As collective knowledge, they are experienced as having an existence "over and above the individuals who happen to embody them at the moment." In this way, institutions come to confront individuals as more or less coercive social facts, but they are still a function of what actors collectively "know." Identities and such collective cognitions do not exist apart from each other; they are "mutually constitutive." On this view, institutionalization is a process of internalizing new identities and interests, not something occurring outside them and affecting only behaviour.

Unless there is an agreed understanding about what the solution is, states cannot produce institutional solutions that meet their expectations in practice. Ideas can help interpret the existing institutional settings and recast new ones as per expectations. While new institutions become the medium through which practical solutions are met, ideas remain a constant causal variable in the form of conventions, protocols or codes of behaviour, that shape and manage those institutions in static and changing situations (Blyth, 2002, pp. 41–44). From the constructivist viewpoint, conventions are the shared ideas about the expectations of people which coordinate or manage their expectations with new institutions to meet common ends and establish stability in the structure (Blyth, 2002, p. 41). The UNFCCC and its subsequent Kyoto Protocol in the 1990s–2000s illustrate the primary conventions that govern the UN climate system. These can be used to support the interests of SIDS throughout present and future climate governance reflecting Paris and post-Paris outcomes (UNFCCC, 2005; Aginam, 2011; Betzold et al., 2012).

4 Shared Ideas and Consensus Building in SIDS' Climate Politics

The early development of SIDS' climate arguments provided a foundation for the ideational aspects of their disproportionate impact on the UN climate governance system. This is observable through their initial engagements in the UN climate negotiations during the late 1980s and early 1990s. The UN climate negotiations became an important phenomenon in the intersubjective structure of climate governance for SIDS. In retrospect, domestic efforts to adapt to changes in climate conditions were made by SIDS, long before the idea or conception of climate change was realised in the international policy context. Adapting to climate change has been at the core of their natural, cultural and policy methods as a way to survive changing environmental conditions (Barnett & Campbell, 2010). It is not within the scope of this chapter to explore those domestic methods in SIDS' responses to climate change. However, it is important to highlight such natural, cultural and policy methods which encompass ideas of adaptation. Ideas about adaptation mirror the ideas generated by SIDS about climate change and solutions they have advocated for. This chapter focuses on the phase beyond national efforts, seeking international efforts to support their existing national capacity to adapt to and curb climate change (Barnett & Campbell, 2010). This outweighing of domestic capacity recognises the engagement of SIDS in international climate negotiations as a meaningful undertaking to reach sustainable solutions, especially with their limited material capabilities.

The climate ideas generated by SIDS were closely linked to the basic principle of equitable responsibilities of states in dealing with broader international environmental issues. The recognition of differentiated treatment for developing countries in the 1972 Stockholm Declaration of the United Nations Conference on the Human Environment made an initial impact on international understanding about the environmental concerns of such vulnerable states (Bodansky, 2001, p. 23). The basic idea entailed 'taking into account the circumstances and particular requirements of developing countries and any costs which may emanate from their incorporating environmental safeguards into their development planning and the need for making available to them, upon their request, additional international technical and financial assistance for this purpose' (Stone, 2004, p. 279). Giving particular attention to developing countries entails the notion of common but differentiated responsibilities of states in dealing with global issues such as environmental issues or climate change. Here, the idea of international cooperation establishes a principle that the rich support or contribute to the poor's efforts in dealing with issues of international significance (Bodansky, 2001; Stone, 2004, p. 79). Such ideas about international cooperation have been embedded in the broader international institutional programme since 1972 (Barnett & Campbell, 2010, p. 36) and has become a fundamental foundation for SIDS' climate debate as they emerged as climate advocates and launched climate foreign policy initiatives to address their climate problems.

It was only in the 1980s that SIDS began international advocacy seeking the international community's cooperation and support for their efforts to address effects of climate change. Markedly, their initial engagement can be understood as a phase of generating ideas and building consensus among SIDS to develop and pursue a common agenda. Agreeing on a 'common problem' was, and still is, one of the key bases of their climate foreign policy (UNFCCC, 2005; AOSIS, 2015). Although the UN climate system has been the main international mechanism through which SIDS have sought climate solutions, their initial ideas were generated and shared among SIDS and international climate negotiators in different multilateral forums. A notable initial event was in 1987 when the Maldives raised the issue of the disproportionate vulnerabilities of small island nations to the effects of climate change at the Commonwealth Heads of Government Meeting (CHOGM) in Vancouver (Sadat,

2009; Barnett & Campbell, 2010, p. 86). Speaking of its own experiences (UNFCCC, 2005, pp. 13–15, 16), the Maldives informed international stakeholders about the devastating effects of climate change and rising sea levels on island nations. Due to inherent vulnerabilities linked to small size, the geographical nature and lack of material resources, such island nations are unable to adapt to increasing harmful climate events by themselves. The Maldives called for international cooperation to give special attention to these states dealing with this global issue. These ideas were later advanced by several Pacific Islands who attended the CHOGM. Pacific island leaders who met in the 1988 annual Pacific Island Forum shared similar views about island vulnerability and sought international cooperation to challenge the effects of climate change (Edwards, 1990, pp. 4–5; Barnett & Campbell, 2010, p. 86). An intergovernmental campaign continued during the same period at regional levels including in the 1987 Summit Meeting of the South Asian Association for Regional Cooperation (SAARC). In 1989, a Commonwealth Group of Experts found that 'there was special need for concern for low-lying island countries (Barnett & Campbell, 2010, p. 86). This was a period of sharing ideas and consensus building among SIDS and international climate negotiators. As similar interests grew among SIDS, in 1987 the Maldives presented the same case before the United Nations General Assembly, making a significant impact on similar states and climate advocates about the special vulnerabilities and need for international cooperation (Edwards, 1990, pp. 4–5).

The 1987 climate foreign policy of the Maldives became one of the SIDS' climate policy initiatives for reaching consensus through the UN system. Due to the mass number of states engaged in UN negotiations, the approach taken by the Maldives also became an inception point for generating ideas and consensus about a common problem faced by SIDS. The Maldives linked island vulnerability to the dangers posed by rising sea levels. For example, it informed the UN that a 'rise of two metres would suffice to virtually submerge the entire country of 1190 small islands, most of which barely rise over two metres above mean sea level' (Maldives Permanent Mission to UN, 1987). The focus was not on mitigation efforts, but on ways to survive any catastrophic events. The mitigation efforts at that time could not have prevented possible increases in sea levels and global warming to the extent that it would have eliminated the climate impacts experienced by island states. The chances of reaching an international agreement to take drastic mitigation measures were slim at best (Betzold, 2015, p. 482). The main focus was on fulfilling the need to create awareness amongst the rest of the world of the reality of climate change and the disproportionate challenges experienced by SIDS (UNFCCC, 2005; Betzold et al., 2012). Therefore, SIDS climate foreign policy was not founded solely on ideas of stopping the causes of climate change, but also based on a claim an equitable right to enjoy life despite their geographic, social, economic and political status. The idea was to primarily focus on 'equity, impacts and adaptation strategies' that could help similar states lacking material resources to combat climate change (Mayer & Arndt, 2009, p. 82; Aginam, 2011). The Maldives informed the 1987 UN meeting that:

> [It is] a time of potential crisis confronting [this] planet and its population [and a] crisis of environmental destruction man has invoked upon himself [engendering] risks of irreversible damage to the human environment that threaten the very life-support systems of the earth–the basis for man's survival and progress (Maldives Permanent Mission to UN, 1987).

Following the statement of the Maldives in the UN meeting in 1987, these ideas about their special vulnerabilities and need for international collective action were mostly agreed upon by other SIDS at regional and international levels. SIDS agreed to come together to share climate concerns and reach international consensus about a common threat posed by climate change. One of the early instances of consensus building was the 1989 Small States Conference on Sea Level Rise held in the capital city of the Maldives, Malé, where more than a dozen small island and low-lying states from the Mediterranean, Caribbean, Indian, and Pacific Ocean put together a regional agreement, the 'Malé Declaration', as a united 'call for action' against the impact of climate change on SIDS (Edwards, 1990). The Declaration stated (UNESCO, 1989, p. 4):

> [The states] decided to . . . seek assistance from the UN, its Agencies and other appropriate institutions in the implementation of the decisions contained in this Declaration. . . . [And] call upon all States of the world family of nations to take immediate and effective measures according to their capabilities and the means at their disposal, to control, limit or reduce the emission of greenhouse gasses, and to consider ways and means of protecting the small states of the world which are most vulnerable to seal level rise.

Antigua and Barbuda, Cyprus, Fiji, Kiribati, Maldives, Malta, Mauritius, Trinidad and Tobago, and Vanuatu made policy statements agreeing with the idea that SIDS are among the states most affected by the adverse effects of climate change while also being the places with the least physical capability and material resources to challenge the devastating impacts from increasing sea level rise and natural disasters linked to global warming. The Maldives, through its then Minister of Transport and Shipping, agreed that experience of sea level rise is a growing common concern with the potential to make small island nations 'innocent victims' of and the 'first countries to suffer' from the 'actions of industrialised nations,' threatening the very existence of the 'small, low-lying and fragile countries' (Ibrahim, 1989). Agreement was further sought about seeking international support including necessary financial and manpower resources (Ibrahim, 1989). Other participants, including Vanuatu, joined the consensus about a common concern for SIDS, stressed the significance of understanding the adverse impacts of sea level rise, and linked the issue to island vulnerabilities to climate change in order to seek the support of industrialised countries to work together with small island nations to reach international climate solutions (Hopa, 1989).

As such, the UN climate governance system for SIDS has been a construction of ideational interplay between SIDS and climate negotiators. Early inceptions of such interplay in both regional and international settings illustrate a period in which ideas generated by a few SIDS were shared among similar states, merging their interests to define the climate change issue as a common problem with disproportionate impacts on small island nations and calling for international collective action. The climate activism of SIDS in the 1980s built on the early developments in international debate on climate change. As the UN became the key multilateral forum for climate debate, SIDS' efforts were almost solely focused on building consensus amongst UN climate negotiators (Bodansky, 2001; UNFCCC, 2005; AOSIS, 2015). Ideas generated and shared among SIDS further developed intersubjective awareness within the UN climate negotiators leading them to perceive SIDS as a case requiring special attention and support. By the time negotiations for a major international framework convention were underway in the early 1990s, the ideas of SIDS had already started to shape international policy interests to ensure equitable treatment for SIDS. In 1992, the UN in Rio de Janeiro formally recognised some 38 small island and low-lying UN member states and 20 non-member states as special cases. Agenda 21 stressed the commitment to 'adopt and implement plans and programmes to support [and] ... enable small island developing States to cope effectively, creatively and sustainably with environmental change [including climate change]' (UN, 1994). In 1994, the Barbados Programme of Action (BPOA) declared the same and stressed the 'vulnerability of SIDS to climate change, climate variability and sea-level rise' (UNGA, 1994, p. 4; UNFCCC, 2005, p. 2). Such early impacts by SIDS on UN-based programmes and conventions portrays an unorthodox behaviour of structural developments in the international system, giving room for micro- and small states to shape climate negotiations. This was mainly possible due to SIDS committing to engage in a constructivist climate dialogue to influence perceptions and expectations of climate negotiators to reduce uncertainties and complexities in dealing with issues affecting them. This ideational consensus building process led these states to form a coalition of small island states, the Alliance of Small Island States (AOSIS) that acted (and still acts) as an institutional mechanism fighting for their common cause in UN climate negotiations.

5 Coalitional Impact and Climate Conventions

Shared ideas have the ability to merge interests and bring states together to reduce the complexity of international norms, structure debates, and engender collective action solutions. Although SIDS share similar physical and material structures in relation to the consequences of climate change, their individual policy objectives in this respect have differed in various aspects (Betzold et al., 2012, p. 594). As observed, despite different individual policy objectives, it appears that SIDS have established climate change as a mutual problem. Ideas about island vulnerability, lack of national or domestic capacity to adapt to and mitigate climate change, and the need for international support for adaptation plans, have been established as the fundamental bases of SIDS' climate agenda (UNFCCC, 2005, pp. 7–9). The mutual understanding generated by such ideas has enabled SIDS to come together from the beginning as a group of states fighting for a common cause. The impact of this intersubjective understanding is seen in the formation and function of AOSIS as an ad hoc coalition seeking climate solutions for SIDS in the UN climate negotiations (Ashe et al., 1999).

In 1990, under the leadership of the Maldives, Vanuatu, and Trinidad and Tobago, 39 UN member-states agreed to form AOSIS, an alliance with the main aim of pursuing the climate foreign policy objectives of SIDS (Betzold et al., 2012, pp. 591, 593). The Alliance was built on common ideas about the threat of climate change, the disproportionate vulnerabilities of SIDS and the need for joint action at national and international levels (Betzold et al., 2012). The Maldives' President, Yameen Gayoom, noted that 'AOSIS would go on to earn a reputation for being the "moral voice" of the UN climate negotiations, by advocating for global action that would protect its most vulnerable members' (Gayoom, 2015). Despite the divergent individual interests among its member states, AOSIS has managed to pursue a common interest based on the concept of island vulnerability. Its function as a coalitional body has been a key enabling force for SIDS to successfully influence the initial UN climate negotiations when the international community sought collective action solutions in the early 1990s (Davis, 1996, p. 18 Betzold et al., 2012, p. 594). While the 1987 Malé Declaration built an ideational base for the formation of AOSIS, it established an intersubjective understanding among SIDS to seek an international framework to ensure equitable responsibilities among states (AOSIS, 2015, p. 3). AOSIS's main objectives in climate negotiations reflected ideas about seeking international cooperation and establishing an international institutional framework that could minimise asymmetries and hence uncertainties in the international system (UNFCCC, 2005; Betzold et al., 2012). Working based on agreed objectives enabled AOSIS to have an impact on climate negotiations as a coalition within the group and in the UN.

By 1990 when the UN decided to run the climate negotiations under the auspices of the UN General Assembly to ensure better representation of both developed and developing states (Bodansky, 2001; Paterson, 1996), SIDS had already established significant representation of small island nations under the AOSIS framework. It became easier for SIDS to influence UN climate negotiators by reducing complexities in their climate foreign policy in relation to the climate crisis and the solutions they sought. The coalitional impact of AOSIS became stronger as interests among members of the group were merged (Starkey et al., 2008). A survey on the number and type of submissions and statements from AOSIS and its members during 1995–2011—which also reflected the early development stage in the 1990s—found that despite the different interests of individual states, AOSIS as a group had made a significant impact on UNFCCC negotiations and prompted an institutional structure that addressed the interests of SIDS (Betzold et al., 2012, p. 605).

The 1992 UN Framework Convention on Climate Change (UNFCCC or the Framework Convention) was the first international institutional framework pronouncing state responsibilities with respect to domestic and international policy responses to climate change (Bodansky, 2001; Paterson, 1996). Although it was not established as an institutional base specific to SIDS, the outcome document adopted in 1992 reflected significant collective action plans catering for their specific interests. Betzold et al. (2012, p. 591) noted that 'despite the relatively small size and lack of political clout of its members, AOSIS has become one of the key players in the UNFCCC negotiations, which in itself is a notable success for island microstates.' As AOSIS took flight in the UN climate negotiations, it utilised the Intergovernmental Negotiating Committee (INC)—the UN negotiating body for the UNFCCC—as the main forum to convince negotiators of the plight of small island states and to ensure their issues were meaningfully addressed in the upcoming framework convention (Ashe et al., 1999). As SIDS engaged in the climate debate, they agreed on fundamental objectives reflecting their concerns and expectations from the framework convention. AOSIS wanted the outcome document to include the following objectives (Ashe et al., 1999, pp. 212–215):

1. Recognise the special case of SIDS: AOSIS was successful in having this requirement included in the UNFCCC document.
2. Recognise that SIDS have special needs that require international support: Articles 3(4), 3(2), 4(4), 4(8), 9 reflected this requirement.
3. Include precautionary principles: Article 3(3) reflected this requirement.
4. Include the obligations of states to stabilised GHG emission: Articles 4(2) and others reflected this requirement.
5. Include requirements for environmental assessments: For example, Articles 4 (1) and 7(2)(d) reflected this requirement.
6. Include monitoring requirements: For example, Articles 4(1)(g) and 4 (8) reflected this requirement.
7. Include requirements for information sharing: For example, Article 6(a) (ii) reflected this requirement.
8. Include priority funding for SIDS: Articles 4(3), 4(3), 11(3) and 21(3) reflected this requirement.
9. Include requirements for transfer of technologies: Articles 4(3), 4(5) and 4 (8) reflected this requirement.
10. Establish institutional mechanisms: For example, Articles 7(2), 9 and 10 reflected this requirement.
11. Include arrangements for protocols to support implementation: Article 17 reflected this requirement.
12. Recognise the polluter pays principle: For example, Article 3(1) and 4 (4) reflected this requirement.

These negotiating points reflected a set of solutions generally agreeable to UN climate negotiators, as evidenced by their articulation in the final document. The general ideas reflected the call for a special case for SIDS and the need for international cooperation and support in their battle against climate change.

These ideas generated conventions that illustrate AOSIS's coalitional impact on UN negotiations. Throughout the 1990s and 2000s, such conventions or shared ideas enabled AOSIS to coordinate the interests of SIDS within the group as well as with other international actors as more and more issues were added to the climate agenda. The addition of other issues to the climate agenda also illustrated the diversity of policy interests among SIDS in areas including land use, land-use change and forestry (LULUCF), and reducing emission from deforestation and forest degradation (REDD). As Betzold et al. (2012, p. 595) noted, 'with a better understanding of climate change and its implications, as well as of the negotiation process, individual states may be better aware of their interests and how they relate to group positions'. Despite the divergent interests, shared ideas about the reality of disproportionate vulnerabilities and the need for international cooperation and support have remained the driving force of the climate foreign policy of SIDS (Betzold et al., 2012, p.592). These ideas as conventions enabled AOSIS to coordinate institutional expectations throughout post-1992 negotiations, especially in defining common grounds to agree on (Ashe et al., 1999). These conventions acted as an interpretive framework and resource-base for AOSIS to combine their interests and pool their resources to convince the broader international audience about the nature of the crisis and types of solutions required to solve it. They have made the coalitional impact of AOSIS on the UN climate governance system more apparent (McMahon, 1993; Chasek, 2005; McNamara & Gibson, 2009; Betzold et al., 2012, p. 594). In a study of AOSIS negotiations between 1995 and 2011, Betzold et al. (2012) also found that the submissions and interventions made by SIDS as a group have decreased overtime; however, the outcome of the negotiations indicated that, despite any decreasing unity among AOSIS members, the objectives pursued by individual members have significantly met group objectives. Although SIDS have individual climate policy interests, shared ideas enabled them to make the maximum impact on international decisions. As Betzold et al. (2012, p. 599) asserted, 'many of these individual submissions have been used to reiterate and reinforce group positions, such as the demand for emissions cuts on the order of 40% compared to 1990 levels.'

The general expectations of SIDS regarding climate solutions have also reflected ideas about special assistance plans. Hence, an effective international climate system is important to meet broader climate foreign policy objectives. As illustrated in recent developments during the Paris negotiations, AOSIS acted as a united front to ensure that the broader UN climate governance system is further strengthened to ensure a meaningful impact on climate solutions. The meaningful impact is made by ensuring the continuity of UN-based climate negotiations (Aginam, 2011; Betzold 2015; Ourbak & Magnan, 2017). For example, emissions reduction has remained at the core of the climate debate within AOSIS and UN forums leading up to the Paris talks, in addition to the need for international assistance in the form of financial resources, technology transfer and building effective national, regional and global policy and governance frameworks (UNFCCC, 2005, p. 10; Betzold et al., 2012; Hoad, 2015). AOSIS has advocated for mitigation efforts in addition to adaptation plans. In the Paris negotiations, AOSIS was acutely committed to convincing climate negotiators to set the temperature goal below 1.5 degrees Celsius (Maldives Permanent Mission to UN, 2015; Thomas et al., 2020). Speaking on behalf of the AOSIS the Maldives reiterated (UNFCCC, 2015b):

> [We] have the honour of delivering this statement on behalf of the [AOSIS], a group of 44 countries most vulnerable to the adverse effects of climate change. . . . limiting the global rise in temperature to 1.5 degrees is a life or death matter for our most vulnerable members and therefore impacts every area of these negotiations.

Commenting on the Paris process, Maldives' President stated:

> It is now evident that certain climate impacts can no longer be managed by cutting emissions or through adaptation, and unfortunately. . . . the Paris agreement must therefore address the full scope of the crisis. This should begin with ambitious commitments from all parties, coupled with a robust process to drive bolder actions year after year, in order to keep temperatures below 1.5 degrees Celsius (Gayoom, 2015).

While the Paris negotiations promised some collective action from some 195 states, subsequent provisions of the final Agreement did accomplish AOSIS' broader objectives of minimising uncertainties in the climate system. For SIDS, consideration to keep global temperature below 2 degrees Celsius above pre-industrial levels has built confidence in the collective action plan (UNFCCC, 2015a). To reiterate Amanda Little (Little, 2015):

> Just before negotiators at the United Nations Climate Change Summit, in Paris, released the final text of their agreement, . . .a group of more than eighty delegates from forty-four low-lying coastal and island countries, through weeping and cheering and bursts of applause, until the chorus of Bob Marley's "Three Little Birds" had been repeated many times over. . . . For all the promise of the agreement, the alliance still has plenty to worry about. . . .And yet the leaders of many nations, including those that the U.N. categorizes as least developed countries (L.D.C.s), described the agreement as a victory.

These outcomes may explain where the UN climate governance system is in resolving adverse effects of climate change. They also explain what the system promises and what needs to be done in order to best address the special circumstances of SIDS. There is general agreement about the likely impacts of climate change on SIDS, and that they are particularly vulnerable to significant adverse effects. There is also general agreement, including among climate negotiators, that SIDS face disproportionate vulnerabilities making responding to the effects of climate change difficult, such that they need international support to do so. There is also general agreement about the value of continued climate negotiations in an attempt to seek international institutional solutions. These intersubjective conventions help to coordinate and manage SIDS' expectations to engage with the UN climate negotiations, and to sustain the collective action plan established by the UNFCCC and its subsequent institutional developments. AOSIS can further help to merge SIDS' expectations under their intersubjective conventions to meet common ends by minimising uncertainty and collective action problem in the international system. In the words of Betzold et al. (2012, p. 595), 'the reduction in uncertainty regarding the reality of climate change may have served to emphasize the overarching common interest: a strong climate change regime in the face of island vulnerability.' Shared climate ideas have predominantly enabled SIDS to ensure international commitment to problematise climate change which is crucial to their climate foreign policy purposes.

6 Conclusion

As places with both physical and economic weaknesses, SIDS have faced and continue to confront an imminent threat to their progress and survival due to climate change. SIDS lack the capacity to respectively fight and challenge the causes and consequences of climate change on their own. This makes them disproportionately vulnerable to the effects of climate change and causes further challenges to their development. Reality has bestowed upon these small island and low-lying coastal states the need to seek cooperation in the international community to support their climate efforts. Over the past 20 years or so, SIDS have managed to ascend international platforms including the UN climate negotiations platform to mark their climate claim and make a disproportionate impact on the design of UN climate governance system. SIDS believe that a stronger international climate system is necessary to meaningfully address their concerns and hence, from the inception of climate debates at UN, SIDS have sought to fulfil this objective by strategically engaging in UN negotiations to influence its climate governance system.

From a constructivist viewpoint, their strategic engagement in UN climate negotiations has been shaped by ideas about climate problems they face and the type of solutions required to best address their special concerns. An observation of the early development of SIDS' climate politics indicates the effective positive impact of their approach in ensuring UN-based climate governance to support their efforts. The first institutional framework, the UN Framework Convention gives special consideration to SIDS in defining the differentiated responsibilities of states. Although the international system is predefined as an anarchical formation of states that share distinctive policy goals, SIDS have managed to influence the design of the broader climate regime especially through constructive engagement in UN climate negotiations. Ideas generated and shared during the 1980s have set interpretive frameworks for SIDS to merge climate policy goals and form coalition forces within UN negotiations platforms, thereby strategically influencing climate foreign policies and the interests of climate negotiators to consider their special case. More importantly, the intersubjective structure of interplay between SIDS climate foreign policy and UN climate negotiations ensured the design and continuity of collective action at international level.

This chapter has argued that a constructivist approach to foreign policy analysis can better explain SIDS' climate politics and how it has impacted the design and continuity of UN climate governance system supporting their climate efforts. As international commitment and cooperation is required to achieve SIDS' policy goals, their approach to addressing climate change meaningfully has been significantly linked to the ideas about international action and shared responsibilities of states. The impacts they have had on the UN process, however disproportionate they have been, are accounted for by the role of ideas generated and shared among climate negotiators. In a state of anarchy, where powerful states such as the USA can avert meaningful collective action, SIDS' climate ideas have played a significant role in shaping collective interests. The early development of the UN climate governance system has shown that ideas have shaped SIDS' climate foreign policy and influenced UN negotiations, starting from the UNFCCC negotiations. The disproportionate impact of SIDS on the UN climate governance system is better understood in the context of an intersubjective structure of international climate politics.

References

Adler, E. (2012). Constructivism in international relations: Sources, contributions, and debates. In W. Carlsnaes, T. Risse, & B. A. Simmons (Eds.), *Handbook of international relations* (pp. 95–118). Sage Publications.

Aginam, O. (2011). *Climate Change Diplomacy and Small Island Developing States*. United Nations University Publications. [online] Retrieved November 1, 2018, from https://unu.edu/publications/articles/climate-change-diplomacy-and-small-island-developing-states.html

Alliance of Small Island States (AOSIS). (2015). *Alliance of Small Island States: 25 Years of Leadership at the United Nations*

Ashe, J. W., Lierop, R. V., & Ashe, A. C. (1999). The role of the Alliance of Small Island states (AOSIS) in the negotiation of the United Nations framework convention on climate change (UNFCCC). *Natural Resources Forum, 23*, 209–222.

Barkdull, J., & Harris, P. G. (2009). Theories of environmental foreign policy. In P. G. Harris (Ed.), *Environmental change and foreign policy: theory and practice* (pp. 19–40). Routledge.

Barnett, J., & Campbell, J. (2010). *Climate change and small island states: Power, knowledge, and the South Pacific*. Earthscan.

Betzold, C. (2015). Adapting to climate change in small Island developing states. *Climatic Change, 133*(3), 481–489.

Betzold, C., Castro, P., & Weiler, F. (2012). AOSIS in the UNFCCC negotiations: from unity to fragmentation? *Climate Policy, 12*, 591–613.

Bishop, M. L. (2012). The political economy of small states: enduring vulnerability? *Review of International Political Economy, 19*, 942–960.

Blyth, M. (2002). *Great transformations: economic ideas and institutional change in the twentieth century*. Cambridge University Press.

Bodansky, D. (2001). The history of the global climate change regime. In U. Luterbacher & D. F. Sprinz (Eds.), *International relations and global climate change (global environmental accord: Strategies for sustainability and institutional innovation)* (p. 505). The MIT Press.

Browning, C. S. (2006). Small, smart and salient? Rethinking identity in the small states literature. *Cambridge Review of International Affairs, 19*(4), 669–684.

Busse, N. (1999). Constructivism and southeast Asian security. *The Pacific Review, 12*(1), 39–60.

Carlsnaes, W. (2008). Actors, structures, and foreign policy analysis. In S. Smith, A. Hadfield, & T. Dunne (Eds.), *Foreign policy: theories, actors, cases* (p. 113). Oxford University Press.

Chasek, P. S. (2005). Margins of power: coalition building and coalition maintenance of the South Pacific Island states and the Alliance of Small Island states. *Review of European Community and International Environmental Law, 14*(2), 125–137.

Chong, A., & Maass, M. (2010). Introduction: The foreign policy power of small states. *Cambridge Review of International Affairs, 23*(3), 381–382.

Coghlan, M. (2002). Prospects and pitfalls of the Kyoto protocol to the United Nations framework convention on climate change. *Melbourne Journal of International Law, 3*, 165.

Constantini, V., Crescenzi, R., Fillipis, F. D., & Salvatici, L. (2007). Bargaining coalitions in the WTO agricultural negotiations. *The World Economy, 30*(5), 863–891.

Danilovic, V. (2002). *When the stakes are high: deterrence and conflict among major powers*. University of Michigan Press.

Davis, W. J. (1996). The Alliance of Small Island states (AOSIS): the international conscience. *Asia–Pacific Magazine, 2*, 17–22.

Doty, R. L. (1993). Foreign policy as social construction. *International Studies Quarterly, 37*(3), 297–320.

Edwards, A. J. (1990). Small states conference on sea level rise. *Marine Pollution Bulletin, 21*(1), 4–5.

Elman, M. F. (1995). The foreign policies of small states: challenging neorealism in its own backyard. *British Journal of Political Science, 25*(2), 171–217.

Flockhart, T. (2016). Constructivism and foreign policy. In S. Smith, A. Hadfield, & T. Dunne (Eds.), *Foreign policy: theories, actors, cases* (p. 79). Oxford University Press.

Gayoom, A. Y. A. (2015). *AOSIS: A history of leadership at UNFCCC*

Goodman, J. S. (1965). The concept of 'system' in international relations theory. *Background, 8*(4), 257–268.

Grecequet, M., Noble, I., & Hellmann, J. (2017). Many Small Island Nations Can Adapt to Climate Change with Global Support. *The Conversation*. [online] Retrieved November 1, 2018, from https://theconversation.com/many-small-island-nations-can-adapt-to-climate-change-with-global-support-86820?utm_medium=email&utm_campaign=Latest%20from%20The%20Conversation%20for%20November%2015%202017%20-%2088027351&utm_content=Latest%20from%20The%20Conversation%20for%20November%2015%202017%20-%2088027351+CID_c2f6e6122e4f00fdc3e0e6f1c19716c4&utm_source=campaign_monitor_us&utm_term=Many%20small%20island%20nations%20can%20adapt%20to%20climate%20change%20with%20global%20support

Gvalia, G., Siroky, D., Lebanidze, B., & Iashvili, Z. (2013). Thinking outside the bloc: Explaining the foreign policies of small states. *Security Studies, 22*(1), 98–131.

Hameiri, S., & Jones, L. (2015). *Governing borderless threats: Non-traditional security and the politics of state transformation*. Cambridge University Press.

Harris, P. G. (2009). *Environmental change and foreign policy: Theory and practice*. Routledge.

Hellmann, G., & Wolf, R. (1993). Neorealism, neoliberal institutionalism, and the future of NATO. *Security Studies, 3*(1), 3–43.

Hey, J. A. K. (2003). *Small states in world politics: Explaining foreign policy behavior*. Lynne Rienner Publishers.

Hoad, D. (2015). Reflections on small island states and the international climate change negotiations (COP21, Paris, 2015). *Island Studies Journal, 10*(2), 259–262.

Hopa, J. T. (1989). *Vanuatu country statement at the small states conference on sea level rise*. Malé.

Houghton, D. P. (2007). Reinvigorating the study of foreign policy decision making: Toward a constructivist approach. *Foreign Policy Analysis, 3*(1), 24–45.

Ibrahim, A. (1989). *Maldives Country Statement*. Paper presented at the Small States Conference on Sea Level Rise held in Malé, 14–18 November 1987.

Keohane, R. O. (1969). *Lilliputians' dilemmas: Small states in international politics*. World Peace Foundation.

Keohane, R. O. (1984). *After hegemony: Cooperation and discord in the political economy*. Princeton.

Keohane, R. O. (1989). *International institutions and state power*. Westview Press.

Keohane, R. O., & Nye, J. S. (1977). *Power and interdependence: World politics in transition*. Little, Brown.

Lee, D., & Smith, N. J. (2010). Small state discourses in the international political economy. *Third World Quarterly, 31*(7), 1091–1105.

Little, A. (2015). What the Paris Climate Agreement Means for Vulnerable Nations. *The New Yorker*. [online] Retrieved November 1, 2018, from https://www.newyorker.com/news/news-desk/what-the-paris-climate-agreement-means-for-vulnerable-nations

Maldives Permanent Mission to the UN. (1987). *Address by His Excellency Mr Maumoon Abdul Gayoom, President of the Republic of Maldives, before the Forty Second Session of the United Nations General Assembly on the Special Debate on Environment and Development*, 19 October 1987. [online] Retrieved December 11, 2020, from http://maldivesmission.com/statements/address_by_his_excellency_mr_maumoon_abdul_gayoom_president_of_the_republic_of_maldives_before_the_forty_second_session_of_the_united_nations_general_assembly_on_the_special_debate_on_environment_and_development_19_october_1987

Maldives Permanent Mission to the UN. (2015). *Statement by Maldives on behalf of the Alliance of Small Island States (AOSIS) at the Meeting of the Parties to the Kyoto Protocol*. 1 December 2015. [online] Retrieved December 11, 2020, from http://maldivesmission.com/statements/statement_by_maldives_on_behalf_of_the_alliance_of_small_island_states_aosis_at_the_meeting_of_the_parties_to_the_kyoto_protocol

Mayer, M., & Arndt, F. J. (2009). The politics of socionatures images of environmental foreign policy. In P. G. Harris (Ed.), *Environmental change and foreign policy: theory and practice* (pp. 94–110). Routledge.

McMahon, V. (1993). Environmental nongovernmental organizations at intergovernmental negotiations. In L. E. Susskind, W. R. Moomaw, & A. Najam (Eds.), *Papers on international environmental negotiation* (Vol. 3, pp. 1–21). The Program on Negotiation at Harvard Law School.

McNamara, K. E., & Gibson, C. (2009). "We do not want to leave our land": Pacific ambassadors at the United Nations resist the category of "climate refugees". *Geoforum, 40*(3), 475–483.

Mead, L. (2020). AOSIS Forum Calls for "Year of Ambition" on Climate Action. *International Institute for Sustainable Development*. [online] Retrieved June 9, 2020, from http://sdg.iisd.org/news/aosis-forum-calls-for-year-of-ambition-on-climate-action/

Mol, P. J. (2014). Small island developing states and international climate change negotiations: The power of moral "leadership". *International Environmental Agreements, 14*, 281–297.

Murphy, D., Tirpak, D., Drexhage, J., & Gagnon-Lebrun, F. (2009). *Encouraging Developing Country Participation in a Future Climate Change Regime by International Institute for Sustainable Development (IISD)*. [online] Retrieved December 11, 2020, from https://www.iisd.org/system/files/publications/developing_country_participation_in_climate.pdf

Organisation for Economic Co-operation and Development (OECD). (2015). *Small island developing states (SIDS) and the post-2015 development finance agenda*. [online] Retrieved December 11, 2020, from https://www.oecd.org/dac/financing-sustainable-development/Addis%20Flyer%20SIDS%20FINAL.pdf

Ourbak, T., & Magnan, A. K. (2017). The Paris agreement and climate change negotiations: small islands, big players. *Regional Environmental Change, 18*(1), 2201–2207.

Panke, D. (2012). Small states in multilateral negotiations. What have we learned? *Cambridge Review of International Affairs, 25*(3), 387–398.

Parry, M., Canziani, O., Palutikof, J., Linden, V. D. P., & Hanson, C. (Eds.). (2007). *Climate change 2007: impacts, adaptation and vulnerability [internet]. Contribution of working group II to the fourth assessment report of the intergovernmental panel on climate change*. Cambridge University Press.

Paterson, M. (1996). *Global warming and global politics*. Routledge.

Paterson, M. (2001). Principles of Justice in the context of global climate change. In U. Uterbacher & D. F. Sprinz (Eds.), *International relations and global climate change (global environmental accord: Strategies for sustainability and institutional innovation)* (pp. 119–126). The MIT Press.

Rasheed, A. A. (2019). Role of small islands in UN climate negotiations: A constructivist viewpoint. *International Studies, 56*(4), 215–235.

Rasheed, A.A. (2020) Climate Ideas as Drivers of Pacific Islands' Regional Politics and Cooperation. *E-International Relations*. [online] Retrieved June 9, 2020, from https://www.e-ir.info/2020/01/15/climate-ideas-as-drivers-of-pacific-islands-regional-politics-and-cooperation/

Rowlands, I. H. (2001). Classical theories of international relations. In U. Luterbacher & D. F. Sprinz (Eds.), *International relations and global climate change (global environmental accord: Strategies for sustainability and institutional innovation)* (pp. 43–64). The MIT Press.

Sadat, N. (2009). Small Islands, Rising Seas. *UN Chronicles*. [online] Retrieved November 1, 2018, from https://unchronicle.un.org/article/small-islands-rising-seas

Simpson, A. W. (2006). Small states in world politics. *Cambridge Review of International Affairs, 19*(4), 649–649.

Smith, S. (2001). Foreign policy is what states make of it. In V. Kubalkova (Ed.), *Foreign policy in a constructed world* (pp. 38–55). M. S. Sharpe.

Smith, S., Hadfield, A., & Dunne, T. (2012). *Foreign policy: Theories actors, cases*. Oxford University Press.

Starkey, B., Boyer, M. A., & Wilkenfeld, J. (2008). *Negotiating a complex world: An introduction to international negotiation*. Rowman & Littlefield Publishers.

Stone, C. D. (2004). Common but differentiated responsibilities in international law. *The American Journal of International Law, 98*(2), 276–301.

Telbami, S. (2002). Kenneth Waltz, neorealism, and foreign policy. *Security Studies, 11*(3), 158–170.

Thomas, A. (2020). Coronavirus crisis underscores small Islands' climate vulnerability. *Climate Home News*. [online] Retrieved June 9, 2020, from https://www.climatechangenews.com/2020/04/17/coronavirus-crisis-underscores-small-islands-climate-vulnerability/

Thomas, A., Baptiste, A., Martyr-Koller, R., Pringle, P., & Rhiney, K. (2020). Climate change and small island developing states. *Annual Review of Environment and Resources, 45*(6), 1–127. https://doi.org/10.1146/annurev-environ-012320-083355

Underdal, A. (2017). Climate change and international relations (after Kyoto). *Annual Review of Political Science, 20*(1), 169–188.

United Nations (UN). (1994). *Programme of Action for the Sustainable Development of Small Island Developing States*. [online] Retrieved December 12, 2020, from https// sustainabledevelopment.un.org/conferences/bpoa1994

United Nations (UN). (2018). *Small Island Developing States*. [online] Retrieved November 1, 2018, from https://sustainabledevelopment.un.org/topics/sids/list

United Nations Educational, Scientific and Cultural Organization (UNESCO). (1989). *Male declaration on global warming and sea level rise*. Retrieved from htp://unesdoc.unesco.org/images/0009/000977/097720eb.pdf

United Nations Framework Convention on Climate Change (UNFCCC). (2005). *Climate change, small island developing States*. [online] Retrieved November 1, 2018, from http://unfccc.int/resource/docs/publications/cc_sids.pdf

United Nations Framework Convention on Climate Change (UNFCCC). (2015a). *Paris Agreement*. [online] Retrieved November 2, 2017, from

United Nations Framework Convention on Climate Change (UNFCCC). (2015b). *Statement delivered by the Maldives on behalf of the Alliance of Small Island States at the Closing SBSTA Plenary in Paris*. [online] Retrieved November 2, 2017, from http://www4.unfccc.int/submissions/Lists/OSPSubmissionUpload/53_152_130937948115158180-AOSIS%20closing%20statement%20SBSTA%2043.pdf

United Nations General Assembly (UNGA). (1994). *Report of the Global Conference on Sustainable Development of Small Island Developing States*. UNGA Resolution A/CONF 167/. [online] Retrieved November 2, 2017, from http://www.un.org/esa/dsd/dsd_aofw_sids/sids_pdfs/BPOA.pdf

United Nations Office of the High Representative for the Least Developed Countries, Landlocked Developing Countries and Small Island Developing States (UN-OHRLLS). (2011). *Small island developing states: Small island big(ger) stake*. Report. New York: UN-OHRLLS. Retrieved from http://unohrlls.org/custom-content/uploads/2013/08/SIDS-Small-Islands-Bigger-Stakes.pdf

Voccia, A. (2012). Climate change: What future for small, vulnerable states? *International Journal of Sustainable Development & World Ecology, 19*(2), 101–115.
Walt, S. M. (1987). *The origins of alliances*. Cornell University Press.
Waltz, K. N. (1979). *Theory of international politics*. Addison-Wesley Pub Co.
Wendt, A. (1992). Anarchy is what states make of it: The social construction of power politics. *International Organization, 46*(2), 391–425.
Wendt, A. (1999). *Social theory of international politics*. Cambridge University Press.
Wicaksana, I. G. W. (2009). The constructivist approach towards foreign policy analysis. *Indonesian Journal of Social Sciences, 1*(1).
Widmaier, W. W. (2005). The democratic peace is what states make of it: a constructivist analysis of the US-Indian 'near-miss' in the 1971 South Asian crisis. *European Journal of International Relations, 11*(3), 431–455.
Widmaier, W. W. (2007). Constructing foreign policy crises: Interpretive leadership in the cold war and war on terrorism. *International Studies Quarterly, 51*(4), 779–794.
Wiegandt, E. (2001). Climate change, equity, and international negotiations. In U. Luterbacher & D. F. Sprinz (Eds.), *International relations and global climate change (global environmental accord: Strategies for sustainability and institutional innovation)* (pp. 128–159). The MIT Press.

Social and Economic Vulnerability to Climate Change: A Gender Dimension for Indian Ocean Islands

Verena Tandrayen-Ragoobur

Abstract The chapter focuses on gender specific socio-economic vulnerability to climate change in four Indian Ocean island economies of Comoros, Madagascar, Maldives and Mauritius. The study provides a systematic analysis of gender differences in environmental changes from vulnerability to adaptability and set out gender-inclusive climate change policies. In analysing the disproportionate bearing of environmental changes on women and their differential coping strategies relative to men, various indicators are first adopted from the INFORM global risk index and EM-DAT database. Second, data from the World Bank Development Indicators, 2018 (World Bank, *World Bank Development Indicators 2018*, 2018a), and World Bank Gender Statistics, 2018 (World Bank, *World Bank Gender Statistics 2018*, 2018b), is analysed for the four islands from 1960 to 2017 using principal component analysis. The principal component analysis extracts the component factors relevant for men and women separately and a Social and Economic Vulnerability Index is developed applying a gender perspective. The results reveal that the human resource dimension (unemployment and education) and demographics have significant bearings on both women's and men's vulnerability to climate change. However, additional key components are observed for women, namely the proportion of women in decision making and health status namely fertility rate. With differences in component factors, gender mainstreaming in climate change policies and strategies is imperative for the Indian Ocean island economies.

Keywords Gender · Climate change · Socio-economic vulnerability · Adaptability · Island economies · Indian Ocean · Gender mainstreaming · Climate risks

---

V. Tandrayen-Ragoobur (✉)
Department of Economics and Statistics, University of Mauritius, Reduit, Mauritius
e-mail: v.tandrayen@uom.ac.mu

S. Moncada et al. (eds.), *Small Island Developing States*, The World of Small States 9, https://doi.org/10.1007/978-3-030-82774-8_9

1 Introduction

Over the past 20 years, the Indian Ocean region has experienced more than 50 natural catastrophic events, amounting to more than US$17 billion in economic costs (IISD, 2016). Island economies like Maldives, Mauritius, Comoros and Madagascar are in the crosshairs of climate change, and the Indian Ocean is the main sphere where natural disasters, geographical isolation, limited resources and sensitive ecosystems add to their vulnerability (IPCC, 2014). Droughts, floods, sea level rise along with soil and coastal erosion are no exception to this part of the Indian Ocean. Coastal erosion, saline intrusion, land-based pollution, and sea flooding are already serious problems in many of these islands. Further, most infrastructure and socioeconomic activities are located along the coastline or close to sea level (Pernetta, 1992; Hay & Kaluwin, 1993).

Changes in climatic conditions and natural hazards are increasingly intertwined with a country's extent of vulnerability and capacity to adapt. The impacts of climate change vary by gender, socioeconomic status, ethnicity, nationality, health, sexual orientation, age, and place, amongst many other factors. Thus, to identify who will be most affected by climate change in planning for climate change responses (O'Brien et al., 2004; Holand et al., 2011), there is a need to measure vulnerability in ways that enable comparison of different groups within a population (Balikoowa et al., 2018). Further, it is important to consider social vulnerability, which comprises the totality of factors that makes people more susceptible to harm from specific threats. It covers the external and/or internal characteristics of a person or group of persons which affect their capacity to anticipate, cope with, resist and recover from the impact of a natural hazard (Blaikie et al., 1994).

Social vulnerability, in particular, highlights the role of gender in shaping vulnerability where women are viewed as being more vulnerable than men. Their higher vulnerability is often linked to existing gender inequalities. Gender inequalities intersect with climate change risks, vulnerabilities and adaptation strategies and in turn climate change magnifies the prevailing patterns of gender disadvantage. Women are more vulnerable to climate changes relative to their male counterparts since they are poorer than men (Arora-Jonsson, 2011; Rahman, 2013) due to low economic opportunities (Johnsson-Latham, 2007), social and cultural constructed roles and responsibilities and also because of unequal access to resources needed to safeguard their livelihoods from climate-related stress (Sharmin & Islam, 2013).

Throughout the limited literature on gender and climate change, two perspectives emerge. First, women in developing countries are more vulnerable than men to the consequences of climate change (Nagarajan et al., 2020). Second, men and women play different roles in coping with climate change and manage climate change risks differently. In effect, women are key actors in numerous areas of mitigation and adaptation. Women are more sensitive to risk, more prepared to change behaviour and more likely to support drastic policies and measures on climate change (Hemmati & Röhr, 2007; Brody et al., 2008). The interest of gender in environment-related research is of increasing interest (Agarwal, 2010;

Arora-Jonsson, 2014; Meinzen-Dick et al., 2014; Leach, 2016). This has evolved from early ecofeminist scholarship arguing that women are more connected and closely linked with nature than men (Shiva, 1988; Mies & Shiva, 1993).

Gender is therefore a critical dimension to be factored in when analysing environmental changes, including integration of gender specific concerns and solutions in climate change policies, programmes and processes. In 2017, during the United Nations Framework Convention on Climate Change (UNFCCC), a Gender Action Plan was adopted and extended the Lima Work Plan on Gender for three years, until 2019. The main thrust of the Gender Action Plan was to ensure the integration of gender into climate change policies and increasing the participation of women at national and global intergovernmental fora. Hence, gender sensitive frameworks are vital to address relevant policy options in mainstreaming gender into climate change responses. Recently there has been a strong case for urgently moving beyond "gender-lite" to gender-responsive implementation across climate risk mapping and response. An important dimension is to re-assess existing national frameworks and plans by designing and promoting women's active participation and leadership at all levels of planning (Nagarajan et al., 2020).

While an emergent literature addresses the specific vulnerability of women within the broader field of disaster and environmental change research (Cutter, 1995; Bolin et al., 1998; Enarson & Morrow, 1998; Enarson, 1998, 2000; Fothergill, 1998; Fordham, 1999, 2000; Bradshaw, 2004; Enarson & Meyreles, 2004; Agarwal, 2010; Arora-Jonsson, 2014; Carr & Thompson, 2014; Leach, 2016), gender analysis of socio-environmental issues still remains understudied, and its incorporation in environmental policies has barely advanced (Ravera et al., 2016).

This chapter focuses on the gendered nature of climate change vulnerability and gender-inclusive climate change policies in the four Indian Ocean island economies of Comoros, Madagascar, Maldives and Mauritius. It provides a systematic, quantitative analysis of gender differences in natural disaster exposure, vulnerability and coping strategies. Different indicators are adopted from the Index for Risk Management (INFORM) global risk index and the Emergency Events Database (EM-DAT) for this purpose. Principal component analysis is applied and a social-vulnerability index across gender is derived by calculating composite factors for the four islands. By accounting for the gender dimension, development strategies can better counter current climate risks and impacts. Whilst national policies need to allow for cultural, traditional, and context-specific factors to facilitate local adaptation and resilience to climate change, regional policies are also of utmost importance to the Indian Ocean islands as they face similar problems. However, trans-boundary adaptation policies are still in their infancy and there is a need to prioritise regional strategies.

The chapter is structured as follows: Sect. 2 links gender and climate change vulnerability via the different channels which explain the reasons for greater vulnerability of women to environmental changes. Section 3 focuses on the social and economic vulnerability index and the different dimensions of the index. Section 4 explains the methods adopted in this study. Section 5 presents the comparative analysis and findings for the four Indian Ocean island economies and finally, relevant policy options are discussed in Sect. 6.

2 Gender, Vulnerability and Adaptability to Climate Change

There are diverse ways in which climate change consequences are driven by gender. Women in the global South are affected more adversely by climate change than men (Arora-Jonsson, 2011). Alber (2009) and Nelson (2011) amongst others postulate that this is because women are the largest proportion of the world's poor and are also more economically dependent on natural resources for their livelihoods and survival. Gender and poverty are two distinct forms of disadvantages (Arora-Jonsson, 2011). Thus, gender unequal access to social and physical goods, gender gaps in education, health, income and time, limited access to resources and human rights violations (Ul Islam et al., 2009) make women more vulnerable to climate change and thus limit the effectiveness of their response and adaptation when a disaster strikes.

Further, socially constructed gender-differentiated roles and responsibilities in the household and community increase women's vulnerability to climate-related natural hazards (Nelson, 2011). Women undertake a disproportionate share of unpaid care work compared to men and, with changes in climate, their burden of unpaid care is likely to rise as women have to travel longer distances to collect water and fuel and also provide additional care for family members affected by climate sensitive diseases (Denton, 2004). These tasks are carried out at the expense of education or income-generating activities (Alber, 2009).

Women are likely to have fewer resources and rights, and less knowledge and time with which to cope with climate change (Cannon, 2002; Nelson et al., 2002; Denton, 2004; Ul Islam et al., 2009; Babagura, 2010). Women's historic disadvantages in terms of their lack of independence, restricted rights and power, and a submissive voice in shaping family and societal decisions, make them highly vulnerable and constrain their ability to adapt to climate change. Women often have limited or no control over family finances and assets and may be subject to cultural restrictions on their mobility, including dress codes and seclusion practices. Cultural restrictions on mobility are especially dangerous during extreme weather events when women may not be able to relocate to safety sites without the consent and/or accompaniment of a male relative (Nelson, 2011). Such gender inequity can be attributed to many interconnected causes, but it is most pronounced where women have lower socioeconomic status and power—and therefore fewer options—so much that their worse outcomes are weighted towards cultural rather than underlying biological or physiological causes (Neumayer & Plümper, 2007).

In addition, women may be less able to escape from catastrophic events due to their smaller average size and physical strength (Cannon, 2002) or most likely, in the case of flooding for example, an inability to swim (Ikeda, 1995). Pregnant and nursing women and those with small children are particularly vulnerable. Women and their children are 14 times more likely to die than men during disasters (Brody et al., 2008). Natural disasters not only kill more women than men but also lower their life expectancy. The stronger the disaster, the stronger the impact on life expectancy (Neumayer & Plümper, 2007). For instance, many women in Sub

Saharan Africa are responsible for producing the food eaten at home and with changing climates, they must work harder for less food and this has a detrimental effect on their health. Moreover, they are usually the first to reduce the amount they eat and sacrifice their diets for other family members.

The cumulative impact of differences in access to resources, income, information, gender divisions of labour, poverty and income inequalities power relations and culturally specific gender norms and roles result in many women being severely affected by climate change. These disparities enhance vulnerability of women and children and create fewer options and capacities to cope with climate variability (Alber, 2009). The more patriarchal the society, the more excluded women and children are from decisions regarding risk mitigation, disaster preparedness, planning and reconstruction, unless specific efforts are made to include them. By not being well represented in decision-making processes, women's ability to meaningfully participate in adaptation and mitigation planning is constrained (UNDP, 2013).

To counteract climate change vulnerability, women adapt their lives either through mitigation or preparedness or build capacities to withstand and cope with hazards or tackle the root causes of vulnerability namely poverty, poor governance, discrimination, inequality and inadequate access to resources and livelihoods. It has been argued that women have an important body of traditional and environmental knowledge and when they are in control of resources, they are more likely than men to use this information for family health and economic stability. Research also shows that women can better change strategies in response to new information and make decisions that minimise risk. When women are effectively empowered, they can be successful agents of adaptation to climate change. Empowering women and achieving gender equality are critical goals in themselves, but represent also critical mechanisms of managing climate change and creating a sustainable future (Yavinsky, 2012).

3 Social and Economic Vulnerability to Climate Change

Women's greater vulnerability to environmental changes is comprised of the specific threats or hazards and the social, cultural and biological factors that make them more vulnerable to that threat. The concept of 'vulnerability' remains complex. An overall understanding of vulnerability is that it is the combination of (1) a 'threat' or 'exposure' to a hazard; (2) the degree of potential for loss (Cutter et al., 2003); and (3) the propensity to be adversely affected, which ascends from historical, social, economic, political, technological, institutional and environmental conditions (Timmerman, 1981; Susman et al., 1983; Cutter, 1996; Weichselgartner, 2001; Bankoff, 2004; Wisner et al., 2004; Allison et al., 2009). However, vulnerabilities vary geographically, over time, space and between social groups with different socio-economic conditions (Cutter et al., 2003; Fuchs et al., 2012). Thus, vulnerability is not static and is more a relative and dynamic concept.

In analysing vulnerability, a hybrid method is often adopted which comprises of risk-hazard approaches, where vulnerability depends on biophysical risk factors and the potential loss of a particular exposed population (Cutter, 1996), as well as political economic-ecological approaches, which emphasise the political, cultural and socio-economic factors explaining differential exposure, impacts and capacities to recover from an event (for instance, the pressure and release model; Blaikie et al., 1994). Although a number of studies have analysed the physical vulnerability component (e.g. Koks et al., 2014; Ocio et al., 2016), the social aspects of vulnerability have often been neglected (Cutter et al., 2003; de Loyola Hummell et al., 2016; Aroca-Jimenez et al., 2017) because of the difficulty of quantifying what are inherently qualitative characteristics.

The concept of vulnerability in the social framework was first studied by O'Keefe et al. (1976) to probe into the role of socioeconomic factors in explaining vulnerability and adaptability to the impacts of hazards. Since, then, various theoretical models have been designed (Haas et al., 1977; Burton et al., 1978; Susman et al., 1983) to explain the specific social inequalities behind vulnerability and the capacity of society to cope with hazards (Blaikie, 1994; Bohle et al., 1994; Cutter, 1996; Cutter et al., 2003). Social vulnerability to natural hazards is most commonly termed as "the differential capacity of groups and individuals to deal with hazards, based on their positions within physical and social worlds" (Dow, 1992, p. 423), or as "the inability to take effective measures to insure against losses" (Bogard, 1988, p. 156). Substantial research has examined social vulnerability to climate change in many international settings (Bohle et al., 1994; Adger, 1999; O'Brien & Leichenko, 2000; O'Brien et al., 2004; Eakin, 2005; Ford et al., 2006; Eakin et al., 2010; Ford & Pearce, 2010; Marshall, 2010).

Measuring social vulnerability is rather complex since it includes numerous differential capacities of groups and individuals which are in turn related to differences in control over resources, risk exposure, awareness and management and the ability to respond. Emphasis is on measuring individual characteristics related to social vulnerability which place people in either a less vulnerable position (such as their socioeconomic status or social connections) or a more vulnerable position (for instance disabilities or the community's limited transport connectivity). Overall, in social vulnerability analysis, a separate assessment is made of vulnerability (Tapsell et al., 2002; Cutter et al., 2003; Nelson et al., 2015) and resilience (Cutter et al., 2008, 2010; Siebeneck et al., 2015). The approach has usually been based on calculating composite indices from socio-demographic and economic characteristics using factor analysis and principal components analysis (Clark et al., 1998; Cutter et al., 2003; Dwyer et al., 2004; Füssel, 2007; de Loyola Hummell et al., 2016; Rogelis et al., 2016).

Although within these socio-economic vulnerability indicators, the gender dimension is factored in as one variable, the different sub-indices are not disaggregated by sex to obtain a more complete picture on the higher vulnerability of women to environmental changes relative to their male counterparts. The main contribution of this chapter is to apply the socio-economic indicators from a gender perspective via the use of sex-disaggregated data wherever available for the four island economies to assess the particular vulnerability and resilience of women to environmental changes and as such develop gender-inclusive climate change policies.

4 Regional Setting: A Brief on the Indian Ocean Islands

In this study, we focus on the four Indian Ocean islands of Maldives, Mauritius, Comoros, and Madagascar, which differ in socio-economic development and susceptibility to environmental changes.

Maldives and Mauritius are upper middle-income economies. They are both small in land area and very vulnerable to climate change impacts and disasters. In particular, the low elevation of Maldives makes it highly vulnerable to sea level rise. Almost half of all settlements and over two thirds of critical infrastructure are located within 100 m of the shoreline and are under immediate threat from rising sea levels. Over the last three decades; the island has also faced additional challenges including extremely high population density, high levels of poverty, communication problems, difficult and expensive transport, and isolation from world markets increasing susceptibility to global influences.

Similar to Maldives, Mauritius encounters multi-faceted environmental challenges, especially in its coastal zones, where a convergence of accelerating sea level rise and increasing frequency and intensity of tropical cyclones result in considerable economic loss, humanitarian stresses and environmental degradation. The Republic of Mauritius is comprised of several islands, such as Mauritius itself, Rodrigues island, Agaléga, Saint Brandon, Tromelin and the Chagos Archipelago. Rodrigues island is located at about 560 km to the north-east of Mauritius. It is part of the Republic of Mauritius with a local population of around 38,500. Its economy is based primarily on fishing, agriculture, farming, handicrafts and a developing tourism sector. The Mauritian economy has expanded at a consistent and moderate pace in recent years, with a gross domestic product (GDP) growth averaging 3.8% in 2016 and 2017. However, despite the recent economic boom, the country is at great risk of climate change and rising sea levels. Mauritius has been facing changes in rainfall patterns both temporally and spatially which affect negatively agricultural production. Both Mauritius and Maldives have developed a Strategic National Action Plan, which integrates disaster risk reduction and climate change adaptation, which is in line with the objectives of the UNFCCC.

The Union of Comoros is classified as a low-income economy, a small archipelago made up of three islands namely Ngazidja (or Grand Comoros), Mohéli, and Anjouan. The Comoros is densely populated, with approximately 400 inhabitants per sq. km, and has relatively young population with more than half of the population (53%) being under the age of 20. The last household survey of 2014 showed almost 18% of the population lives under the international poverty line US $1.9 per capita per day. There is also considerable inequality, with a Gini index of 44.9 in 2017 and high infant and maternal mortality rates (World Bank, 2018a,

2018b). Further, the population is highly dependent on natural resources with the production of three main export crops (vanilla, cloves, and ylang-ylang) contributing about 95% of export earnings. Climate change impacts are experienced on agriculture and food security, coastal zones and marine ecosystems, water resources and therefore public health. In addition, sea level rise is a major concern as one hundred per cent of the population of Comoros live within 10 km of the coastline. To reduce its vulnerability and adapt to a changing climate, Comoros is working on adaptation strategies namely, agricultural diversification of production, beach rehabilitation, irrigation efficiency, improving water management, and research on potential epidemic diseases related to climate change.

Finally, Madagascar is the world's fourth largest island and has one of the highest poverty rates in Africa, with 81% of the population living on less than $1.25 per day. Madagascar is distinctively vulnerable to environmental impacts and natural disasters due to its extensive coastline. Its leading climate change stressors are increased temperatures; extended drought periods and increased variability of rainfall; intensification of cyclones and floods and rising sea level and sea surface temperatures. The driving sectors of Madagascar's economy include agriculture (predominantly rainfed), fisheries and livestock production, all of which rely on climate-sensitive natural resources. Food security is a major concern, with 25% of the country's rural population classified as food insecure. Madagascar launched its National Adaptation Plan process in 2012 aiming to reduce climate vulnerability in the medium and long term, and to integrate climate-related risks and opportunities into development planning and budgeting systems.

Each of these four island nations is highly vulnerable to climate change. A better understanding of their degree of vulnerability and capacity to adapt to specific hazards is needed, including a gender dimension given the greater vulnerability of women to climate change.

5 Methods

Generic indices of national level vulnerability are often used (Cardona, 2007) but only a minority focused on islands (see for instance Blancard & Hoarau, 2013). The island-specific indicators that exist often suffer from lack of data (Peduzzi et al., 2009) and there have been recent moves towards qualitative analysis with participatory approaches that link existing data with local visions of vulnerability (Park et al., 2012). This chapter uses standard indicators of socio-economic vulnerability by disaggregating the indices by gender to assess the higher susceptibility of women to climate change.

5.1 The Inform Risk Index and the Emergency Events Data

Thus, the chapter provides an analysis of the different degree of vulnerabilities faced by the four island economies in the context of climate change by first using the INFORM Risk Index 2018.[1] The index summarises multiple factors contributing to the risk for humanitarian crises and disasters and as such creates a risk profile for every country. An overall risk score out of ten is made available for each country. It covers three dimensions of risk, namely: hazards and exposure (events that may occur and exposure to such hazards); vulnerability (the susceptibility of communities to those hazards); and lack of coping capacity (lack of resources available that can alleviate the impact). The aspects of physical exposure and physical vulnerability are integrated in the hazard and exposure dimension. The characteristic of fragility of the socio-economic system becomes INFORM's vulnerability dimension while lack of resilience to cope and recover is treated under the lack of coping capacity dimension. The split of vulnerability in three components is particularly useful for tracking the results of disaster reduction strategies over time.

In addition, data from the Emergency Events Database—Université Catholique de Louvain (UCL)—CRED, D. Guha-Sapir (EM-DAT) is used for the four island economies.[2] However, these two data sources do not allow for differences between women and men. Hence, the socio-economic vulnerability index within a gender perspective is computed using principal components analyses of data from the World Bank Development Indicators, 2018 (World Bank, 2018a), and World Bank Gender Statistics, 2018 (World Bank, 2018b), for each of the four islands from 1960 to 2017.

5.2 Building the Socio-Economic Vulnerability Index via a Gender Lens

The objective of the study is first to construct a socio-economic index of overall vulnerability from a set of indicators for the four island economies. The vulnerability indicator is a useful tool for identifying and monitoring vulnerability over time and also for developing and prioritising strategies to reduce vulnerability (Rygel et al., 2006). Second, different sub-indices are split between data pertaining to males and females to construct a socio-economic vulnerability measure through a gender lens. The index is calculated for each nation over time subject to availability of data.

Based on existing literature (Cutter et al., 2003; Oulahen et al., 2015; de Loyola Hummell et al., 2016), a set of 40 variables was initially characterised for each country. An indicator based approach is adopted with an array of indicators to measure the socio-economic vulnerability of the island states. These variables are

[1] Data on the index can be obtained at http://www.inform-index.org/.

[2] The data can be accessed at https://www.emdat.be/database.

country-specific and due to limited data, more indicators could not be added to the list. The indicators were then split into seven dimensions: (1) Socio-economic (GDP per capita, Gross National Income for male and female, the Human Development Index (HDI) for male and female and the Gender Development Index); (2). Human Resources (employment by gender (including sectoral and vulnerable employment as well as unemployment rate) and education); (3) Demographics (age dependency ratio, urban and rural population by gender and female migrants); (4) Health status (life expectancy, fertility rate, population growth and health expenditure per capita); (5) Infrastructural development (telephone mainlines per 100 people); (6) Built environment (access to water in rural and urban areas and population density); and (7) Social Capital (proportion of women in ministerial positions as it explains the network of relationships among people who live and work in a society). The gender aspect and the urban-rural dichotomy are well covered within the seven different dimensions of the index.

Table 1 below provides an overview of the main socio-economic vulnerability indicators and it can be observed that Gross National Income per capita for all countries is twice as much for males as for females. There is also a discrepancy between the female and male HDI index with the latter being higher. Across all four economies, male employment rate exceeds female employment rate and a greater percentage of women than men tend to be in the agricultural sector. Vulnerable employment also tends to be on the higher side for women compared to men in all islands except Mauritius. Similarly, female unemployment rate exceeds that of male in all countries. In terms of life expectancy, women tend to live longer than men in all island economies and they also have higher primary education completion rates. The proportion of women in ministerial level positions is highest in Madagascar followed by Mauritius.

6 Data Analysis: The Hazards and Exposure Dimension: A Regional Perspective

Under the INFORM model, the hazards and exposure dimension of countries is rated on a scale from 0 to 10. Hazard-dependent factors are treated in the hazard and exposure dimension, while hazard-independent factors are divided among the two dimensions: the vulnerability dimension that considers the strength of the individuals and households relative to a crisis situation, and the lack of coping capacity dimension that considers factors of institutional strength. Within the hazard and exposure dimension, both natural and human hazards are depicted in Table 2.

It can be observed that the overall score of the Hazard and Exposure Index for Madagascar is the highest at 3.9 followed by Maldives 2.1 and Mauritius 1.9. Madagascar stands out in terms of greater exposure to both natural and human hazards and that exposure is likely to be higher due to the occurrence of floods, followed by tropical cyclones, tsunamis and droughts. Madagascar has had

Table 1 Socio-economic vulnerability indicators of the four island economies (Sources: Compiled from the World Development Indicators 2018 and Human Development Reports 2017. Adapted from Cutter et al. (2003), Holand et al. (2011), Holand and Lujala (2013), Bergstrand et al. (2015) and UNDP (2017))

Sub-indices	Comoros	Madagascar	Maldives	Mauritius
SOCIOECONOMIC				
Gross Domestic Product per capita (USD)	775.1	401.7	9875.3	9630.9
Gross National Income per capita (Female)	715	1091	7155	10,540
Gross National Income per capita (Male)	1945	1549	13,591	25,539
Human Development Index	0.49	0.52	0.70	0.78
Human Development Index (Female)	0.44	0.50	0.68	0.76
Human Development Index (Male)	0.54	0.53	0.72	0.80
Gender Development Index	0.82	0.95	0.94	0.95
HUMAN RESOURCES				
Employment Level				
Employment to population ratio, 15+, female (%)	27.2	81.3	54.8	41.23
Employment to population ratio, 15+, male (%)	64.8	87.5	80.1	70.43
Sectoral Employment				
Employment in agriculture, female (%)	76	76.4	3.7	5.7
Employment in agriculture, male (%)	56.2	72.1	10.6	8.3
Employment in industry, female (%)	0.6	4.5	20.4	16.5
Employment in industry, male (%)	7.1	13.9	24.4	30.5
Employment in services, female (%)	23.3	19.1	75.81	77.7
Employment in services, male (%)	36.7	14	65.0	61.3
Wage and salaried workers, female (%)	28.7	8.4	72.3	84.4
Wage and salaried workers, male (%)	41.4	13.4	76.8	76.5
Vulnerable Employment				
Vulnerable employment, female (%)	70.4	89.6	26.4	13.96
Vulnerable employment, male (%)	54.7	81.0	12.8	18.26
Unemployment level				
Unemployment, female (%)	23.9	3	5.8	11.3
Unemployment, male (%)	18.4	1.7	3.1	4.9
Education				
Primary Completion Rate, female (%)	79.2	71.4	89.0	103.1
Primary Completion Rate, male (%)	73.8	67.3	95.1	99.2
DEMOGRAPHICS				
Age dependency ratio (% of working-age population)	75.1	80.1	38.0	41.5
Urban population, female (% of total)	14.0	17.9	22.7	20.3
Urban population, male (% of total)	14.3	17.2	23.0	19.5
Female migrants (% of international migrant stock)	51.6	43.0	29.7	44.6
Rural population, female (% of total)	35.6	32.3	26.9	30.5

(continued)

Table 1 (continued)

Sub-indices	Comoros	Madagascar	Maldives	Mauritius
Rural population, male (% of total)	36.1	32.6	27.5	29.8
HEALTH STATUS				
Life expectancy at birth, female (years)	65.2	67.1	78.2	77.8
Life expectancy at birth, male (years)	61.8	64.0	76.1	71.1
Population growth (%)	2.3	2.7	2.0	0.1
Fertility rate	4.4	4.2	2.1	1.4
Health expenditure per capita	56.8	13.7	1165.1	482.5
INFRASTRUCTURE				
Telephone lines (per 100 people)	1.7	0.6	5.8	30.73
BUILT ENVIRONMENT VULNERABILITY				
Population density	427.5	42.8	1391.6	622.4
Improved water source (% of population with access)	90.1	51.5	98.6	99.9
Improved water source, rural population (%)	89.1	35.3	97.9	99.8
Improved water source, urban population (%)	92.6	81.6	99.5	99.9
SOCIAL CAPITAL				
Proportion of women in ministerial level positions (%)	6.1	19.2	5.9	11.6

35 cyclones, eight floods and five periods of severe droughts (a three-fold increase) over the past 20 years. These have caused $1 billion in damages and affected food security, drinking water supply and irrigation, public health systems, environmental management and quality of life (USAID, 2016). Mauritius is the next hazardous country, followed by Maldives with tropical cyclones more prominent in the former and tsunamis in the latter. In terms of human hazard, the Projected Risk of Conflicts is estimated using the Global Conflict Risk Index. The index covers structural indicators such as socioeconomic, political, geographic and security variables to determine a given country's risk for conflict. The projected risk of conflict is 1.7 for Madagascar and 1.3 for Maldives.

In addition, data from EM-DAT is analysed for all four economies in terms of total deaths, total number of people affected by natural disasters over the last decades and the total damage to the country. The information is provided in Table 3 and supports the analysis that Madagascar followed by Mauritius has been more exposed and more affected by natural disasters relative to the other island states in this region.

Table 2 Hazards and exposure across four Indian Ocean Islands (Source: Data retrieved from the INFORM Risk Index, 2018)

Country	Earthquake (0–10)	Flood (0–10)	Tsunami (0–10)	Tropical cyclone (0–10)	Drought (0–10)	Natural (0–10)	Projected conflict risk (0–10)	Current highly violent conflict intensity (0–10)	Human (0–10)	Hazard and exposure (0–10)
Comoros	0.1	0.1	6.6	2.8	1.0	**2.6**	0.6	0.0	**0.4**	**1.6**
Madagascar	0.1	7.7	7.2	7.4	4.3	**5.9**	1.7	0.0	**1.2**	**3.9**
Maldives	0.1	0.1	8.9	0.0	0.0	**3.1**	1.3	0.0	**0.9**	**2.1**
Mauritius	0.1	0.1	5.9	6.8	1.3	**3.4**	0.1	0.0	**0.1**	**1.9**

Table 3 Number of people affected, died and total damage to the country (1980–2017) (Source: Data retrieved from the EM-DAT Database, 2017)

Country	Events count	Total deaths	Total affected	Total damage ('000 US$)
Comoros	23	670	519,262	47,804
Madagascar	78	4998	17,381,188	2,321,801
Maldives	6	325	65,559	506,100
Mauritius	22	81	1,032,006	803,373

7 Findings

7.1 The Socio-Economic Vulnerability Index: A Gender Approach

The second dimension of the INFORM Risk index analyses the vulnerability or susceptibility of communities to those hazards. From Table 4, it can be argued that Comoros is more vulnerable with an index of 4.5 compared to 4.2 for Madagascar and 1.7 and 1.5 for Mauritius and Maldives, respectively. Comoros is highly vulnerable to natural disasters not only because of its geographical position and climatic factors but also due to human-induced pressure on natural resources. Unsustainable land use practices, including deforestation and expanding agriculture give rise to the underlying vulnerability of the Comorian population. Local communities living in vulnerable areas within proximity of the sea are also exposed to coastal erosion as a result of heavy rainfall, tides or sand removal and are often cut off from food, water and medical supplies as well as emergency services during such climate-related natural disasters (UNDP, 2017). Similarly, Madagascar faces significant risks imposed by an increasingly variable and changing climate. Further, development challenges loom large for Madagascar, with rampant and persistent poverty, poor rural female-headed families, high illiteracy rate, low connectivity and an uncertain political situation (Government of Madagascar, 2017).

In addition, the broad indicators outlined in Table 2 are included in the principal component analysis of social economic vulnerability. The use of both percentages and densities is not problematic to the analysis since the main objective of the principal component analysis is to reduce a complex set of many correlated variables into a set of fewer, uncorrelated components (Rygel et al., 2006). The variables were separated into male and female and entered into a correlation matrix and the Varimax orthogonal rotation and Kaiser normalisation was applied to the solution (Kaiser, 1974). The Varimax orthogonal rotation minimises the number of variables that have high loadings on each factor and simplifies the interpretation of the factors. Each original variable tends to be associated with one (or a small number) of factors, and each factor denotes only a small number of variables. After rotation, the loadings are rescaled back to the proper size using Kaiser normalisation which ensures the stability of solutions across samples. The factors can be inferred from the opposition of few variables with positive loadings to few variables with negative loadings. For the variables focusing specifically on women, three components with eigenvalues

Table 4 Vulnerability across four Indian Ocean Islands (Source: Data retrieved from the INFORM Risk Index, 2018)

Country	Development & deprivation	Inequality	Aid dependency	Socio-economic vulnerability	Uprooted people	*Health conditions*	*Children U5*	*Recent shocks*	*Food security*	Other vulnerable groups	Vulnerable groups	Vulnerability
	(0–10)	(0–10)	(0–10)	(0–10)	(0–10)	(0–10)	(0–10)	(0–10)	(0–10)	(0–10)	(0–10)	(0–10)
Comoros	5.2	7.7	5.7	**6.0**	0.0	3.2	4.8	0.0	7.7	4.5	**2.5**	**4.5**
Madagascar	7.5	3.9	3.0	**5.5**	0.0	2.8	3.8	4.2	7.4	4.8	**2.7**	**4.2**
Maldives	2.1	3.6	1.5	**2.3**	0.0	0.6	2.4	0.0	2.1	1.3	**0.7**	**1.5**
Mauritius	2.6	3.9	1.2	**2.6**	0.0	1.1	1.0	0.0	2.5	1.2	**0.6**	**1.7**

Table 5 Rotated component matrix rotated component matrix. Extraction method: Principal Component Analysis. Rotation Method: Varimax with Kaiser Normalisation. Threshold for exclusion = low correlations (<0.2). (Source: Author's computation)

	Female			Male	
Social economic vulnerability	Component 1	Component 2	Component 3	Component 1	Component 2
GNI per capita	−0.234	−0.328		−0.283	
Employment in agriculture	**0.332**			**0.328**	
Employment in industry	**−0.323**	−0.221		−0.273	
Employment in services	−0.307			**−0.326**	
Vulnerable employment	0.278		0.249	**0.325**	
Wage and salaried workers	−0.280		−0.253	**−0.328**	
Unemployment rate			**0.794**		**0.893**
Primary education completion rate	**−0.310**	−0.323	−0.342	−0.255	
Proportion of women in ministerial level positions		**−0.751**	−0.205		
Life expectancy at birth	**0.315**			0.295	0.208
Fertility rate	0.308				
Population density	**0.328**	0.283		0.301	
Age dependency ratio	0.286			0.313	
Health expenditure per capita				−0.279	

greater than 1.0 were retained and they accounted for 93.4% of the social vulnerability variation for women in the data. For those variables explaining male social vulnerability to climate change, two components with eigenvalues greater than 1.0 were generated and they accounted for 87.1% of the variation in the data.

The factorial reduction procedure indicated statistical significance and showed that data were suitable for factorial analysis for both the male and female samples. For the latter, the Kaiser–Meyer–Olkin measure of sampling adequacy is 0.805 and the highly significant Bartlett's test of sphericity (Chi-square = 1948.672) suggests the strength of the correlation of the variables. For male, the Kaiser–Meyer–Olkin measure of sampling adequacy stands at 0.817 and the Bartlett test of sphericity (Chi-square = 1959.472) is also highly significant. The next step is to gain a maximum variation of factors and functions of combined variables, by applying the Varimax rotation. This is shown in Table 5 below.

For women, the first principal component is a measure of employment in agriculture and industry, population density, life expectancy and primary education completion rate. It is therefore crucial to understand that women critically contribute to specific sectors like agriculture (Ampaire et al., 2020; Huyer & Partey, 2020), and their absence exacerbated due to climate change can lead to sectoral economic losses. Hence, framing cross-sectoral solutions for women is imperative to reduce the higher female vulnerability to environmental changes. Further, education and demographics also correlate well with the first component. The second principal component seems to be a measure of the proportion of women in ministerial level positions hence women can participate in decision making processes, for instance, in climate change decisions. Access to participation in decision-making processes and access to leading positions/power are not gender neutral in many developing economies. When women's voices are not equally represented in the political and other decision-making structures, it negatively impacts on climate policies and practices. Gender mainstreaming efforts in mitigation and adaptation policies and programmes are vital; ignoring gender differences within these decision-making processes can increase existing gender inequalities. Female unemployment rate also loads strongly on the third component. It can be noted that demographics and human resources are strongly associated with socio-economic vulnerability for both females and males. One major difference between them across the factor loadings is that for women, health status and their participation in decision making are important components.

Finally, factor scores are obtained and the social economic vulnerability index is computed for each country. For Comoros, the male and female socio-economic vulnerability indices range from 2.4 to 3.3 with a higher value for women. Socio-economic vulnerability has increased over the years for the Comoros with the figure being more pronounced for women. The same holds for Madagascar with the index ranging between 3.1 and 4.4 and women appear to be more vulnerable relative to men. Similar to Comoros, Madagascar's socio-economic vulnerability to climate-related shocks over the years has increased especially with several droughts and cyclones over the past decades. For Maldives, the socio-economic vulnerability index ranges from 1.8 to 3.3 and in contrast with Madagascar and Comoros the female vulnerability index is lower than that for males. A similar picture is depicted for Mauritius whereby the index is between 2.1 and 3.1 and that of women is slightly lower than for men. For both Mauritius and Maldives, susceptibility to environmental changes has been rising over the last decades. Across the Indian Ocean islands, Madagascar leads in terms of socio-economic vulnerability with higher vulnerability for women, followed by Comoros, Mauritius and Maldives. The differences across countries in terms of the gender dimension of the vulnerability index can be attributed to a mixture of factors namely the development level, the percentage of women in the agricultural sector, the female primary education completion rate, the number of women in vulnerable employment, and the fertility rate.

7.2 Lack of Coping Capacity

The last dimension which is analysed pertains to the lack of coping capacity measured by governance structure, communication, physical infrastructure, access to health care and infrastructural development. These are key components of climate resilient development covering the generation and communication of climate risk and vulnerability information, institutional capacity and coordination, community level activities and financing of disaster risk reduction and adaptation to climate change. From Table 6, Madagascar ranks the highest followed by Comoros, Maldives and Mauritius. Hence in terms of the overall risk, Madagascar tops the list among these economies and ranks high in terms of climate change risk.

Women's vulnerability to disaster in these four islands is a combined effect of both climate change and social factors. The first factor includes greater exposure to climate induced disasters with higher probabilities of cyclones, floods and droughts. The second factor is related to status of women in terms of their education, employment situation (unemployed or being in vulnerable jobs), livelihood, fertility and existing social and cultural barriers which hinder their full participation in all phases of disaster management activities. Although women across the four island economies seem to be worse off relative to men, they play a vital role in coping with and preparing for prevention, response, recovery and adaptation activities.

8 Conclusion and Gender-Inclusive Climate Change Policy Options

From the INFORM Risk indicators, there is evidence of an elevated climate change exposure of the four Indian Ocean island states included in this study. Using data for over more than five decades for the four countries, the findings from the principal component analysis, indicate that the factors leading to greater socio-economic vulnerability differ across islands. However, the components which stand out relate to human resources (employment, sectoral employment and education), demographics (population density) for both men and women but for women in particular, health (fertility rate) and the proportion of women in ministerial level positions.

In addition, the Socio-Economic Vulnerability Index computed from the principal component analysis varies by gender over the period 1960 to 2017 for all four islands. For instance, for Comoros, the male and female socio-economic vulnerability indices range from 2.4 to 3.3 with a higher value for women. The same holds for Madagascar with the index ranging between 3.1 and 4.4 and women appear to be more vulnerable relative to men. For Maldives, the socio-economic vulnerability index ranges from 1.8 to 3.3 whilst for Mauritius, the index is between 2.1 and 3.1. In contrast with Madagascar and Comoros, the female vulnerability indices are lower than the indices for males in the two upper middle-income economies. Hence, the socio-economic vulnerability index obtained from the principal component analysis

Table 6 Lack of Coping Capacity Index and Inform Risk Index (Source: Data retrieved from the INFORM Risk Index, 2018)

Country	Disaster risk reduction	Governance	Institutional	Communication	Physical infrastructure	Access to health care	Infrastructure	Lack of coping capacity	Inform risk	Risk class
	(0–10)	(0–10)	(0–10)	(0–10)	(0–10)	(0–10)	(0–10)	(0–10)	(0–10)	(Very low-Very high)
Comoros	7.8	7.8	**7.8**	5.9	5.2	4.5	**5.2**	**6.7**	**3.6**	Medium
Madagascar	4.7	7.5	**6.1**	8.1	9.6	8.4	**8.7**	**7.6**	**5.0**	High
Maldives	5.8	6.1	**6.0**	1.2	0.2	2.7	**1.4**	**4.1**	**2.3**	Low
Mauritius	3.3	3.8	**3.6**	2.5	0.3	3.2	**2.0**	**2.8**	**2.1**	Low

reveals that women in these islands, in particular in Madagascar and Comoros are more vulnerable to changes in climate compared to men. Further, Madagascar leads in terms of socio-economic vulnerability with higher vulnerability for women, followed by Comoros, Mauritius and Maldives. Gender sensitive measures are thus needed for disaster preparedness and adaptation options. Clearer gender focused strategies will allow policy-makers to move past the simple recognition that climate change affects men and women differently and push for policy solutions which substantially close gender gaps.

First, policy options must address the current gender constraints in these island economies. It is important to understand the existing gender differentials at the community/micro levels rather than at the macroeconomic level as characteristics of communities, regions and villages differ in shaping their existing vulnerabilities and capacity to adapt. An assessment of the differences in men's and women's roles, responsibilities, access to and management of strategic resources, and participation in decision making needs to be undertaken at the level of local communities. This will generate quantitative and qualitative gender disaggregated data and enable analysis of potential loss and damage for men and women and the capacity to adapt to climate change. It will also provide opportunities for communities to articulate their concerns and propose solutions, as well as to assist government and other private bodies to take into account the complex local-level impacts. Deep and wide consultations with communities in these countries represent therefore, a necessary ingredient in the development of gender-sensitive, participatory, equitable and transparent strategies.

This study shows that women operating in the agricultural sector are highly vulnerable to environmental changes. Hence, the inclusion of gender considerations in agriculture and natural resource management policies is of paramount importance for these island-developing states. In particular, it supports the need for an increase in women's access to productive resources and opportunities such as land and assets, agro advisory information, new agricultural techniques, credit facilities, technology development and knowledge transfer, finance, fertilisers and seeds (Partey et al., 2020). Women can be encouraged and trained to develop sustainable agribusinesses, retain control of their assets, contribute to food security and participate meaningfully in household and community-level decision making and leadership. Further, recognising women's responsibilities in providing for the family's food security and income needs is vital. The governments of island economies must also adopt a holistic approach to promote greater investments in initiatives that enable the empowerment of women, advancement in technology, renewable energy, land and water use and management and sustainable livelihoods in agriculture. Climate-resilient approaches including climate-smart agriculture can promote gender equality and women's empowerment (Huyer & Partey, 2020).

This study also shows that education plays a significant role in addressing vulnerabilities and as such builds women's resilience and adaptive capacity to risks associated with climate change. Improved access to formal education and advancement in technology and innovation can lead to major changes in societal expectations and gender relations. Empowering girls and women to learn how to integrate traditional knowledge and science through quality education can help communities better understand the impact of climate change on their livelihoods. More opportunities, including access to grants and scholarships from childhood must be provided to get women and girls into higher education.

In addition, there is a clear link between higher levels of female education and lower rates of fertility. Teaching girls about their rights, specifically about their reproductive rights, helps improve their health. This can ultimately lead to the stabilisation of poverty levels and enable girls to stay in school for a longer period of time. Hence, empowering girls and women via a combination of education and family planning may lessen their vulnerability to climate change. Quality education with much needed information on female reproductive health may open up more options, provide skills and greater financial independence for young women who will thereafter invest in the education of their own children. For instance, Kwauk and Braga (2017) show that for every additional year of schooling a girl receives on average, her country's resilience to climate disasters, is expected to improve by 3.2 points.

Along the same lines, equipping girls with leadership and life skills through education can generate a significant positive social impact. From the results above, the proportion of women in ministerial positions is an important factor explaining climate change vulnerability of women. Likewise, in 2017, the first ever UNFCCC Gender Action Plan was adopted at COP23 and aimed at integrating gender equality and human rights into climate action. Its overall goal is to support the implementation of the gender-related decisions and mandates in the UNFCCC process. The Gender Action Plan proposes a 50% representation of women in all party delegations and constituted bodies under the UNFCCC by 2019. However, the number of women involved in decision making at national levels remains limited. In developing countries, less gender equality in participation in formal decision-making roles is associated with greater negative impacts on women in climate change decisions. Female leaders are extremely effective in conservation and protection efforts, and are more likely to pursue better sustainable futures for their communities. Hence, the promotion and inclusion of more women in leadership position and in climate change decisions can help reduce their vulnerability and improve their adaptability to climate risks and disasters. Mainstreaming gender in climate change policies is a must, especially for island nations where climate change vulnerability is particularly high.

References

Adger, W. N. (1999). Social vulnerability to climate change and extremes in coastal Vietnam. *World Development, 27*(2), 249–269.

Agarwal, B. (2010). *Gender and green governance: The political economy of women's presence within and beyond community forestry*. Oxford University Press.

Alber, G. (2009). Gender and climate change policy. In (n.a.) *Population dynamics and climate change* (pp. 149–163). UNFPA & IIED.

Allison, E. H., Perry, A. L., Badjeck, M. C., Adger, W. N., Brown, K., Conway, D., Halls, A. S., Pilling, G. M., Reynolds, J. D., Andrew, N. L., & Dulvy, N. K. (2009). Vulnerability of national economies to the impacts of climate change on fisheries. *Fish and Fisheries, 10*(2), 173–196.

Ampaire, E. L., Acosta, M., Huyer, S., Kigonya, R., Muchunguzi, P., Muna, R., & Jassogne, L. (2020). Gender in climate change, agriculture, and natural resource policies: Insights from East Africa. *Climatic Change, 158*(1), 43–60. https://doi.org/10.1007/s10584-019-02447-0

Aroca-Jimenez, E., Bodoque, J. M., Garcia, J. A., & Diez-Herrero, A. (2017). Construction of an integrated social vulnerability index in urban areas prone to flash flooding. *Natural Hazards and Earth System Sciences, 17*(9), 1541–1557.

Arora-Jonsson, S. (2011). Virtue and vulnerability: Discourses on women, gender and climate change. *Global Environmental Change, 21*(2), 744–751.

Arora-Jonsson, S. (2014). Forty years of gender research and environmental policy: Where do we stand? *Women's Studies International Forum, 47*(B), 295–308.

Babagura, S. (2010). *Gender and climate change: South Africa case study*. Heinrich-Böll-Stiftung.

Balikoowa, K., Nabanoga, G., & Tumusiime, D. M. (2018). Gender stereotyping: Evidence from gender differentiated household vulnerability to climate change in Eastern Uganda. *Cogent Environmental Science, 4*(1), 1–15.

Bankoff, G. (2004). Time is of the essence: Disasters, vulnerability and history. *International Journal of Mass Emergencies and Disasters, 22*(3), 23–42.

Bergstrand, K., Mayer, B., Brumback, B., & Zhang, Y. (2015). Assessing the relationship between social vulnerability and community resilience to hazards. *Social Indicators Research, 122*(2), 391–409.

Blaikie, P., Cannon, T., Davis, I., & Wisner, B. (1994). *At risk: Natural hazards, people's vulnerability and disasters*. Routledge.

Blaikie, P. M. (1994). *Political ecology in the 1990s: An evolving view of nature and society*. Michigan State University, Center for Advanced Study of International Development.

Blancard, S., & Hoarau, J. F. (2013). A new sustainable human development indicator for small island developing states: A reappraisal from data envelopment analysis. *Economic Modelling, 30*, 623–635.

Bogard, W. C. (1988). Bringing social theory to hazards research: Conditions and consequences of the mitigation of environmental hazards. *Sociological Perspectives, 31*(2), 147–168.

Bohle, H. G., Downing, T. E., & Watts, M. J. (1994). Climate change and social vulnerability: Toward a sociology and geography of food insecurity. *Global Environmental Change, 4*(1), 37–48.

Bolin, R., Jackson, M., & Crist, A. (1998). Gender inequality, vulnerability, and disaster: Issues in theory and research. In E. Enarson & B. H. Morrow (Eds.), *The gendered terrain of disaster: Through women's eyes* (pp. 27–44). Praeger.

Bradshaw, S. (2004). *Socio-economic impacts of natural disasters: A gender analysis* (Vol. 32). United Nations Publications.

Brody, A., Demetriades, J., & Esplen, E. (2008). *Gender and climate change: Mapping the linkages. A scoping study on knowledge and gaps*. BRIDGE, Institute of Development Studies (IDS).

Burton, I., Kates, R. W., & White, G. F. (1978). *The environment as hazard*. Oxford University Press.

Cannon, T. (2002). Gender and climate hazards in Bangladesh. *Gender & Development, 10*(2), 45–50.

Cardona, O. D. (2007). *Indicators of disaster risk and risk management*. Washington, D.C.: Inter-American Development Bank. [online] Retrieved December 14, 2020, from https://www.ipcc.ch/apps/njlite/ar5wg2/njlite_download2.php?id=9430

Carr, E. R., & Thompson, M. C. (2014). Gender and climate change adaptation in agrarian settings: Current thinking, new directions, and research frontiers. *Geography Compass, 8*(3), 182–197.

Clark, G. E., Moser, S. C., Ratick, S. J., Dow, K., Meyer, W. B., Emani, S., Jin, W., Kasperson, J. X., & Kasperson,& and Schwarz, H.E. (1998). Assessing the vulnerability of coastal communities to extreme storms: The case of Revere, MA., USA. *Mitigation and Adaptation Strategies for Global Change, 3*(1), 59–82.

Cutter, S. L. (1995). The forgotten casualties: Women, children, and environmental change. *Global Environmental Change, 5*(3), 181–194.

Cutter, S. L. (1996). Vulnerability to environmental hazards. *Progress in Human Geography, 20*(4), 529–539.

Cutter, S. L., Boruff, B. J., & Shirley, W. L. (2003). Social vulnerability to environmental hazards. *Social Science Quarterly, 84*(2), 242–261.

Cutter, S. L., Barnes, L., Berry, M., Burton, C., Evans, E., Tate, E., & Webb, J. (2008). A place-based model for understanding community resilience to natural disasters. *Global Environmental Change, 18*(4), 598–606.

Cutter, S. L., Burton, C. G., & Emrich, C. T. (2010). Disaster resilience indicators for benchmarking baseline conditions. *Journal of Homeland Security and Emergency Management, 7*(1).

de Loyola Hummell, B. M., Cutter, S. L., & Emrich, C. T. (2016). Social vulnerability to natural hazards in Brazil. *International Journal of Disaster Risk Science, 7*(2), 111–122.

Denton, F. (2004). Gender and climate change: Giving the "latecomer" a head start. *IDS Bulletin, 35* (3), 42–49.

Dow, K. (1992). Exploring differences in our common future (s): The meaning of vulnerability to global environmental change. *Geoforum, 23*(3), 417–436.

Dwyer, A., Zoppou, C., Nielsen, O., Day, S., & Roberts, S. (2004). *Quantifying social vulnerability: A methodology for identifying those at risk to natural hazards* (pp. 2–3). Geoscience Australia Record 2004/14.

Eakin, H. (2005). Institutional change, climate risk, and rural vulnerability: Cases from Central Mexico. *World Development, 33*(11), 1923–1938.

Eakin, H., Lerner, A. M., & Murtinho, F. (2010). Adaptive capacity in evolving peri-urban spaces: Responses to flood risk in the Upper Lerma River Valley, Mexico. *Global Environmental Change, 20*(1), 14–22.

EMDAT Database. (2018). *The International Disaster Database*. Centre for Research on the Epidemiology of Disasters – CRED. [online] Retrieved January 5, 2018, from https://www.emdat.be/database

Enarson, E. (1998). Through women's eyes: A gendered research agenda for disaster social science. *Disasters, 22*(2), 157–173.

Enarson, E. P. (2000). *Gender and natural disasters*. International Labour Organization.

Enarson, E. P., & Meyreles, L. (2004). International perspectives on gender and disaster: Differences and possibilities. *International Journal of Sociology and Social Policy, 24*(10/11), 49–93.

Enarson, E. P., & Morrow, B. H. (1998). *The gendered terrain of disaster*. Praeger Press.

Ford, J. D., & Pearce, T. (2010). What we know, do not know, and need to know about climate change vulnerability in the western Canadian Arctic: A systematic literature review. *Environmental Research Letters, 5*(1), 14008.

Ford, J. D., Smit, B., & Wandel, J. (2006). Vulnerability to climate change in the Arctic: A case study from Arctic Bay, Canada. *Global Environmental Change, 16*(2), 145–160.

Fordham, M. H. (1999). Balancing resilience and vulnerability. *International Journal of Mass Emergencies and Disasters, 17*, 15–36.

Fordham, M. H. (2000). *The place of gender in earthquake vulnerability and mitigation*. Anglia Polytechnic University.

Fothergill, A. (1998). The neglect of gender in disaster work: An overview of the literature. In E. Enarson & B. H. Morrow (Eds.), *The gendered terrain of disaster: Through women's eyes* (pp. 11–25). Praeger.

Fuchs, S., Birkmann, J., & Glade, T. (2012). Vulnerability assessment in natural hazard and risk analysis: Current approaches and future challenges. *Natural Hazards, 64*(3), 1969–1975.

Füssel, H.-M. (2007). Vulnerability: A generally applicable conceptual framework for climate change research. *Global Environmental Challenge, 17*, 155–167.

Government of Madagascar. (2017). *Madagascar's Strategic Programme for Climate Resilience 2017*. National Bureau for Coordination of Climate Change, PPCR Sub-Committee Meeting of 12th December 2017, Washington DC. [online] Retrieved December 14, 2020, from https://www.climateinvestmentfunds.org/sites/cif_enc/files/meeting-documents/madagascar_spcr_presentation_ppcr_subcommittee_meeting_dec_12_2017.pdf

Haas, E., Kates, R. W., & Bowden, M. J. (1977). *Reconstruction following disaster*. Massachusetts Institute of Technology.

Hay, J. E., & Kaluwin, C. (1993). Climate change and sea level rise in the south Pacific region. In *Proceedings of the Second SPREP Meeting*, Nomea.

Hemmati, M., & Röhr, U. (2007). *A huge challenge and a narrow discourse. Women and environments*. Women and Gender Studies Institute, New College, University of Toronto.

Holand, I. S., & Lujala, P. (2013). Replicating and adapting an index of social vulnerability to a new context: A comparison study for Norway. *The Professional Geographer, 65*(2), 312–328.

Holand, I. S., Lujala, P., & Röd, J. K. (2011). Social vulnerability assessment for Norway: A quantitative approach. *Norsk Geografisk Tidsskrift - Norwegian Journal of Geography, 65*(1), 1–17.

Huyer, S., & Partey, S. (2020). Weathering the storm or storming the norms? Moving gender equality forward in climate-resilient agriculture. *Climatic Change, 158*, 1–12.

Ikeda, K. (1995). Gender differences in human loss and vulnerability in natural disasters: A case study from Bangladesh. *Indian Journal of Gender Studies, 2*(2), 171–193.

INFORM. (2018). *The Risk Management Index*. [online] Retrieved January 10, 2018, from http://www.inform-index.org/

Intergovernmental Panel on Climate Change (IPCC). (2014). *Climate Change 2014 impacts, adaptation, and vulnerability. Part A: Global and sectoral aspects*. Cambridge University Press.

International Institute for Sustainable Development (IISD). (2016). *PR: Indian Ocean Commission Develops Climate Change Adaptation Plan*. IISD. [online] Retrieved March 25, 2018, from http://sdg.iisd.org/news/indian-ocean-commission-develops-climate-change-adaptation-plan/

Johnsson-Latham, G. (2007). *A study on gender equality as a prerequisite for sustainable development*. Report to the Environment Advisory Council, The Environment Advisory Council, Ministry of the Environment, Sweden.

Kaiser, H. F. (1974). An index of factorial simplicity. *Psychometrika, 39*, 31–36.

Koks, E. E., de Moel, H., Aerts, J. C., & Bouwer, L. M. (2014). Effect of spatial adaptation measures on flood risk: Study of coastal floods in Belgium. *Regional Environmental Change, 14* (1), 413–425.

Kwauk, C., & Braga, A. (2017). *Three platforms for girls' education in climate change strategies*. Brookings Institution.

Leach, B. (2016). *Reshaping gender and class in rural spaces*. Routledge.

Marshall, N. A. (2010). Understanding social resilience to climate variability in primary enterprises and industries. *Global Environmental Change, 20*(1), 36–43.

Meinzen-Dick, R., Kovarik, C., & Quisumbing, A. R. (2014). Gender and sustainability. *Annual Review of Environment and Resources, 39*, 29–55.

Mies, M., & Shiva, V. (1993). *Ecofeminism*. Zed Books.

Nagarajan, C., Le Masson, V., Benoudji, C., Reyes, S. S., Ensor, M. O., Kellogg, M., Tanyag, M., Csevar, S., Sawas, A., Anwar, N., & Anjum, G. (2020). *Gender, climate & security: Sustaining inclusive peace on the frontlines of climate change*. UNEP, UN Women, UNDP & UNDPPA/PBSO.

Nelson, K. S., Abkowitz, M. D., & Camp, J. V. (2015). A method for creating high resolution maps of social vulnerability in the context of environmental hazards. *Applied Geography, 63*, 89–100.

Nelson, V. (2011). *Gender, generations, social protection & climate change: A thematic review*. Overseas Development Institute.

Nelson, V., Meadows, K., Cannon, T., Morton, J., & Martin, A. (2002). Uncertain predictions, invisible impacts, and the need to mainstream gender in climate change adaptations. *Gender & Development, 10*(2), 51–59.

Neumayer, E., & Plümper, T. (2007). The gendered nature of natural disasters: The impact of catastrophic events on the gender gap in life expectancy, 1981–2002. *Annals of the Association of American Geographers, 97*(3), 551–566.

O'Brien, K. L., Leichenko, R., Kelkar, U., Venema, H., Aandahl, G., Tompkins, H., Javed, A., Bhadwal, S., Barg, S., Nygaard, L., & West, J. (2004). Mapping vulnerability to multiple stressors: Climate change and globalization in India. *Global Environmental Change, 14*(4), 303–313.

O'Brien, K. L., & Leichenko, R. M. (2000). Double exposure: Assessing the impacts of climate change within the context of economic globalization. *Global Environmental Change, 10*(3), 221–232.

O'Keefe, P., Westgate, K., & Wisner, B. (1976). Taking the naturalness out of natural disasters. *Nature, 260*, 566–567.

Ocio, D., Stocker, C., Eraso, Á., Martínez, A., & de Galdeano, J. M. S. (2016). Towards a reliable and cost-efficient flood risk management: The case of the Basque Country (Spain). *Natural Hazards, 81*(1), 617–639.

Oulahen, G., Mortsch, L., Tang, K., & Harford, D. (2015). Unequal vulnerability to flood hazards: "ground truthing" a social vulnerability index of five municipalities in Metro Vancouver, Canada. *Annals of the Association of American Geographers, 105*(3), 473–495.

Park, S. E., Marshall, N. A., Jakku, E., Dowd, A. M., Howden, S. M., Mendham, E., & Fleming, A. (2012). Informing adaptation responses to climate change through theories of transformation. *Global Environmental Change, 22*(1), 115–126.

Partey, S. T., Dakorah, A. D., Zougmoré, R. B., Ouédraogo, M., Nyasimi, M., Nikoi, G. K., & Huyer, S. (2020). Gender and climate risk management: Evidence of climate information use in Ghana. *Climatic Change, 158*(1), 61–75. https://doi.org/10.1007/s10584-018-2239-6

Peduzzi, P., Dao, H., Herold, C., & Mouton, F. (2009). Assessing global exposure and vulnerability towards natural hazards: The Disaster Risk Index. *Natural Hazards and Earth System Sciences, 9*(4), 1149–1159.

Pernetta, J. C. (1992). Impacts of climate change and sea-level rise on small island states: National and international responses. *Global Environmental Change, 2*(1), 19–31.

Rahman, M. S. (2013). Climate change, disaster and gender vulnerability: A study on two divisions of Bangladesh. *American Journal of Human Ecology, 2*(2), 72–82.

Ravera, F., Iniesta-Arandia, I., Martín-López, B., Pascual, U., & Bose, P. (2016). Gender perspectives in resilience, vulnerability and adaptation to global environmental change. *Ambio, 45*(3), 235–247.

Rogelis, M. C., Werner, M., Obregón, N., & Wright, N. (2016). Regional prioritisation of flood risk in mountainous areas. *Natural Hazards and Earth System Sciences, 16*(3), 833–853.

Rygel, L., O'Sullivan, D., & Yarnal, B. (2006). A method for constructing a social vulnerability index: An application to hurricane storm surges in a developed country. *Mitigation and Adaptation Strategies for global Change, 11*(3), 741–764.

Sharmin, Z., & Islam, M. (2013). Consequences of climate change and gender vulnerability: Bangladesh perspective. *Bangladesh Development Research Working Paper Series (BDRWPS), 16*, 1–14.

Shiva, V. (1988). *Staying alive: Women, ecology, and survival in India*. Kali for Women

Siebeneck, L., Arlikatti, S., & Andrew, S. A. (2015). Using provincial baseline indicators to model geographic variations of disaster resilience in Thailand. *Natural Hazards, 79*(2), 955–975.

Susman, P., O'Keefe, P., & Wisner, B. (1983). Global disasters, a radical interpretation. In K. Hewitt (Ed.), *Interpretations of calamity. From the viewpoint of human ecology* (pp. 263–283). Routledge

Tapsell, S. M., Penning-Rowsell, E. C., Tunstall, S. M., & Wilson, T. L. (2002). Vulnerability to flooding: Health and social dimensions. *Philosophical Transactions of the Royal Society of London, A: Mathematical, Physical and Engineering Sciences, 360*(1796), 1511–1525.

Timmerman, P. (1981). *Vulnerability, resilience and the collapse of society. Environmental Monograph 1*. Institute for Environmental Studies, Toronto University.

Ul Islam, S., Rehman, N., & Sheikh, M. M. (2009). Future change in the frequency of warm and cold spells over Pakistan simulated by the PRECIS regional climate model. *Climatic Change, 94* (1–2), 35–45.

United Nations Development Programme (UNDP). (2013). *Overview of linkages between gender and climate change*. United Nations Development Programme.

United Nations Development Programme (UNDP). (2017). *Social Vulnerability Assessment Tools for Climate Change and DRR Programming*. [online] Retrieved January 20, 2018, from http://www.adaptation-undp.org/Social-Vulnerability-Assessment-Tools

USAID. (2016). *Climate Change Risk Profile Madagascar*. Country Fact Sheet. https://pdf.usaid.gov/pdf_docs/pa00mtz8.pdf

Weichselgartner, J. (2001). Disaster mitigation: The concept of vulnerability revisited. *Disaster Prevention and Management: An International Journal, 10*(2), 85–95.

Wisner, B., Blaikie, P., Cannon, T., & Davis, I. (2004). *At risk: Natural hazards, people's vulnerability and disasters* (2nd ed.). Routledge.

World Bank. (2018a). *World Bank Development Indicators 2018*. [online] Retrieved February 24, 2018, from https://data.worldbank.org/products/wdi

World Bank. (2018b). *World Bank Gender Statistics 2018*. [online] Retrieved January 15, 2018, from https://datacatalog.worldbank.org/dataset/gender-statistics

Yavinsky, R. W. (2012). *Women more vulnerable than men to climate change*. Population Reference Bureau. [online] Retrieved December 14, 2020, from https://www.prb.org/women-vulnerable-climate-change/

Vulnerability of Jamaica's South Coast Fishing Communities to Coastal Erosion and Flooding

Donovan Campbell and Simone Lee

Abstract Several fishing communities across the Caribbean are experiencing rapid changes in the coastal ecosystem services that underpin their livelihood security. In Jamaica, these changes are partly reflected in the collapse of the local capture fishery, which has exacerbated ecological, social and food security challenges. In this paper, coastal and community-based vulnerability approaches are combined to assess the climate change vulnerability of three major fishing communities (Old Harbour Bay, Rocky Point, and Alligator Pond) along the southern coast of Jamaica. These communities are at the frontline of some of the most serious threats from climate change and have been badly affected by rapid changes in coastal and marine ecosystems. The impacts of coastal erosion on fishing businesses, as well as the vulnerability of fishing infrastructure and assets to coastal flooding are examined in the three sites. Coastal flooding and erosion directly impact household and personal property, fishery business assets, public infrastructure, and disrupt opportunities to reduce intergenerational poverty in the targeted communities. The paper concludes by highlighting four key strategies for reducing livelihood vulnerabilities to coastal erosion and flooding. These include creating economic alternatives, improving ecosystem resilience, strengthening co-management approaches, and developing gender-sensitive indicators to assess vulnerability in fishing communities. The results demonstrate the value of integrative social and coastal vulnerability assessments in characterizing nuanced climate impacts on fishery-related businesses.

Keywords Climate change · Fisheries · Jamaica · Vulnerability · Capacity assessment · Climate change adaptation · Coastal erosion · Coastal flooding · Livelihoods

---

D. Campbell (✉)
Department of Geography and Geology, University of the West Indies (Mona Campus), Kingston, Jamaica
e-mail: donovan.campbell@uwimona.edu.jm

S. Lee
Centre for Environmental Management, University of the West Indies (Mona Campus), Kingston, Jamaica

S. Moncada et al. (eds.), *Small Island Developing States*, The World of Small States 9, https://doi.org/10.1007/978-3-030-82774-8_10

1 Introduction

Caribbean fisheries are of crucial importance for human wellbeing as it provides food, employment, and enjoyment for thousands of coastal inhabitants. Fisheries employ nearly 200,000 people in the Caribbean community, earning US$5 billion to $6 billion per year in foreign exchange, and providing about 10 percent of the region's protein intake (Singh-Renton & McIvor, 2015). In recent times, the fisheries sector has been badly affected by a series of natural disasters, which resulted in adverse impacts on the livelihoods of fishers across the region (Monnereau et al., 2017). The problem is amplified by overfishing and coastal development, including human settlements and tourism, which often result in the conversion and destruction of habitats such as coastal wetlands (mangroves), coral reef and seagrass beds (IPCC, 2019; World Bank, 2019). The loss of these habitats reduce fisheries productivity and resilience to environmental impacts such as those caused by climate change (Robinson et al., 2019). The destruction of these habitats in the region pose great challenges for coastal livelihoods, food security and the conservation of marine biodiversity.

In Jamaica, fisheries play an important economic and cultural role. Fisheries provide multiple socioeconomic contributions, such as income generation, food security, and livelihood diversification opportunities (MICAF, 2018). It forms the backbone of the local economy in many coastal communities across the island. Additionally, fisheries form part of the cultural identity in these communities (Chan et al. 2019). Participation in fisheries is a tradition, and the knowledge, practices, and way of life associated with fisheries have transcended many generations. Jamaican fisheries resources are already precarious, and further unsustainable exploitation could have grave consequences for those most dependent on the sector (Marschke et al., 2020). The challenges to fisheries posed by climate change are numerous, and include ocean acidification, degrading reefs, changing species' distributions, increasing uncertainty about fish stocks, and extreme events (e.g. hurricanes) affecting fishers' equipment and infrastructure (Marschke et al., 2020; McWilliams et al., 2005; Pulwarty et al., 2010). Recent studies suggest that the net loss of local species (e.g., warm water corals) and profound changes in ecosystem services will be a key outcome of climate warming in many marine systems, and notably those closer to equatorial regions (Turner et al. 2020; Beaugrand et al., 2015; Gattuso et al., 2015). Strategies to promote sustainable fisheries are urgently needed to secure institutional and governance mechanisms capable of responding to current and future climate risks.

Jamaica's fisheries sector is categorized into (i) inland, freshwater capture; (ii) marine capture; and (iii) aquaculture. The marine capture fishery comprises both artisanal and industrial operators and provides employment directly and indirectly to some 40,000 fisherfolk, and contribute to the livelihoods of over 200,000 Jamaicans (MICAF, 2018). Jamaica has one of the highest levels of fish consumption per capita in the Americas (30.6 kg per year), but is highly dependent on imports, which accounted for about 79% of all fishery products consumed domestically in 2014 (Singh-Renton & McIvor, 2015). In 2013, fisheries imports in Jamaica were valued at US$61 million while exports were estimated at US$14 million (Department of Economic and Social Affairs/United Nations Statistics Division, 2018). At the end of 2015, there were 23,631 registered fisherfolk and 7133 registered boats operating from 187 fishing beaches and two cays located at the Pedro Bank (PIOJ, 2019). The marine fishery resource in Jamaica includes those within the territorial sea and archipelagic waters which are approximately 17,995 and 22,000 km^2 respectively. This implies that upwards of 50,000 individuals (registered and unregistered) may be engaged in some form of fishing activity in Jamaica's maritime space, which is estimated at 274,000 km^2

The majority of these fishers are considered 'small-scale or 'artisanal', often fishing from small, open canoes power by outboard motors. They target demersal and pelagic fish, lobster, shrimp and conch within a significant portion of Jamaica's maritime waters. In addition to the artisanal fishers, industrial operations also target Queen Conch and Caribbean Spiny Lobster around the Pedro Bank. Fisheries exports were valued at approximately USD $12 million in 2018. Almost 80% of local fish consumed was imported in 2018, and which was valued at USD $109.85 million (PIOJ, 2019). Imports are driven by decreasing production, which results in part from fisheries governance challenges associated with overfishing, habitat degradation, and illegal fishing (MOAF, 2015). Fisheries in Jamaica have been in decline over recent years, which poses ecological, social and food security challenges that are being exacerbated by the effects of climate change.

1.1 Climate Change Implications

Climate change is seriously impacting fisheries production and marine ecosystems across the Caribbean (CSGM, 2020; IPCC, 2018; Brander, 2010; Cheung et al., 2010; Monnereau et al., 2013; Mora et al., 2013). The Near-Term Climate Scenarios (2014) report for Jamaica highlights that observed sea level rise from 1950 to 2000 shows that the rise in the Caribbean is near the global mean. In Jamaica, sea level measurements at Port Royal between 1955 and 1971 indicate a 0.9 mm per year rising trend (CSGM, 2012). Warming trends were also found to be consistent with a warming pattern across the globe. Data from the airport stations indicate historical warming of 0.20–0.31 °C per decade, with greatest warming occurring between June and August. This trend is validated by recent analysis of sea surface temperatures around Jamaica where a statistically significant annual increase of +0.15 °C per decade from 1980 to 2015 was observed (CCCCC, 2015). The seasonal trends offer a better indication of the degree to which the waters surrounding Jamaica are warming. The seasonal trends in sea surface temperature (SST) show that warming is occurring at rate of +0.09 °C per decade for January SST and + 0.17 °C per decade for August SST. It is believed that more extreme weather will occur with increased SST. The annual SST is expected to increase locally by 2 °C, which would warm the waters surrounding Jamaica to an annual average of +28 °C with summer averages exceeding 29 °C (CCCCC, 2015).

The past 100 years have demonstrated that climate change is likely to alter the quality and quantity of Jamaica's natural resources, which could have significant negative impacts on ecosystem services and associated livelihoods (Chan et al. 2019; GOJ, 2013; Monnereau et al., 2013). Specifically for fisheries, the increasing frequency and/or intensity of extreme climatic events can affect fish habitat, productivity or distribution, as well as have direct impacts on fishing operations and the physical infrastructure of coastal communities (Turner et al. 2020; GOJ, 2013). Small changes in environmental conditions, such as temperature, salinity, wind and ocean currents, can alter the abundance, distribution and availability of fish populations (IPCC, 2019; McIlgorm et al., 2010). Fishing communities in the Caribbean are expected to be severely impacted by climate change as risks of increased exposure and sensitivity are high while adaptive capacity is low (IPCC, 2018; Robinson & Wren, 2020). Many fisheries-dependent communities across the region precariously positioned due to poverty, lack of social services, and critical infrastructure (IPCC, 2019; Singh-Renton & McIvor, 2015).

The Jamaican fisheries sector is already experiencing the impacts of climate change. The majority of fishery livelihood assets are located along the shoreline, with many of the fishing beaches in Jamaica experiencing severe erosion and damage from storm events. In recent times, the sector has been badly affected by a series of natural disasters, which resulted in adverse impacts on the livelihoods of fishers across the island. For example between 2002 and 2010, the country was affected by six major storms and several flood events amounting to JM $118.67 billion in losses (PIOJ, 2008). Additionally, there was an estimated $90 million in damage to the fishing industry from Hurricane Sandy in 2012 (JM $76.8 m was accounted for by the Marine Fishing Industry while the remaining JMD $13.5 million was attributed to the Aquaculture Industry) (PIOJ, 2013).

2 Methodology

The vulnerability and capacity assessment (VCA) methodology used in this study is guided by a framework developed by the Caribbean Community Climate Change Centre (CCCCC, 2015) for conducting and mainstreaming of climate change assessments. The VCA methodology was developed to provide usable decision support information and tools to assist civic and business leaders in making critical decisions to reduce climate impacts across the region. The methodology not only facilitates vulnerability assessment, but is also used to evaluate adaptation interventions. The VCA helps ensure that the adaptation measures being considered are likely to achieve the desired outcomes by capturing drivers of vulnerability at the community-level and supporting more targeted adaptation interventions. In this paper we combine empirical coastal and social vulnerability data in a single study to characterize the impacts of climate change on Jamaica's south coast fishing communities.

Key drivers of sensitivity to climate impacts are determined by the degree of poverty, health conditions, access to water and food. Geographical factors such as proximity to coast and land elevation are key drivers of exposure in fishing communities. Adaptive capacity at different levels is determined by the institutional architecture (governance structure), and economic, social, technological and cultural factors. The approach used in this study facilitates the integration of local knowledge, especially individual and community perceptions of climate change and local coping strategies. Figure 1 shows the location of the three target communities (Old Harbour Bay, Rocky Point and Alligator Pond). With an estimated population of 7388, Old Harbour Bay is the largest fishing community on the island. Rocky Point and Alligator Pond have populations of approximately 4240 and 2152 respectively (STATIN, 2016). According to the latest census report, more than 80 per cent of the livelihoods in these communities are dependent on fisheries.

A socio-ecological vulnerability assessment developed by Marshall et al. (2010) is used as a conceptual guide for the assessment. Marshall et al. (2010) propose a heuristic framework for assessing linked social-ecological vulnerability that posit "in the ecological domain, ecological exposure and ecological sensitivity create impact potential. The impact and ecological recovery potential forms the ecological vulnerability or exposure in the social domain. This ecological vulnerability combined with the sensitivity of people from the impact potential for society. The social adaptive capacity and the impact potential together create social-ecological vulnerability" (p.6). For this assessment, the framework was modified to include socioeconomic stressors as a major influence on exposure conditions in the social domain. This is important for Small Island Developing States (SIDS) in general, and the fisheries sector in particular, where dual processes of economic and ecological change have profound impacts on vulnerability conditions.

2.1 Coastal Vulnerability Assessment

The coastal vulnerability assessment was performed in two stages; the first involved a detailed site survey and geomorphic assessment and informal interviews with local experts. The second involved a Geographic Information System (GIS) based flood/erosion vulnerability assessment. This assessment was led by Van Proosdij and Baker (2015) who outlined the following steps. All data were integrated within ArcGIS 10.2 and used to assess the spatial variability in vulnerability alongshore within all study sites. Prior to the field deployment, a coastline was selected from existing datasets and edited with the most current satellite imagery as needed to best reflect the current shoreline position. The shoreline was characterized based on a dynamic segmentation model for changes in foreshore, backshore and nearshore zones. Lines are segmented at changes in the backshore, foreshore and nearshore. The characterization of the shorezone was facilitated by decision tree flow charts,

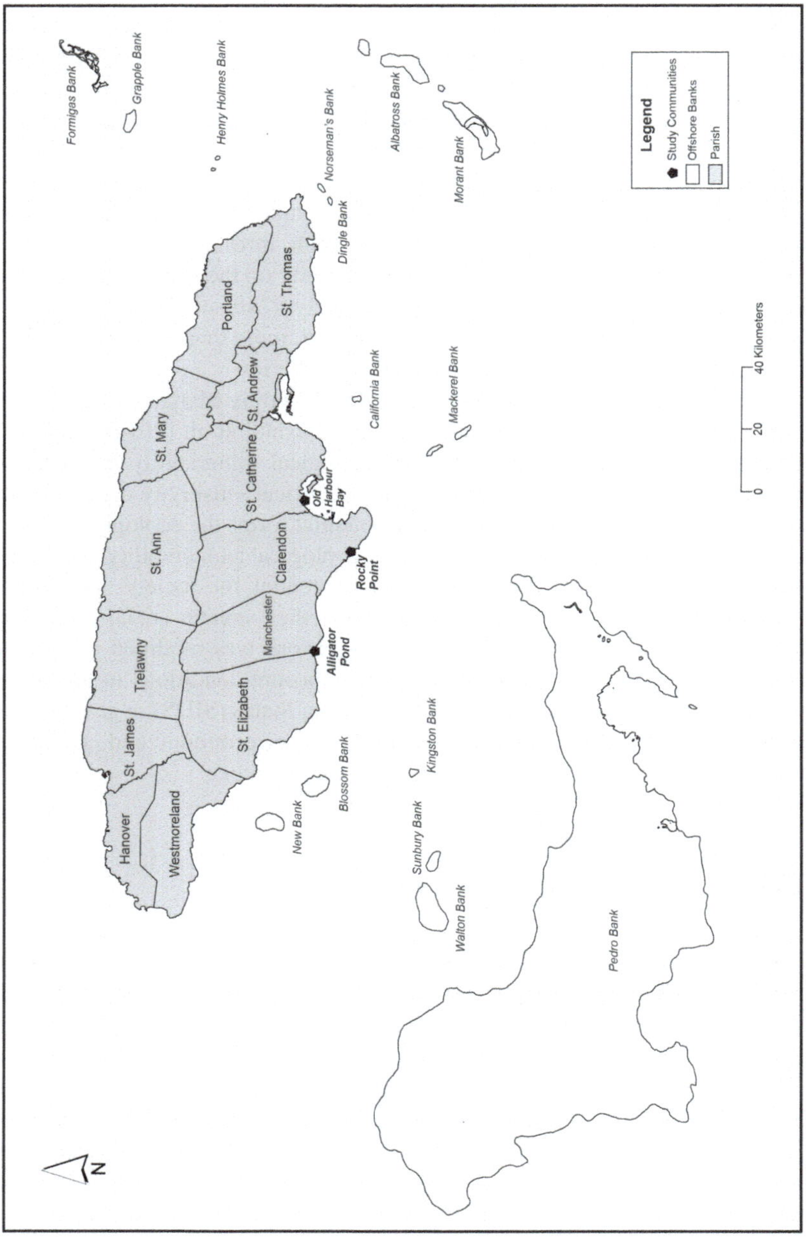

Fig. 1 Study of communities and offshore banks (*Source: Author*)

with a different chart for each shoreline to account for varying forms and features associated with different parts of the coast. It captures information such as landform type, material, height, slope, vegetation type, observed erosion and anthropogenic structures.

In the field, the coastline was photographed with GPS enabled camera and segmented within ArcGIS 10.0 (using a YUMA field tablet) at changes in backshore (furthest extent on land of highest tides), foreshore (at waterline) or nearshore (below water line or intertidal zone) characteristics. Land inundation models were created from Digital Elevation Models (DEMs). NextMap5 elevation models created by Intermap Technologies were used for all sites. The DEMs used a calculated vertical datum which approximates the equipotential surface of the sea, but does not necessarily represent the mean level of water. Inundation modelling utilized Geographic Resources Analysis Support System (GRASS) GIS to calculate flooded regions only where there is a direct connection to the sea.

The mean 60 year rate of sea-level rise in the Caribbean region was found to be similar to the global average of approximately 1.8 $mm{\cdot}yr^{-1}$ (Palanisamy et al., 2012). Forbes et al. (2013) estimated relative sea level rates (up to 2100) between 1 to 1.34 m for select Caribbean islands (e.g. Trinidad and Tobago, Bermuda, Barbados) and accounted for local vertical crustal motion. Given the complexity of tectonic plate processes (some area uplifting while others are subsiding) within Jamaica, it is recommended that future work incorporate measured rates of vertical crustal motion in predicting relative sea level rates for the island. For this study, however, given that the vertical resolution of the digital elevation model currently available is 1 m, finer resolution predictions of flooding are not possible. As a result, inundation levels are provided for a range of scenarios. Levels 1–3 are considered as highly likely given previous hurricane-induced storm surge extents and current estimates of global sea-level rise (Church et al., 2013; Forbes et al., 2013). Water level elevations up to 7 m were included in corresponding with the coastal hazard zones identified in Robinson et al. (2008).

Historical coastal erosion was calculated from shoreline features digitized from historical aerial photography, as well as historical and contemporary satellite imagery. Each site was covered by historical aerial photography flown in either 1953 or 1954. These images were scanned at the Jamaica Mines and Geology division at 600dpi and georeferenced to modern satellite imagery using ArcGIS. A mixture of Worldview-1, Worldview-2, Worldview-3 and GeoEye-1 satellite imagery was used to provide modern satellite coverage of all sites. The latest cloud-free satellite images from the Digital Globe imagery archive was used to cover each site. All sites are covered by imagery captured either in 2014 or 2015. Satellite imagery from the Jamaica-wide Ikonos 2001 satellite mosaic was used as the source of an additional shoreline position for the year 2001.

2.2 Community-Based Vulnerability Assessment

Coastal vulnerability assessment, climate models and projections are only parts of the climate change picture, which is characterized by unique and complex interactions between the biophysical environment and human systems. In the fisheries sector, issues related to social vulnerability and resilience have become dominant themes in the climate change discourse. Incorporating the human dimensions of fisheries management in the climate change adaptation processes can enhance the effective implementation of community-based programmes. Because vulnerable fishing communities face inherent adaptation challenges and are expected to respond to negative climate change impacts, it is important that their priorities are clearly understood and incorporated in adaptation solutions. The livelihood of small-scale fishers is situated in the context of multiple socioeconomic, environmental and institutional stressors that are being exacerbated by climate change. The combined and interactive effects of these stressors create complex trade-offs that retard the uptake of adaptation interventions.

The relationship of people to impacted physical environments and ecosystems and their capacity to cope with and adjust to the new situation play a fundamental role in the level of vulnerability to climate impacts. In other words, communities with varying capacity to respond to climate events will likely yield different levels of vulnerability. It is therefore important that biophysical and socioeconomic assessments be integrated and complements each other, to provide a holistic understanding of vulnerability to climate change (Wongbusarakum & Loper, 2011).

In this paper, vulnerability is conceptualized as a function of the exposure of the system to a given impact, the sensitivity of the system to that impact, and the adaptive capacity of that system to recover from impacts and evolve to mitigate future impacts and take advantage of new opportunities (Barsley et al., 2013). The extent to which people's livelihoods are vulnerable to the impacts of climate change is dependent on: their exposure to climate impacts (i.e. if impacts are felt in their location); their sensitivity (i.e. the extent to which their livelihood is affected by an impact); and their capacity to adapt to the likely impacts (Adger, 2006). Vulnerability to specific impacts of climate change will be most severe when and where they are experienced together with stresses from other sources, such as changes in the ecosystem, demography and the socio-political and economic environments.

Vulnerability assessments can be carried out at different scales. In this assessment we adopt a multi-scale qualitative approach to assess vulnerability in the targeted fishing communities. Two gender-specific focus group discussions were conducted in each of the three target communities. The discussions were guided by participatory community engagement tools and focused on issues related to the climate change exposure, sensitivity and adaptive capacity. The selection of participatory tools was guided by the type of livelihood/hazard vulnerability information they are designed to capture. Seasonal calendars were used to capture community exposure to coastal erosion, flooding, changes in climatic and oceanographic variables (and other interacting hazards); hazard ranking and impact to determine the degree of sensitivity (issues related to food and livelihood security are explored here); and vulnerability and capacity matrix to determine adaptive capacity. The vulnerability matrix was used to determine which livelihoods groups and resources are most vulnerable and to identify coping strategies currently used to address the hazards. The Focus Group Discussions (FGDs) were recorded using a digital tape recorder and transcribed verbatim. The data analysis involved documenting interpretations, developing seasonal calendars, hazard ranking, vulnerability and capacity matrix and existing coping and adaptation strategies assessment. The FGD transcripts were coded based on the domains of social vulnerability (exposure, sensitivity and adaptive capacity) and analysed using NVIVO qualitative data analysis software. Emerging themes were highlighted and used to support findings from the quantitative survey.

Adaptive capacity is the ability or potential of a system to respond successfully to climate variability and change, and includes adjustments in both behaviour and in resources and technologies (IPCC, 2014).

Exposure: The presence of people, livelihoods, species or ecosystems, environmental functions, services, and resources, infrastructure, or economic, social, or cultural assets in places and settings that could be adversely affected (IPCC, 2014).

Sensitivity is the degree to which a system will be affected by, or responsive to climate stimuli (Smith et al., 2001). Sensitivity is basically the biophysical effect of climate change; but sensitivity can be altered by socio-economic changes. For example, new crop varieties could be either more or less sensitive to climate change (IPCC, 2014).

Vulnerability: The propensity or predisposition to be adversely affected. Vulnerability encompasses a variety of concepts and elements including sensitivity or susceptibility to harm and lack of capacity to cope and adapt (IPCC, 2014).

3 Results

In the target sites, three major coastal changes were found to be heavily impacting fishing activities, namely—shoreline erosion, coastal flooding and sea level rise. These changes and the resulting impacts have increased the livelihood vulnerability of hundreds of fishers and their dependent families.

3.1 Coastal Erosion and Fisheries Businesses

Coastal erosion was identified as the main hazard affecting fisheries livelihoods in Alligator Pond and Old Harbour Bay where (relatively) rapid changes were

Table 1 Historical (1953–2001) and contemporary (2001–2015) change in shoreline position (in metres) at intersection for field profiles

Community	Length of shoreline assessed	Station	1953–2001 Change	2001–2014 Change
Alligator pond	2.58	1	+58	+20
		2	+50	+5
		3	+24	−7
		4	+45	−35
		5	+13	−24
Old Harbour Bay	1.2	1	−9	−19
		2	+86	−6
		3	+26	+11
Rocky point	1.45	1	−22	+49
		2	−3	+28
		3	−43	+40

Notes: Measurement accuracy ±5 m. Erosion indicated by negative values

experienced at the main landing beach in the past 10 years (Table 1). Both these sites primarily accreted between 1953 and 2001 (a 48-year period), as much as 58 m at Alligator Pond and 86 m at Old Harbour Bay. Contrastingly, between 2001 and 2014 (a 13-year period), both sites displayed comparatively rapid rates of erosion (35 m at Alligator Pond and 19 m at Old Harbour Bay). A series of storm events combined with anthropogenic pressure (e.g. increased number of fishers) occurred during the period of rapid change, it is likely that cumulative impacts associated with anthropogenic changes to the coastline are partly to blame for the changes observed.

A closer analysis of Alligator Pond revealed that the system experienced varying beach changes at western and eastern ends. In the western end, the beach system remained progradational between the 2 study periods, changing from 1.23 to 1.33 m·yr.$^{-1}$, and settling at a relatively stable beach slope of 16.8^{0}. The eastern end however, experienced severe erosion between the study periods, changing from 0.28 to −2.33 m·yr.$^{-1}$ (see Fig. 2 below). Of note is the tourist area known as 'Little Ochi' which has eroded approximately 35 m since 2001, with an additional 7–10 m experienced between 2014 and 2015 alone.

Between 1953 and 2001, coastal changes (including infrastructure building) resulted in the merging of an isolated sandy barrier with the main shoreline (supported by an 86 m and 26 m progradation of sections of downdrift shoreline). Between 2001 and 2015, the mainland coast retreated between 6 to 19 m, feeding sections of the beach further east through west to east longshore drift, resulting in a progradation rate of 0.73 m·yr.$^{-1}$ (see Fig. 2 below). Unlike Alligator Pond, attempts at shoreline stabilization have occurred through the installation of nearshore armour units in 2014, to mitigate some of the wave energy entering the site and encouraging sand deposition, and are from local, anecdotal information appearing to be effective.

The 2015 shoreline characterizations support the changes observed during the two study periods. Both foreshores at Alligator Pond and Old Harbour Bay are a

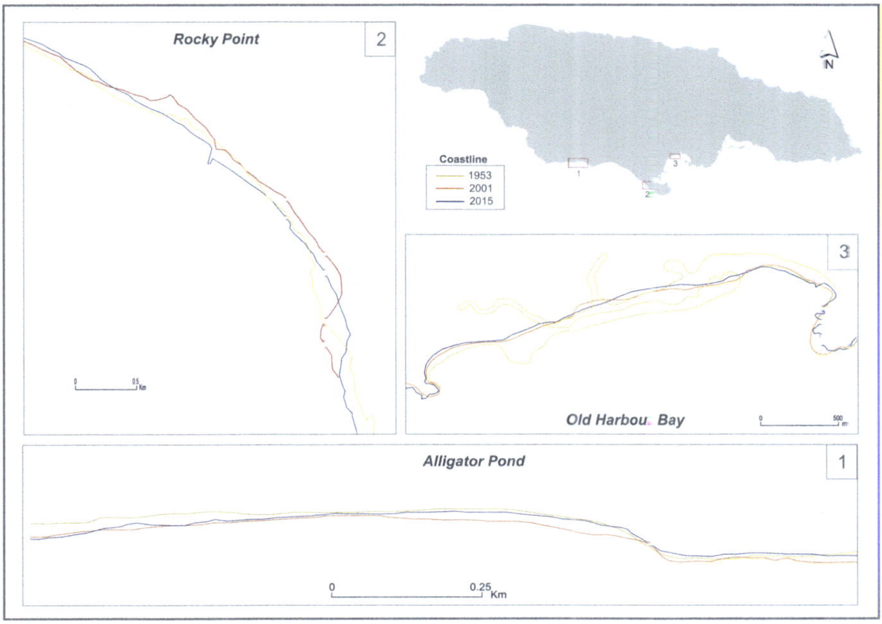

Fig. 2 Comparison of position of shorelines digitized from historical aerial photography from 1953 and satellite imagery from 2001 and 2014 (*Notes: 2015 shoreline based on GPS points collected in the field on July 18, 2015. Results shown for Alligator Pond (left), Rocky Point (centre) and Old Harbour Bay (right). Results for Rocky Point shown for comparison.*) (*Source:* modified from Van Proosdij & Baker, 2015)

mixture of chronically eroding (32 per cent and 16 per cent of the beach respectively), and partially stabilized (60 per cent and 84 per cent respectively). To put these changes into context with social vulnerability, it is important to consider the associated shoreline characterization of each target site, with focus on the backshore. Less than 10 per cent of Alligator Pond's backshore is anthropogenic structures, while these make up 56 per cent of Old Harbour's backshore. The remaining percentage of backshore at Alligator Pond is primarily dune (47 per cent), clastic slope (32 per cent) and waterbody (13 per cent), while the remaining percentage of backshore at Old Harbour Bay is clastic slope (31 per cent) and mangrove (13 per cent) (see Fig. 3).

Further, Fig. 3 shows the condition of the anthropogenic structures and natural features located on the backshore of the target sites. At Alligator Pond the majority (6 per cent) of the anthropogenic structures are failing, while none are considered 'intact'. Similarly, the majority (54 per cent) of Alligator Pond's natural features are considered 'not stabilized'. Old Harbour Bay exhibits similar patterns, though not to the same extent. The majority (40 per cent) of the anthropogenic structures are considered 'damaged', and none considered 'intact'. The majority (25 per cent) of the backshore's natural features are considered 'partially stabilized.'

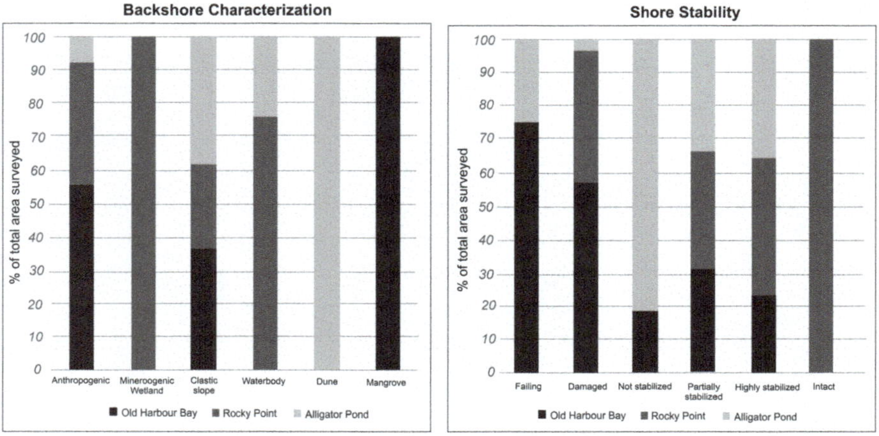

Fig. 3 Field backshore characterization conducted in July 2015 and assessment of shore stability. *'Failing, damaged, intact' refer to anthropogenic structures and the remainder associated with the stability of a natural feature.* (*Source:* modified from Van Proosdij & Baker, 2015)

While a relatively small percentage of Alligator Pond's backshore is anthropogenic, there is significant erosion as one transition to the central area, including houses in water (including toilet facilities), collapsing buildings and no foreshore. Attempts at shore protection using sandbags and rocks have been largely unsuccessful. More systematic use of armourstone has been used near Little Ochi and the main village where the beach used to extend beyond the restaurants. This use however, can be seen as short-term solution to the problem and is exacerbating erosion at either end. While the use of tires does reduce wave energy reaching the toe of the slope, it could be better stabilized with paired plantings of vegetation. Further east, erosion rate decreases initially as beach is protected by a nearshore platform; on the other hand, backshore rates increase significantly, exposing layers of garbage and debris within the old dune field.

The coastal landforms observed in the field indicate that the system is not sediment starved, demonstrating strong potential for a range of adaptation options. Efforts need to concentrate on decreasing wave energy reaching the shore and trapping sand within the coastal zone before it is removed from the system inland by strong winds or alongshore by longshore currents. This may be achieved by a combination of methods including nearshore armourstone breakwater (designed to provide benthic and fish habitat) similar to the ones observed in Old Harbour Bay. Further to the west, embryo dunes are forming naturally on top of old dunes or at the base of eroding, scarped dunes indicating potential for recovery. Planting of dune vegetation is recommended to assist this process. These results demonstrate the overall degraded state of the shoreline—a fishing communities' primary area of onshore activity—the impacts of which are examined in a later section.

A decrease in the productivity of local fisheries is likely to lead to an increase in the importation of fish and fish products thereby increasing competition. A sustained increase in the number of windy days as fishers complain that they are dealing with presently will further reduce the number of days per year that fishers can go to sea especially as the majority of artisanal fishers lack the type of gear and equipment necessary to combat heavy winds. The competitiveness of fisheries businesses can be enhanced through small and medium-scale enterprises (SMEs) development. This can serve as an important strategy for improving livelihood security, while simultaneously reducing pressures on the marine resources. Possible areas for SME development include renewable energy, cold storage, value-added operations, marketing and financing for environmentally-friendly fishing gear. Figure 4 highlights some of the key stressors impacting the fisheries businesses across the three study sites.

Existing economic and social challenges make the operators particularly sensitive to climate change impacts. In Alligator Pond, fishing is central to the local economy and helps to support a population of 190,800 (STATIN, 2013). During the focus group discussion, participants estimated that between 75 to 80 per cent of local businesses are dependent fishing. This includes a mix of nearshore and offshore pelagic fishing. The fishing methods most commonly used are: hook and line, pots and diving/spear shooting. Participants in the focus group discussions also noted that the fishing population in the community has increased in recent times. This has resulted in increased competition and greater levels of investment in fishing operations to remain viable. Fishers in Alligator Pond also indicated that beach erosion and strong winds are major factors currently affecting their operations.

Tourism and fishing activities have been affected due to flooding that occurs from higher elevations during heavy rainfall. Some private businesses have implemented measures (e.g. the use of tyres and other materials) to reduce the rate of erosion that may be considered as maladaptive and could exacerbate the erosion problem. So much of the fishing beach has been lost to erosion that fishers have resorted to ad hoc measures to secure their boats ashore, while fish vendors have been relegated to operating from the roadside. Some fishers are of the view that the customer being able to buy the freshly caught fish boat or beachside was part of the Alligator Pond experience. They believe that the rapid loss of the beach area is one reason why some customers have stayed away—which has impacted their earnings. Participants in the FGDs also reported that they have experienced significant increases in the intensity of winds in recent times which has reduced the number of fishing days. Some of the fishers explained that due to strong winds they had not been able to go out for up to 6 months.

Over the period of 2009–2010, the beach located within the Alligator Pond fishing village is estimated to have eroded by approximately 12.2 m (Myers, 2010). Tourism (especially at Little Ochi, a prominent restaurant/retreat spot) and fishing activities have been affected due to flooding that occurs from higher elevations during heavy rainfall. Robinson and Khan (2011) reported that the eastern and western edge of the Little Ochi restaurant complex has a huge, ilmenite-rich sand dune. This provides the 'black' material of the sand.

The material wellbeing of fishers is dependent on the profitability of his/her fishing operation. Per McClanahan (2010), 'profitability of fishing equals total revenue minus total costs, and total revenue equals price times the quantity of

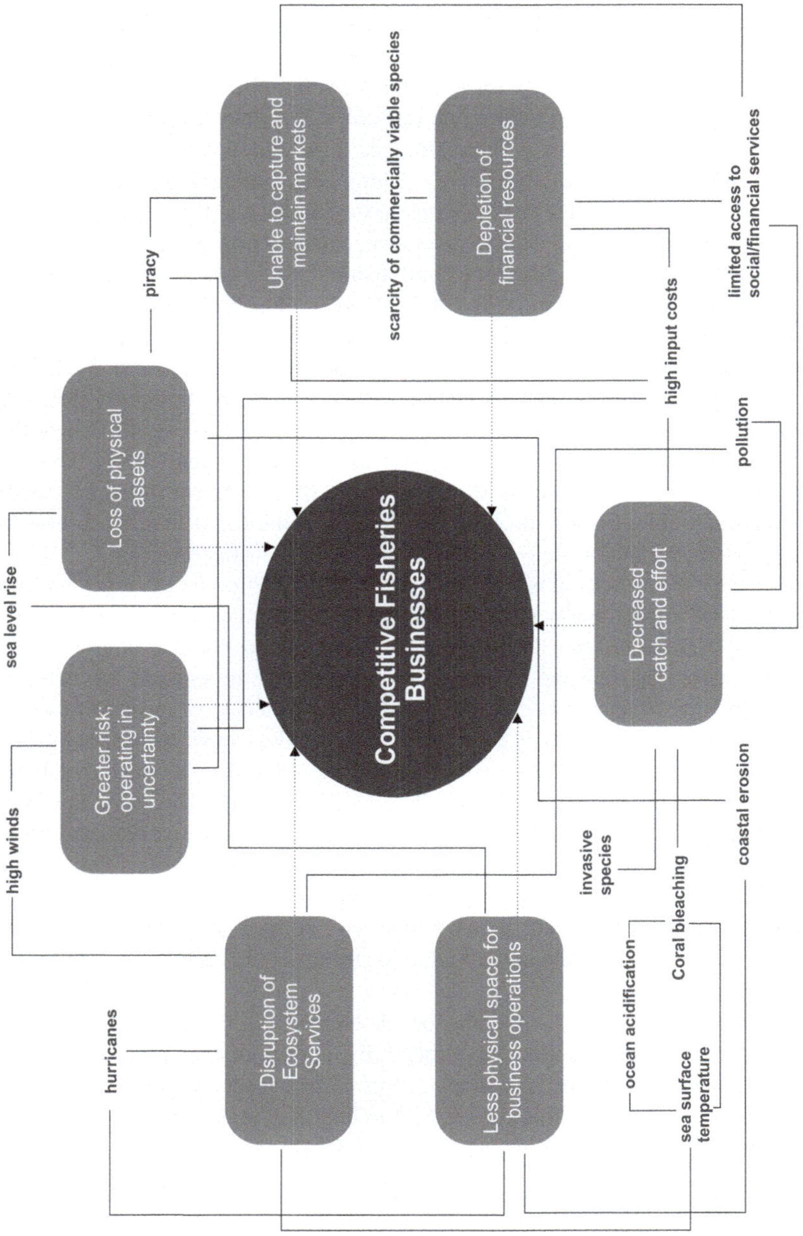

Fig. 4 Stressors and shocks to marine fisheries businesses in the study area (*Source: Author*)

catch summed across all price groups'. Prices are a function of supply and demand and are influenced by fish size, species, consumer preferences and fish quality, and the catch quantity-demand function. The income of fishers can be highly variable due to variation in the productivity and distribution of fish stocks, changes in prices, and fishers' luck at finding and catching fish (Holland et al., 2017). Understanding the seasonality of fishers' catch and composition is crucial to understanding the social wellbeing of fishers. Not only does seasonal catch variability impact the material wellbeing of fishers it can also affect subjective and relational aspects of their wellbeing. The analysis also shows how fish catch and potential income varies per the primary fishing technique employed by fishers in the study. Keeping in mind that the income generated from a catch does not necessarily depend on its market value since fishers make individual decisions at times to sell fish at lower than market prices. Reef fish species are the most landed fish species across the three communities. The quality of a catch is determined by the ratio of "trash" to "clean" fish caught. Aiken and Kong (2000), placed fish species into three market categories; Quality fish, Common fish, and Trash fish, based on the market value of the species.

3.2 Vulnerability to Coastal Flooding

Flooding was identified by participants in the FGDs as the most persistent hazard in Rocky Point and Old Harbour Bay. Rocky Point is located on a low-lying alluvial plain which increases its exposure to both inland and coastal flooding. Inundation modelling indicates that 80 buildings are within the zone of risk with 2 m of flooding with this number jumping to 805 at 3 m including 10 km of road (see Fig. 5); which constitutes a significant risk. In addition, while the prograding beach system has increased the resilience of the beach system to coastal erosion, it has also introduced challenges regarding appropriate drainage of floodwaters. Planting of mangroves could also be considered as an adaptation response to absorb some of the storm surge wave energy. Because of its coastal location, hurricane threats and storm surges also threaten fishing livelihoods. The Planning Institute of Jamaica (PIOJ) (2008) estimates that sea level rise by 0.18 m may cause loss of land in the area of 101.9 km in areas such as Rocky Point/Portland Cottage. Hurricane Ivan (2004) related storm surge affected Rocky Point, completely debilitating fishing livelihoods in the aftermath. A section of the fishing community was destroyed and houses further inland were damaged by the rising water from the sea (PIOJ, 2008).

The Old Harbour Bay community is currently at significant risk of flooding, which could increase with climate change (IPCC, 2013) and require eventual relocation of residents some sections of the community. Actual measurements from storm events, specifically Hurricane Dean (2007) where marine inundation extended beyond 150 m and the 3.5 m surge was compounded by the town gully. For example, a one-metre flood depth was recorded 380 m from the sea in the area west of the town gully. However, to demonstrate using models, close to 200 buildings are within 1 m above sea level and numerous old foundations were evident in the field of

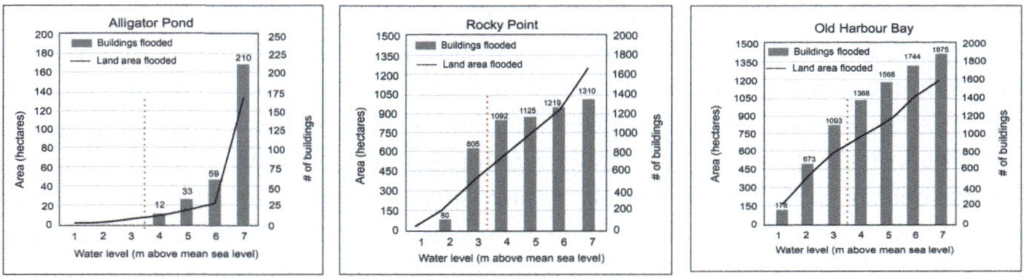

Fig. 5 Flooding scenarios based on contiguous flood models for 1 m flood intervals at the three sites Pond. *Floods less than 3 m represent combined contemporary sea level rise and storm surge. Results shown for Alligator Pond (left), Rocky Point (centre) and Old Harbour Bay (right). Results for Alligator Pond shown for comparison.* (*Source:* modified from Van Proosdij & Baker, 2015)

homes that were destroyed during previous hurricanes. Enforcement of the no-build zone could be important in the future. A total of 1093 buildings and close to 850 hectares of land will flood with water levels up to 3 m. Robinson and Khan (2011) reported that extensive damage to the housing by high winds and waves occurred from Hurricane Dean. The overall development challenges in the study communities include: high levels of unemployment (particularly amongst the youth), limited opportunity to access technical and vocational training, poor drainage facilities, poor roads and poor representation by elected leaders.

Old Harbour Bay fishing community is exposed to overland flooding, liquefaction, storm surges, tropical systems and tsunamis. An analysis of reported disaster events within this region identified flooding as the most persistent natural hazard. Observed events were collated from a number of different sources, namely the Disaster Catalogue and from damage assessment reports prepared by the Office of Disaster Preparedness and Emergency Management (ODPEM). Recurring incidences of flooding (both coastal and riverine) and wind damage, in addition to a growing population, has prompted the authorities to investigate the extent of the area's vulnerability to the above-mentioned hazards and its effect on the population.

3.3 *Community Wellbeing and Adaptation Response Options*

The livelihood of fishers in the study areas is situated in the context of multiple socioeconomic, environmental and institutional stressors that are being exacerbated by climate change. The combined and interactive effects of these stressors create complex trade-offs that can retard the uptake of adaptation solutions. The cumulative impact of shocks and stress can exacerbates community vulnerability leading to food and livelihood insecurities. Table 2 shows the results of the participatory vulnerability and capacity matrix for fisher folk in Old Harbour Bay.

Climate change impacts on the fisheries sub-sector was found to be different for men when compared to women. The findings of the FDGS indicate that women are primarily involved in the postharvesting activities such as fish vending,

Table 2 Participatory vulnerability and capacity matrix, old harbour fisherfolk

	Hurricane	Strong Winds	Coastal Erosion	Changes in Coral Reefs	Changes in Currents	Piracy
Fishing equipment	3	3	1	2	2	3
Labour/value chain	3	3	2	3	3	3
Infrastructure	3	2	2	2	1	0
Small business	3	3	1	3	3	3
Fish stock	2	1	2	3	3	3
Fish catch	3	3	2	3	3	3

0 = no threat; 1 = low threat; 2 = somewhat of a threat; 4 = high threat

fish cleaning while men are fishers, builders and mechanics. Women are also heavy investors in the sub-sector and are more likely to own fishing vessels than their counterparts. Effectively tackling multiple stressors and building climate-resilient development pathways is dependent on the success of efforts to redress power asymmetries and intergenerational inequity.

4 Conclusions

The socioeconomic fragility of Jamaica's fisheries sector has been highlighted in the past two decades by multiple economic and environmental shocks and stressors. Increased production costs, global competition, and economic recessions in key markets have resulted in decreased profitability of fisheries business operations. Coastal flooding and erosion have directly impacted fishing assets, critical coastal infrastructure, and the wellbeing of fishers in the three study communities. Loss of coastal infrastructures such as vending areas and docking bays affect fishers' capacity to sustain business operations. Based on the findings from this research we recommend four key strategies/response options for reducing livelihood vulnerabilities across the three study communities.

(1) Promote Economic Alternatives Response options to promote economic alternatives in the three communities include: (i) supporting new fisheries opportunities (e.g. pelagic or aquaculture fisheries), market (e.g. export), products (e.g. fish oil), species or livelihoods (e.g. sports fishing, ecotourism, algal farming); (ii) promote climate-smart fishing practices (through more efficient gear, equipment and fishing methods, increased regulation for sustainable fisheries; tackle over-exploitation of species to meet quotas) that can have negative long-term impacts and; (iii) increased access and use of technology (e.g. the use of improved fish pots with biodegradable panels to prevent ghost fishing), greater involvement of fishers in the management of the new protected area and, coral restoration program as a strategy to enhance fish stock.

(2) Improve Ecosystem Resilience Planting (or restoring) mangrove other vegetations between armour units and the fishing beach zones can reduce current rates of erosion. Fishers expressed more interest in nature-based solutions (NBS) (e.g. mangrove and coral restoration, and Special Fishery Conservation Areas—SFCAs) than hard infrastructure (e.g. breakwaters). Interventions to improve ecosystems' ability to adapt or recover from impacts from ocean acidification, disease outbreaks, and more frequent storm damage are urgently required. Additionally, resilience gains can be achieved through ecosystem restoration, reducing anthropogenic stresses (such as habitat destruction, pollution, destructive fishing practices), reducing carbon footprint through more efficient gear/equipment and practices, and improve the ability of ecosystems such as wetlands to act as a carbon sink (blue carbon). Special emphasis should also be placed on increasing the area and health of seagrass beds at the Pedro Cays, which is a key conch fishing ground and supports the livelihood of many fishers from the study communities.

(3) Strengthen Fisheries Co-Management Multi-species and area-based fisheries management approaches provide unique opportunities for multi-stakeholder inclusion and participation. The presence of autonomous and diverse community-based institutions in the study areas (e.g. the Caribbean Coastal Area Management Foundation, the Old Harbour Bay and Rocky Point Fishermen Cooperatives) are well-positioned to contribute more significantly to fisheries management in general, and SFCAs in particular. The value of co-management and ecosystems-based fisheries management approaches as an area of strategic investment for climate change adaptation is enhanced by, among other things, application of data from ecological monitoring of climate-sensitive parameters. Strengthening the institutional capacity of fishing cooperatives provides multiple co-benefits toward to effective stewardship and co-management of fishery resources. Training on how to manage cooperatives effectively and awareness of the role and value of inclusive membership emerged as key areas for capacity building in the study areas.

(4) Develop Gender-Sensitive Vulnerability Indicators While female fishers are predominantly engaged in the postharvest side of production, they tend to own and manage a significant amount of assets in the study communities and the impact of disasters on their livelihood are often underestimated. Greater involvement of female fishers in disaster risk management and community development planning at the community-level could help to address this problem. Across the study sites, female fisherfolk are struggling with the dual burden of managing their business and domestic labour. A baseline of the youth dynamics in the fisheries in Jamaica would be valuable for understanding how participation in the fisheries sub-sector varies by age (e.g. fishing methods, frequency of fishing trips, risk appetite, livelihood diversification strategies). A focus on gender and age-sensitive approaches to climate change adaptation would be particularly useful for ensuring that appropriate indicators are developed to capture the vulnerabilities of fishery-based livelihoods.

Developing a climate-sensitive Gender Strategy for the fisheries sub-sector would be a positive step towards this objective.

References

Adger, N. (2006). Vulnerability. *Global Environmental Change, 16*(3), 268–281.

Aiken, K., & Kong, G. (2000). *The marine fisheries of Jamaica*. Naga.

Barsley, W., De Young, C., & Brugère, C. (2013). *Vulnerability assessment methodologies: an annotated bibliography for climate change and the fisheries and aquaculture sector*. Food and Agriculture Organization.

Beaugrand, G., Edwards, M., Raybaud, V., Goberville, E., & Kirby, R. (2015). Future vulnerability of marine biodiversity compared with contemporary and past changes. *Nature Climate Change, 5*(7). https://doi.org/10.1038/nclimate2650

Brander, K. (2010). Impacts of climate change on fisheries. *Journal of Marine Systems. 79*(3), 389–402.

Caribbean Community Climate Change Centre (CCCCC). (2015). *National Adaptation Strategy and action plan for Jamaica's fisheries sub-sector*. GCCA – ACP Project.

Chan, C., Armitage, D., Alexander, S., & Campbell, D. (2019). Examining linkages between ecosystem services and social wellbeing to improve governance for coastal conservation in Jamaica. *Ecosystem Services, 39*, 100997. https://doi.org/10.1016/j.ecoser.2019.100997

Cheung, W. W. L., Lam, V. W. Y., Sarmiento, J. L., Kearny, K., Watson, R., Zeller, D., & Pauly, D. (2010). Large-scale redistribution of maximum fisheries catch potential in the global ocean under climate change. *Global Change Biology, 16*(1), 24–35. https://doi.org/10.1111/j.1365-2486.2009.01995.x

Church, J. A., Clark, P. U., Cazenave, A., Gregory, J. M., Jevrejeva, S., Levermann, A., Merrifield, M. A., Milne, G. A., Nerem, R. S., Nunn, P. D., Payne, A. J., Pfeffer, W. T., Stammer, D., & Unnikrishnan, A. S. (2013). Sea level change. In T. F. Stocker, D. Qin, G.-K. Plattner, M. Tignor, S. K. Allen, J. Boschung, A. Nauels, Y. Xia, V. Bex, & P. M. Midgley (Eds.). *Climate change 2013: The physical science basis. Contribution of working group I to the fifth assessment report of the intergovernmental panel on climate change* (pp. 1137–1216). Cambridge University Press. https://doi.org/10.1017/CBO9781107415324.026

Climate Studies Group Mona (CSGM). (2012). *State of the Jamaican climate 2012: Information for resilience building (summary for policymakers)*. Planning Institute of Jamaica.

Climate Studies Group Mona (Eds.). (2020). *The state of the Caribbean Climate*. Produced for the Caribbean Development Bank.

Department of Economic and Social Affairs/United Nations Statistics Division. (2018). *United Nations 2018 Comtrade database*. [online] Available from: https://comtrade.un.org/

Forbes, D. L., James, T. S., Sutherland, M., & Nichols, S. E. (2013). Physical basis of coastal adaptation on tropical small islands. *Sustain Sci, 8*(3), 327–344.

Gattuso, J.-P., Magnan, A., Billé, R., Cheung, W. W. L., Howes, E. L., Joos, F., Allemand, D., Bopp, L., Cooley, S. R., Eakin, C. M., Hoegh-Guldberg, O., Kelly, R. P., Pörtner, H.-O., Rogers, A. D., Baxter, J. M., Laffoley, D., Osborn, D., Rankovic, A., Rochette, J., Sumaila, U. R., Treyer, S., & Turley, C. (2015). Contrasting futures for ocean and society from different anthropogenic CO2 emissions scenarios. *Science, 349*(6243). https://doi.org/10.1126/science.aac4722

Government of Jamaica (GOJ). (2013). *Green paper: climate change policy framework and action plan. Ministry of Water, land, environment and climate change*. Kingston. [online] Available from: http://www.japarliament.gov.jm/attachments/440_Climate%20Change.pdf

Holland, D. S., Speir, C., Agar, J., Crosson, S., DePiper, G., Kasperski, S., & Perruso, L. (2017). Impact of catch shares on diversification of fishers' income and risk. *Proc Natl Acad Sci, 114* (35), 9302–9307. https://doi.org/10.1073/pnas.1702382114

Intergovernmental Panel on Climate Change (IPCC). (2014). *Climate change 2014: Impacts, adaptation, and vulnerability*. Part B: *regional aspects*. In V. R. Barros, C. B. Field, D. J. Dokken, M. D. Mastrandrea, K. J. Mach, T. E. Bilir, M. Chatterjee, K. L. Ebi, Y. O. Estrada, R. C. Genova, B. Girma, E. S. Kissel, A. N. Levy, S. MacCracken, P. R. Mastrandrea, & L. L. White (Eds.), *Contribution of working group II to the fifth assessment report of the intergovernmental panel on climate change* (p. 688). Cambridge University Press.

IPCC. (2018). Global warming of 1.5°C. An IPCC Special Report on the impacts of global warming of 1.5°C above pre-industrial levels and related global greenhouse gas emission pathways, in the context of strengthening the global response to the threat of climate change, sustainable development, and efforts to eradicate poverty. In V. Masson-Delmotte, P. Zhai, H. O. Pörtner, D. Roberts, J. Skea, P. R. Shukla, A. Pirani, W. Moufouma-Okia, C. Péan, R. Pidcock, S. Connors, J. B. R. Matthews, Y. Chen, X. Zhou, M. I. Gomis, E. Lonnoy, T. Maycock, M. Tignor, & T. Waterfield (Eds.), (in press).

IPCC. (2019). IPCC Special Report on the Ocean and Cryosphere in a Changing Climate. In H.-O. Pörtner, D. C. Roberts, V. Masson-Delmotte, P. Zhai, M. Tignor, E. Poloczanska, K. Mintenbeck, A. Alegriá, M. Nicolai, A. Okem, J. Petzold, B. Rama, & N. M. Weyer (Eds.), (in press).

Marschke, M., Campbell, D., & Armitage, D. (2020). Precarious livelihoods: Examining the intersection of fish work and ecological change in coastal Jamaica. *People and Nature, 2*(1), 152–162. https://doi.org/10.1002/pan3.10061

Marshall, N. A., Marshall, P. A., Tamelander, J., Obura, D., Malleret-King, D., & Cinner, J. E. (2010). *A framework for social adaptation to climate change; Sustaining tropical coastal communities and industries*. IUCN. [online] Available from: https://www.iucn.org/content/a-framework-social-adaptation-climate-change-sustaining-tropical-coastal-communities-and-industries

McClanahan, T. R. (2010). Effects of fisheries closures and gear restrictions on fishing income in a Kenyan coral reef. *Conservation Biology, 24*(6), 1519–1528. https://doi.org/10.1111/j.1523-1739.2010.01530.x

McIlgorm, A., Hanna, S., Knapp, G., Le Floc, P., Millerd, H. F., & Pan, M. (2010). How will climate change alter fishery governance? Insights from seven international case studies. *Marine Policy, 34*(1), 170–177.

McWilliams, J., Côté, I., Gill, J., Sutherland, W., & Watkinson, A. (2005). Accelerating impacts of temperature induced coral bleaching in the Caribbean. *Ecology, 86*(8), 2055–2060. https://doi.org/10.1890/04-1657

Ministry of Agriculture and Fisheries (MOAF). (2015). *National Fisheries and aquaculture policy*. Kingston.

Ministry of Industry, Commerce, Agriculture and Fisheries (MICAF). (2018). *Promoting community-based climate resilience in the fisheries sector*. Environmental and Social Management Framework, Kingston, Jamaica, Fisheries Division, Ministry of Industry, Commerce, Agriculture and Fisheries (MICAF).

Monnereau, I., Mahon, R., McConney, P. & Nurse, L. (2013), Vulnerability of the fisheries sector to climate change impacts in Small Island developing states and the wider Caribbean: Early findings, , Centre for Resource Management and Environmental Studies (CERMES)

Monnereau, I., Mahon, R., McConney, P., Nurse, L., Turner, R., & Vallès, H. (2017). The impact of methodological choices on the outcome of national-level climate change vulnerability assessments: An example from the global fisheries sector. *Fish and Fisheries, 18*, 717–731. https://doi.org/10.1111/faf.12299

Mora, C., Frazier, A., Longman, R., Dacks, R., Walton, M., Tong, E., Sanchez, J., Kaiser, L., Stender, Y., Anderson, J., Ambrosino, C., Fernandez-Silva, I., Giuseffi, L., & Giambelluca,

T. (2013). The projected timing of climate departure from recent variability. *Nature, 502*, 183–187.

Myers, G. (2010). *Fishers alarmed at erosion in Alligator Pond*. [online]. Retrieved July 2001, from http://www.jamaicaobserver.com/news/Alarming-beach-erosion-at-Alligator-Pond_7789235

Nurse, L. (2011). The implications of global climate change for fisheries management in the Caribbean. *Climate and Development, 3*(3), 228–241.

Palanisamy, H., Becker, M., Meyssignac, B., Henry, O., & Cazenave, A. (2012). Regional Sea level change and variability in the Caribbean Sea since 1950. *Journal of Geodetic Science, 2*(2), 125–133. https://doi.org/10.2478/v10156-011-0029-4

Planning Institute of Jamaica (PIOJ). (2008). *Jamaica macro socio-economic and environmental assessment of the damage and loss caused by hurricane Ivan*. Kingston.

Planning Institute of Jamaica (PIOJ) (2013). *Jamaica macro socio-economic and environmental assessment of the damage and loss caused by hurricane sandy*. Planning Institute of Jamaica, Kingston. [online] Available from: https://www.pioj.gov.jm/product/damage-and-loss-assessments-2013-pioj-report-hurricane-sandy/

Planning Institute of Jamaica (PIOJ). (2019). *Economic and social survey Jamaica (2014)*. Kingston, Jamaica, Planning Institute of Jamaica.

Pulwarty, R. S., Nurse, L., & Trotz, U. O. (2010). Caribbean Islands in a changing climate. Environment. *Science and Policy for Sustainable Development, 52*(6), 16–27.

Robinson, S., & Wren, C. (2020). Geographies of vulnerability: A research note on human system adaptations to climate change in the Caribbean. *Geografisk Tidsskrift-Danish Journal of Geography, 120*(1), 79–86. https://doi.org/10.1080/00167223.2020.1733432

Robinson, E., Khan, S., & Coutou, R. (2008). *Beach erosion and other coastal hazards: Ensuring safety at the community and management levels*. Kingston, Jamaica, Environmental Foundation of Jamaica. [online] Available from: http://www.mona.uwi.edu/geoggeol/MGU/Past projects/BEACHES Phase two report 2008.pdf

Robinson, E., & Khan, S. A. (2011). *Impacts on Jamaica's coastline from recent hurricanes*. Kingston, Jamaica, Environmental Foundation of Jamaica. [online] Available from: https://www.mona.uwi.edu/geoggeol/MGU/Publications/2007%20Workshops/Post_Hurricane_DEANWorkshop.pdf

Robinson, J., Wilson, S., Robinson, J., Gerry, C., Lucas, J., Assan, C., Govinden, R., Jennings, S., & Graham, N. (2019). Productive instability of coral reef fisheries after climate-driven regime shifts. *Nature Ecology & Evolution, 3*(2), 183–190. https://doi.org/10.1038/s41559-018-0715-z

Singh-Renton, S., & McIvor, I. (2015). *Review of current fisheries management performance and conservation measures in the WECAFC area*. (FAO Fisheries and Aquaculture Technical Paper 587). Food and Agricultural Organization.

Statistical Institute of Jamaica (STATIN). (2013). *Population census*. Kingston, Jamaica, Government of Jamaica. [online] Available from: http://statinja.gov.jm/Demo_SocialStats/populationbyparish.aspx

Statistical Institute of Jamaica (STATIN). (2016). *Population census*. Kingston, Jamaica, Government of Jamaica.

Turner, R., McConney, P., & Monnereau, I. (2020). Climate change adaptation and extreme weather in the small-scale fisheries of Dominica. *Coastal Management, 48*(5), 435–455. https://doi.org/10.1080/08920753.2020.1795970

Van Proosdij, D., & Baker, G. (2015). *Coastal vulnerability assessment of fishing communities in Jamaica*. (Technical Report). Kingston, Jamaica, Caribsave Partnership.

Wongbusarakum, S., & Loper, C. (2011). *Indicators to assess community-level climate change vulnerability: An addendum to SocMon and SEM-Pasifika regional socioeconomic monitoring guidelines*. Secretariat for the Pacific Environment Programme through the Coral Reef InitiativeS for the Pacific (CRISP) and IUCN.

World Bank. (2019), *Forces of Nature: Assessment and Economic Valuation of Coastal Protection Services Provided by Mangroves in Jamaica*. [online] Available at: https://www.nepa.gov.jm/new/projects/docs/WorldBank2019_ForcesOfNature.pdf

Fisheries Sector Vulnerabilities to Climate Change in Small Island Developing States

Iris Monnereau, Robin Mahon, Patrick McConney, Leonard Nurse, Rachel Turner, and Henri Vallès

Abstract Small Island Developing States (SIDS) are considered to be highly vulnerable to climate change. Our ability to identify where adaptation is most needed is constrained by the outputs of vulnerability assessments to date, which typically produce single aggregate rankings, providing limited information on the underlying factors of vulnerability. SIDS have often been banded together to address common sustainability challenges. However, despite their similarities they are not homogenous and differences between the three SIDS groups (Caribbean, Pacific and Atlantic, Indian Ocean, Mediterranean and South China Sea (AIMS)) can be expected. Using a diverse range of relevant indicators and a well-established conceptual framework, we compare the vulnerability of the national fisheries sectors of Caribbean, Pacific, and AIMS SIDS at three levels of decreasing data aggregation: (1) overall vulnerability, (2) across the three components making up overall vulnerability, i.e. exposure, sensitivity, and adaptive capacity, and (3) across groups of correlated indicators making up each of the aforementioned components. The underlying components and subcomponents show distinct differences between the country groups. The vulnerability of the fisheries sector in Pacific SIDS is associated with high sensitivity and low adaptive capacity whereas, in contrast, the vulnerability of Caribbean SIDS is driven by high exposure. Our results highlight that what drives overall vulnerability differs fundamentally among SIDS and that the use of a single aggregate score to assess overall vulnerability masks much of the information on specific underlying factors that is needed to guide adequate adaptation strategies to climate change in SIDS.

I. Monnereau (✉)
Food and Agriculture Organization of the United Nations (FAO), Subregional Office for the Caribbean, UN House Barbados, Bridgetown, Barbados
e-mail: Iris.monnereau@fao.org

R. Mahon · P. McConney · L. Nurse · H. Vallès
Centre for Resource Management and Environmental Studies, University of the West Indies, Bridgetown, Barbados

R. Turner
Penryn Campus, University of Exeter, Exeter, UK
e-mail: R.Turner@exeter.ac.uk

S. Moncada et al. (eds.), *Small Island Developing States*, The World of Small States 9, https://doi.org/10.1007/978-3-030-82774-8_11

Keywords Vulnerability assessments · Climate change · Fisheries management · Ocean health index · Small Islands Developing States · Adaptive capacity · Climate Change Adaptation

1 Introduction

Although Small Island Developing States (SIDS) produce only 0.6%[1] of global greenhouse gases, they are expected to be disproportionally affected by the threats of climate change (Mimura et al., 2007; Nurse et al., 2014; Thomas et al., 2020). SIDS are small island or low-lying coastal countries located in the tropical and subtropical regions (partly) surrounded by oceans (Boto & Biasca, 2012), which have been banded together under the United Nations to address common sustainability challenges. They share similar sustainability challenges related to their specific characteristics such as small size, susceptibility to natural disasters and climate change, vulnerability to external shocks, concentration of population and infrastructure in the coastal zone, high dependence on limited resources including marine resources; fragile environments, and excessive dependence on international trade (Guillotreau et al., 2012; Mimura et al., 2007; Monnereau et al., 2017; Nurse et al., 2014; Polido et al., 2014). Despite longstanding recognition of these issues facing SIDS (in e.g. Rio Earth Summit in 1992, Mauritius Conference in 2005), recent global conferences have highlighted ongoing and emerging challenges (see e.g. Rio+20 Summit in 2012 and the SAMOA Pathway in 2014 and the Paris Agreement).

It is widely recognised that SIDS are among the most vulnerable to the effects of climate change (Betzold, 2015; Thomas et al., 2020). Climate change impacts can be expected to exacerbate this vulnerability because SIDS are mostly low-lying islands with vulnerable coastlines, with a high percentage of population and vital infrastructure located in the coastal zone. In SIDS, 85% of the population lives within 10 km of the coastline in comparison to 26% for all other coastal and island countries. Small islands will be most at risk from climate change as a result of sea-level rise, tropical and extra-tropical cyclones, increasing air and sea surface temperatures, and changing rainfall patterns (Nurse et al., 2014). It has been suggested that the very existence of some atoll nations is threatened by rising sea levels associated with climate change. Although this scenario is not applicable to all SIDS, it is without a doubt that climate change impacts overall will have serious negative effects on SIDS especially on socio-economic conditions and bio-physical resources, although impacts will be shaped by the extent and effectiveness of adaptation (Nurse et al., 2014).

[1] This percentage is based on our calculations of SIDS' carbon production in 2009 from the Carbon Dioxide Information Analysis Center. See http://cdiac.ornl.gov/trends/emis/meth_reg.html (accessed 23 July 2013). All SIDS are included in this analysis except for: American Samoa; Guam; Puerto Rico; Tuvalu; and the US Virgin Islands (due to lack of data).

There is increasing concern over the direct and indirect impacts of climate change and climate variability on marine capture fisheries (Brander, 2010; Cheung et al., 2010; Mora, 2013; Oremus et al., 2020). Climate change impacts such as sea surface temperature increases, ocean acidification, increased intensity of storms, and sea level rise are expected to trigger a series of biophysical and socio-economic impacts on national fisheries (Allison et al., 2009; Brander, 2007; Cheung et al., 2010; Mahon, 2002; Mora, 2013; Nurse, 2011; Pörtner et al., 2014). SIDS' Exclusive Economic Zones (EEZ) are on average 2,133 times larger than their landmass while for all other coastal or island countries in the world their EEZ is only six times larger than their landmass. It is therefore not surprising SIDS are also referred to as "large ocean states" (UN-OHRLLS, UNESCO and UN-DOALOS, 2014). Their large EEZs relative to their landmass underscores the importance of their coastline and maritime space for SIDS and the importance of understanding climate change vulnerability of the fisheries sector in SIDS.

Negative impacts on the fisheries sector from climate change that are already evident in SIDS include; coral bleaching, increasing intensity of storms and hurricanes, increased sea level, and coastal erosion and storm surges affecting landing sites. Smaller catches can be expected to have significant socio-economic implications for those working in the harvest and post-harvest sectors (Oxenford & Monnereau, 2018). This will also have implications for national governments such as reduced domestic productivity in the fishing sector, food security and international trade and foreign currency earnings, especially for those countries with export-oriented fisheries (Oxenford & Monnereau, 2018).

Fish resources are typically shared by multiple nations and climate-induced changes to their distribution may require changes to multi-lateral or international agreements, whilst changes in genetic connectivity of stocks may impact the effectiveness of marine reserve networks in protecting source populations (Oxenford & Monnereau, 2018). Climate change stressors such as sea level rise and increased frequency of severe hurricanes in the region will continue to have significant negative impacts on safety of fishers, fisheries infrastructure, boats and fishing equipment, and coastal fishing communities (Oxenford & Monnereau, 2018).

Vulnerability assessments based on a range of biophysical and socio-economic indicators have become the dominant method to establish who and what is vulnerable to the negative effects of climate change (Klein, 2009; Tschakert et al., 2013). They are considered to be particularly relevant now that the impacts of climate change are increasingly being observed (Hinkel, 2011). Climate change vulnerability assessments have, as a result, been receiving increasing attention in policy and academic circles (Khazai et al., 2014). A comparative approach using the country or state as the unit of analysis can be used to identify particularly vulnerable groups of countries. These national level vulnerability assessments can help guide appropriate climate change adaptation policies (Allison et al., 2009; Blasiak et al., 2017; Brooks et al., 2005). It has been argued that on a strategic level, SIDS have had to demonstrate their vulnerabilities within the United Nations Framework Convention on Climate Change (UNFCCC) in order to remain relevant and compete with other groups for these resources (Oculi & Stephenson, 2018).

Understanding where the impacts of climate change on the fisheries sector have the greatest social and economic significance is crucial, as fisheries are important for food security, livelihoods and employment and for the generation of foreign exchange for national governments globally (Allison, 2011; Allison et al., 2009). Vulnerability assessments and the ranking of countries can have both political and practical consequences. National level fisheries sector climate change vulnerability assessments as well as other overall vulnerability assessments to date have indicated Least Developed Countries (LDCs) are most vulnerable in comparison to SIDS and other coastal countries (see e.g. Allison et al., 2009; Blasiak et al., 2017). Monnereau et al. (2017) have demonstrated that this is partly due to methodological choices in the implementation of these assessments, which can substantially change the analysis of which country groups (SIDS, LDCs and other coastal countries) are most vulnerable. Monnereau et al. (2017) carried out an assessment of climate change vulnerability of the fisheries sector at the national level across 173 countries, based on 35 indicators across the components of exposure, sensitivity and adaptive capacity. The results suggested that SIDS are most vulnerable in the sensitivity component, while for LDCs the area of greatest vulnerability lies in their low adaptive capacity (Monnereau et al., 2017). However, the results also indicated high variability in vulnerability scores within the SIDS group. Given the serious implications of vulnerability assessment outcomes for adaptation, in this study we argue that the special characteristics of SIDS need to be further explored. The UN recognizes 51 SIDS, spread over three geographical regions: (1) the Caribbean, (2) the Pacific, and (3) the Atlantic, Indian Ocean, Mediterranean and South China Sea (AIMS). Here we systematically compare the vulnerability outcome for the three groups of SIDS: Caribbean, Pacific and AIMS.

Understanding the differences in vulnerability between the three SIDS groups is important as this can help determine adequate pathways for fisheries sector adaptation in the face of climate change. We build on the assessment carried out by Monnereau et al. (2017), expanding on the 35 indicators used there by adding 11 more indicators that are particularly tailored to SIDS climate change vulnerability. This paper compares the vulnerability scores across the three categories and final vulnerability score for the three different SIDS groups and assesses which subcomponents (group of indicators) explain the largest vulnerability in terms of vulnerability between the three groups. Understanding the differences in vulnerability between the three SIDS groups is important as this can help determine adequate pathways for fisheries sector adaptation in the face of climate change.

2 Differences and Similarities in Vulnerability of SIDS to Climate Change

There is no common accepted definition of what constitutes SIDS (Boto & Biasca, 2012; Polido et al., 2014) and the exact countries will vary depending on the criteria used (Boto & Biasca, 2012; Crowards, 2002).[2] In this study we follow the classification of the United Nations Office of the High Representative for the Least Developed Countries, Landlocked Developing Countries and Small Island Developing States (UN-OHRLLS) and included 50 of the 51 SIDS with an additional seven overseas territories (Table 1). We follow the UN-OHRLLS as this organisation will be responsible for creating reports on the progress of SIDS in the context of both the SAMOA Pathway and the 2030 agenda for sustainable development (which includes the set of 17 Sustainable Development Goals (SDGs)).

SIDS share several common characteristics such as small size, susceptibility to natural disasters and climate change, vulnerability to external shocks, concentration of population and infrastructure in the coastal zone, high dependence on limited resources including marine resources; fragile environments, and excessive dependence on international trade. There has been a tendency to generalize about the potential impacts of climate change on SIDS and their adaptive capacity (Nurse et al., 2014). Table 2 presents a number of vulnerability characteristics of SIDS found in the literature and the additional vulnerability pressures that climate change is expected to exacerbate.

Despite these common characteristics, SIDS are by no means homogenous and they vary by geography, physical, climatic, social, political, cultural, and ethnic character as well as by level of economic development. Table 3 shows some differences between the three SIDS groups and all other coastal and island countries. The average population per island and gross domestic product (GDP) is highest in the Caribbean, although closely followed by AIMS SIDS. Pacific SIDS have the smallest total population on average and the lowest GDP per capita. AIMS are by far the smallest islands in landmass in comparison to Caribbean and Pacific SIDS with the Pacific SIDS having the largest landmass. The average EEZ is highest for Pacific SIDS with an average of 1,372,131 sq. km, while Caribbean SIDS have a small EEZ

[2]Even United Nations (UN) agencies have never agreed a common definition of SIDS and six different UN agencies use a different compilation of SIDS (Lisowska, 2016). Currently there are as many as eight different SIDS lists depending on which institution has produced it, ranging from 31 to 52 SIDS (Lisowska, 2016). This lack of definition and the grouping of SIDS is important, however, because of two factors: a) the international support and assistance available to states that are classified as SIDS; b) the reliability of the aggregated statistics that show the overall economic progress and health of these countries (Lisowska, 2016). The World Bank and the International Monetary Fund, for example, use different compilation of SIDS affecting funding streams for small islands as only those on their lists are eligible states for financial assistance such as grants and concessional lending or access to the Multi-Donor Trust Fund but those not on the respective lists are not (Lisowska, 2016). A classification of SIDS based on size of population, land or size of economy will each create a different set of countries (although with overlaps) (Crowards, 2002).

Table 1 The countries per SIDS groups included in this study

Caribbean region (29)	Pacific Region (19)	Africa Indian Ocean Mediterranean and South China Sea (AIMS) region (9)
Antigua and Barbuda	Cook Islands	Cape Verde
Bahamas	Federated States of Micronesia	Comoros*
Barbados	French Polynesia	Guinea-Bissau*
Belize	Fiji	Maldives
Cuba	Guam	Mauritius
Dominica	Kiribati*	Sao Tomé and Principe*
Dominican Republic	Marshall Islands	Seychelles
Grenada	Nauru	Singapore
Guyana	New Caledonia	Réunion
Haiti*	Niue	
Jamaica	Palau	
Puerto Rico	Papua New Guinea	
St. Kitts and Nevis	Samoa*	
Saint Lucia	Solomon Islands	
St. Vincent and the Grenadines	Timor-Leste	
Suriname	Tonga	
Trinidad and Tobago	Tuvalu*	
US Virgin Islands	Vanuatu*	
Anguilla (UK Overseas Territory (OT))	Tokelau (OT)	
Bermuda (UK OT)		
British Virgin Islands (UK OT)		
Cayman Islands (UK OT)		
Montserrat (UK OT)		
Turks and Caicos (UK OT)		
French Guiana (FR OT)		
Guadeloupe (FR OT)		
Martinique (FR OT)		
Aruba (NL OT)		
Netherlands Antilles (NL OT)		

*(Note: * indicates SIDS is also a least developed country (LDC))*

of only 135,580 km^2 and AIMS SIDS an EEZ of 563,322 sq. km. The EEZ/landmass ration is highest for Pacific SIDS whom have an EEZ 5,346 times larger than their landmass, followed by AIMS with an EEZ 796 times larger and Caribbean SIDS 443 times larger. The three different SIDS groups (Caribbean, Pacific and AIMS SIDS) can thus be expected to score differently in vulnerability assessments.

Table 2 Vulnerability characteristics of SIDS

Vulnerability categories of SIDS	General vulnerability characteristics	Climate change vulnerability
Smallness	• High competition between land use, spatial concentration of population, infrastructure and other productive assets in coastal zone • Large % population concentrated in one centre in the coastal zone • SIDS are often comprised of a number of small islands	• Sea level rise and increased intensity and frequency of El Niño-Southern Oscillation (ENSO) events can lead to flooding in the coastal zones where most population infrastructure and other productive assets are located • Often SIDS are single-town islands where flooding and damage to the main city will shut down the majority of civil services • Small population often means limited institutional capacity in the public and private sectors. Government departments and private firms are relatively small, affecting their capacity and competitiveness. Governments have less capacity to deal with climate change issues and provision of public services is limited. As government capacity is limited, accessing large funds and writing grants can prove more difficult than for larger countries. • Sources of revenue are limited with many small islands relying heavily on tariffs for a substantial proportion of public revenue • They tend to largely depending on trade. Yet external capital is limited and many small states are dependent on Official Development Assistance (ODA)
Insularity and remoteness	• High external transport costs and time delays • High costs in accessing imported goods delays and reduceds quality in information flows • High freight costs and reduced competitiveness • Geopolitically weak • Heavy reliance on limited number of natural resources (high reliance on coastal and marine resources)	• Accessing material needed to improve adaptation (e.g. building hard and soft sea defences) are limited and costly • Remoteness entails dependence on transport providers and high transport costs, which are already higher for smaller volumes • As SIDS are divided by large ocean spaces, organisation and collaboration between the states can be limited. As a result, negotiation power of SIDS (groups) at the global fora for climate change can be limited

(continued)

Table 2 (continued)

Vulnerability categories of SIDS	General vulnerability characteristics	Climate change vulnerability
Environmental factors	• Located in tropical and sub-tropical regions • High reliance on marine resources to meet protein needs • High dependence on fish for livelihoods, employment and exports • Small size, isolation and fragility of island ecosystems creates biodiversity threats	• High level of exposure to natural disasters and climate change impacts such as sea level rise, increased intensity and frequency of ENSO events, sea surface temperature rise and ocean acidification due to geographical location of SIDS in tropical and subtropical regions • Climate change impacts expected to; increase threat to biodiversity and important ecosystems SIDS depend on for ecosystem services; cause coral reef degradation (e.g. coral reef bleaching); mangrove destruction (e.g. salt-water intrusion); result in negative impacts on pelagic species due to increases sea surface temperatures and coastal erosion
Economic factors	• Small economies (small internal market) • Limited resource base • Dependence on external finance • Dependence on narrow range of export products • High import content (particularly of strategic imports such as food and fuel) • Susceptible to volatility of international trade • High use of foreign exchange for energy imports • High energy costs inhibit development of particular new economic sectors • GDP per capita often higher than cut-off for ODA so limited access to concessionary resources • High dependence remittances (not secure and is volatile)	• Loss of land and flooding will lead to destruction of existing economic infrastructure and human settlements • Limited number of natural resources available on which the population depends to a high degree (e.g. coral reefs) are expected to be disproportionally affected by climate change impacts • Exports and imports depend on ports which can be affected by climate change impacts such as increased frequency and intensity of storms • High dependence on foreign aid and remittances which makes economy volatile while demand for financial aid for climate change adaptation will only increase this dependency • Limited resource base makes SIDS vulnerable for natural disasters as there is the chance it will wipe out/affect the supply of key resources for the whole country
Demographic factors	• Small population (limited human resource base; costly public administration and infrastructure; limited institutional capacity) • High population density which increases pressure on already limited resources • Small economies (high per capita costs for infrastructure and services)	• Limited human capacity for climate change adaptation • High costs of infrastructural adaptation

(continued)

Table 2 (continued)

Vulnerability categories of SIDS	General vulnerability characteristics	Climate change vulnerability
Governance	• Limited human, financial and technical capacity to adapt	• Climate change adaptation requires high level of human, financial and technical capacity and expertise which is often not or only available to a limited degree in SIDS • Loss of land and disappearance of low-lying island and atolls can lead to loss of EEZs as maritime boundaries will shift • Large EEZ to manage in comparison to land area and population makes enforcement and implementation of rules and regulations difficult putting increasing pressure on fisheries sustainability
Disaster mitigation capacity	• Limited hazard forecasting ability, complacency, little insurance cover	• Increased intensity and frequency of natural disasters such as floods, hurricanes and storms will be further intensified by the limited forecasting ability of SIDS • Limited preparedness for disaster and lack of early warning systems for the fisheries sector

Briguglio (1995, 2003), Von Tigerstrom (2005), Guillaumont (2009), Guillotreau et al. (2012), Nurse et al. (2014) and authors

Table 3 General characteristics of the three SIDS groups and other coastal nations)

	Region			
Characteristics	AIMS	Caribbean	Pacific	All other countries
Population (sum total of all in region)[a]	1,244,488	1,500,122	562,988	5,387,1253
GDP per capita (mean)[b]	14,044	17,562	6,708	17,411
Land surface (sq. km)[c]	4,554	23,629	29,336	965,043
EEZ (sq. km)[d]	563,322	135,580	1,372,131	816,718
EEZ/landmass (sw. km)[e]	796	443	5,346	6
% population living <5m above sea level[f]	22	21	34	9
% population living within 10 km of coastline[f]	86	81	90	26

[a]CIA Factbook 2015
[b]World Bank (2011a, b)
[c]World Bank (2011a, b)
[d]www.seaaroundus.org
[e]Use of sources of [c] and [d]
[f]CIESIN 2010

3 Methodology

3.1 Vulnerability Framework

In this paper we build on the work of Monnereau et al. (2017) whom in turn build on to the framework developed by Allison et al. (2009) (see Fig. 1), which was based on the commonly applied definition of vulnerability used in the Third Assessment Report (TAR) of the Intergovernmental Panel on Climate Change (IPCC) (2001). In this interpretation, the vulnerability of any sector to climate variability or change is a function of (a) the degree of exposure to the threat; (b) the sector's sensitivity - the degree to which a system is affected (either adversely or beneficially); and (c) the capacity of the sector to cope with or adapt to the threat, to take advantage of or create opportunities, or to cope with the consequences (Smit & Wandel, 2006). We have adapted the original framework as discussed in the TAR (2001) and AR4 (2007), which has also been used by a number of other vulnerability assessments (Bell et al., 2011; Cinner et al., 2012, 2013; Blasiak et al., 2017).

We have built on the framework developed by Monnereau et al. (2017) with 35 indicators and have added another 11 indicators. This analysis thus involves a set of 46 indicators (5 for exposure, 18 for sensitivity and 23 for adaptive capacity). Exposure is defined as the degree of climate stress upon a particular unit of analysis; it may be represented as either long-term change in climate conditions, or by changes in climate variability, including the magnitude and frequency of extreme events (IPCC, 2001). Both slow-onset changes (e.g. sea surface temperatures, ocean acidification) and an increased number of extreme-weather events and intensity thereof are expected to impact fisheries worldwide. Sensitivity is usually defined as the

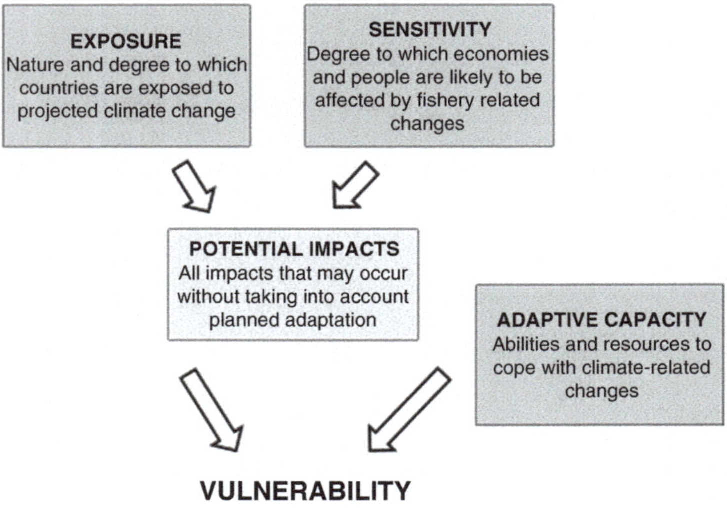

Fig. 1 Vulnerability framework. (Allison et al., 2009)

degree to which biophysical, social and economic conditions are likely to be influenced by extrinsic stresses or hazards due to climate change, including beneficial and harmful effects. Adaptive capacity relates to the capacity of a community or country to cope with, and adapt to, a variety of climate change impacts and is strongly influenced by several factors related to economic vulnerability, governance, education, and health. A number of indicators within exposure are based on projected data (e.g. sea level rise and maximum potential yield change in fisheries by 2050) while all the indicators for sensitivity and adaptive capacity are based on the current state of affairs (see Appendix A for a list of the 46 indicators).

3.2 Data Analysis

In this analysis we have accounted for potential redundancy among indicators within each vulnerability component (exposure, sensitivity, adaptive capacity) by means of Principal Component Analysis (PCA) on the ranked-transformed indicator data This data is derived from the sources found in Appendix A. A separate PCA was undertaken on each vulnerability component, allowing the identification of groups of redundant (correlated) indicators and ensuring an equal weighting of each of these groups within each vulnerability component. Each PCA was based on a correlation matrix and was followed by varimax rotation of the Principal Components (PCs) to help interpret indicator loadings. Only principal components (PC) corresponding to eigenvalues ≥ 1 were retained (Legendre and Legendre, 2011). To help interpret each PC, only indicators with relatively high loadings (≥ 0.6) on that PC were considered. The country scores on the retained PCs of each vulnerability component (exposure, sensitivity, adaptive capacity) were extracted and rank-transformed. The rank-transformed country scores on the retained PCs were averaged across PCs within each vulnerability component to yield a single average score rank for each country for each vulnerability component. This implies that each retained PC contributed equally to the final country score rank for a given vulnerability component, even though such PCs differed in the amount of total variance (and number of high loading indicators) that they captured. In order to integrate vulnerability component scores into overall vulnerability, the final average score ranks for each country for each vulnerability component were averaged to yield a single final vulnerability score.

The PCAs identified in total 11 subcomponents across the three components: two for exposure, four for sensitivity, and five in adaptive capacity. These 11 subcomponents represented coherent and interpretable groups of indicators that explained relatively high amounts of variation in the data. For each component (exposure, sensitivity and adaptive capacity), we compared the country scores of the PCA subcomponents among the three SIDS groups. Boxplots were used to visually assess the median score and distribution of values within each vulnerability component (exposure, sensitivity, and adaptive capacity), and for the final vulnerability score across the three SIDS groups.

4 Results

4.1 Component Comparison of the Three SIDS Groups

The three components and final vulnerability rankings show differences among the three SIDS groups. The results suggest that Caribbean SIDS are more vulnerable in terms of exposure, with the Pacific and AIMS SIDS having lower levels of exposure (see Fig. 2). The values for sensitivity for all three SIDS groups are more similar with Pacific SIDS being the most vulnerable in sensitivity. For adaptive capacity, Pacific and AIMS SIDS are the most vulnerable with Caribbean SIDS the least vulnerable. As a result of these differences across the three categories per SIDS group for the final vulnerability rankings, Caribbean SIDS are the most vulnerable based on median rank, followed very closely by Pacific SIDS, while AIMS SIDS were the least vulnerable overall.

4.2 Subcomponents of the Three Categories

Each component (exposure, sensitivity, and adaptive capacity) has several subcomponents that contribute to variance in the data. We used the first two main PCA

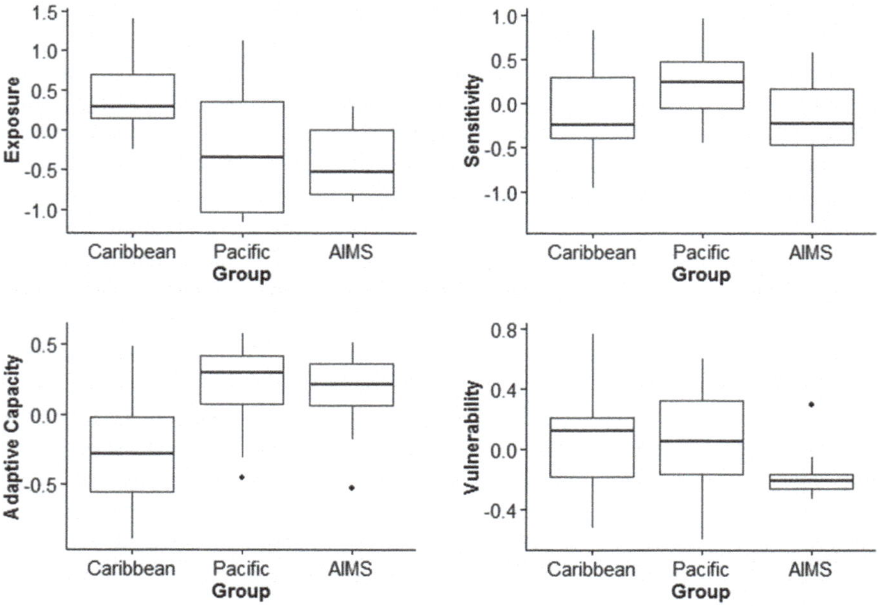

Fig. 2 Scores per component and final vulnerability of the three SIDS groups. (Based on analysis of authors) (Sources in Appendix A)

subcomponents of each component to look at the scoring of the different country groups, which explained 59% of variation in exposure, 44% of variation in sensitivity, and 40% of variation in adaptive capacity (Table 1).

For exposure, dimension one represents high values for sea level rise but low scores for future thermal stress, while dimension two reflects low scores for ocean acidification and high scores for sea surface temperature change. Variability in sensitivity indicators was driven mainly by coastal vulnerability, coastal health and fisheries dependency. Variation in adaptive capacity scores is strongly influenced by the first dimension of socio-economic vulnerability (31%) with remaining dimensions all equal at 9%. Dimension one in adaptive capacity relates to socio-economic indicators such as GDP per capita, education, governance and health. The second dimension in adaptive capacity relates to the level of fisheries management and magnitude of MPAs in the EEZ of a country. In our results, except for fisheries management and MPAs, other fisheries sector related indicators (e.g. fisher folk as percentage of Economic Active Population, fish nutrition, fish exports, and fish catch/per population) did not load strongly onto the first two components in these PCAs (Table 4).

Scoring for the three SIDS groups across the first two dimensions of exposure shows distinct patterns. Caribbean SIDS score highly on both subcomponents of exposure (sea level rise and thermal stress (E1), and sea surface temperature and ocean acidification (E2)) and are relatively clustered, while Pacific and AIMS SIDS show lower and more diverse scores (Fig. 3).

In respect of sensitivity, all three groups show high variability on both the first two dimensions, suggesting there are no clear patterns in coastal vulnerability or coastal health among the three SIDS groups.

In respect of adaptive capacity, Caribbean SIDS show comparatively low scores on both dimension 1 and dimension 2. Pacific SIDS show high variability on both dimensions. AIMS SIDS show high variability although less variability for PC1 but tend towards higher scores on dimension 2 (Figs. 4 and 5).

5 Discussion and Conclusion

As climate change is expected to impede the ability of nations to alleviate poverty and achieve sustainable development, adaptation to climate change in SIDS has been highlighted as a high priority. Given the scarcity of funds currently available for adaptation, the UNFCCC Secretariat has suggested that prioritization among eligible countries should be based on their vulnerability to climate change (Hinkel, 2011; Klein, 2009). Article 4.4 of the UNFCCC calls on the developed country parties to "assist the developing country parties that are particularly vulnerable to the adverse effects of climate change in meeting costs of adaptation" (United Nations, 1992). This has triggered a significant amount of research that assesses the vulnerability of different regions, countries, sectors, communities and groups of people (Hinkel, 2011).

Table 4 Scoring of indicators on the first two subcomponents of exposure, sensitivity and adaptive capacity

Exposure	*Dimension 1: Sea level rise and thermal stress (30%)*	*Dimension 2: sea surface temperature and ocean acidification (29%)*
Sea level rise	0.81	
Future thermal stress	−0.78	
Sea surface temperature change		0.82
Ocean acidification		−0.76
Sensitivity	*Dimension 1: coastal vulnerability (23%)*	*Dimension 2: coastal health (21%)*
% land 10 km coastline	0.91	
% pop 10 km coastline	0.91	
% cities in coastal zone	0.67	
% land 5 m above sea level	0.65	
Habitat status		0.91
Biodiversity status		0.87
Natural coastal protection		0.82
Coral reef health		0.65
Adaptive Capacity	*Dimension 1: Socio-economic vulnerability (31%)*	*Dimension 2: Fisheries management and MPAs (9%)*
GDP per capita	0.90	
Agriculture % GDP	0.90	
Governance	0.83	
Infant mortality	0.80	
Healthy life expectancy	0.77	
Night Light Development Index	0.77	
Sanitation	0.77	
Fisheries resilience	0.68	
Fisheries management		0.77
MPA (% of EEZ)		0.73

Note: only factors > 0.6 are shown

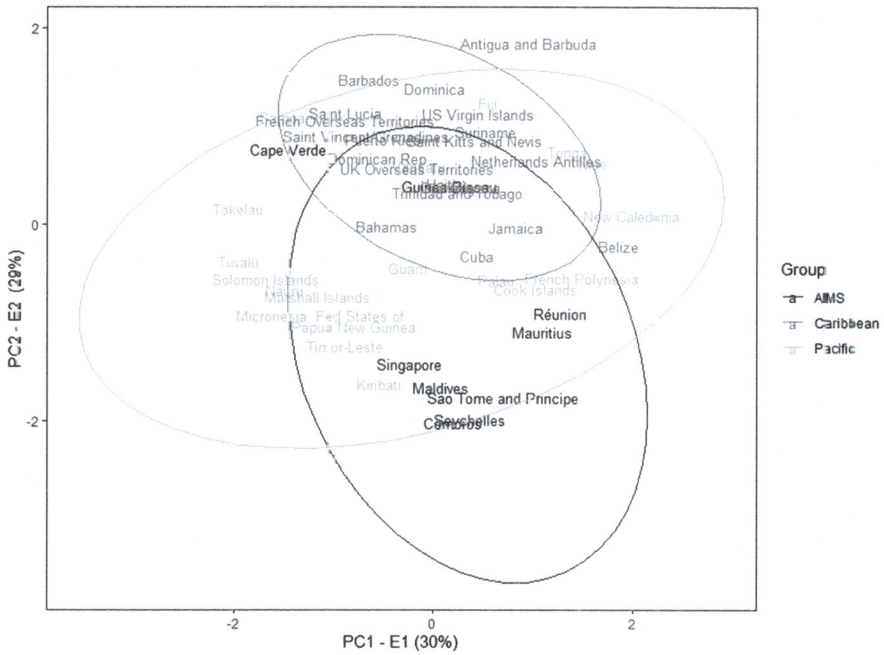

Fig. 3 Plot of the three SIDS groups along first two principal components for exposure. (E1 = sea level rise and thermal stress, E2 = sea surface temperature and ocean acidification. Ellipses show 95% confidence interval)

In this paper we examine the vulnerability of the fisheries sector to climate change in the three SIDS groups- Caribbean, Pacific, and AIMS. This is the first study as far as we are aware to carry out a rigorous in-depth statistical quantitative comparison of vulnerability scoring of the fisheries sector between the three SIDS groups. The comparison of vulnerability of the fisheries sector in the three SIDS groups of this study showed that single vulnerability scores are not adequate in comparing countries' vulnerability, and an examination of the underlying subcomponents of their vulnerability can highlight important differences with implications for policy. The Caribbean SIDS vulnerability scores were on average marginally higher than those of Pacific SIDS yet both groups had variable scores. However, on closer examination of the ranking of country groups on vulnerability components or subcomponents, we see larger differences. Examining these underlying subcomponents of vulnerability provides a richer explanation of their vulnerability and is thus considered to be valuable in identifying barriers to climate change adaptation needed for successful adaptation.

When we take a closer look at the differences between the three components, we find contrasting results for exposure and adaptive capacity. The final result is the outcome of Caribbean SIDS having the highest median vulnerability score overall, driven mainly by high scores on exposure, while being less vulnerable in adaptive

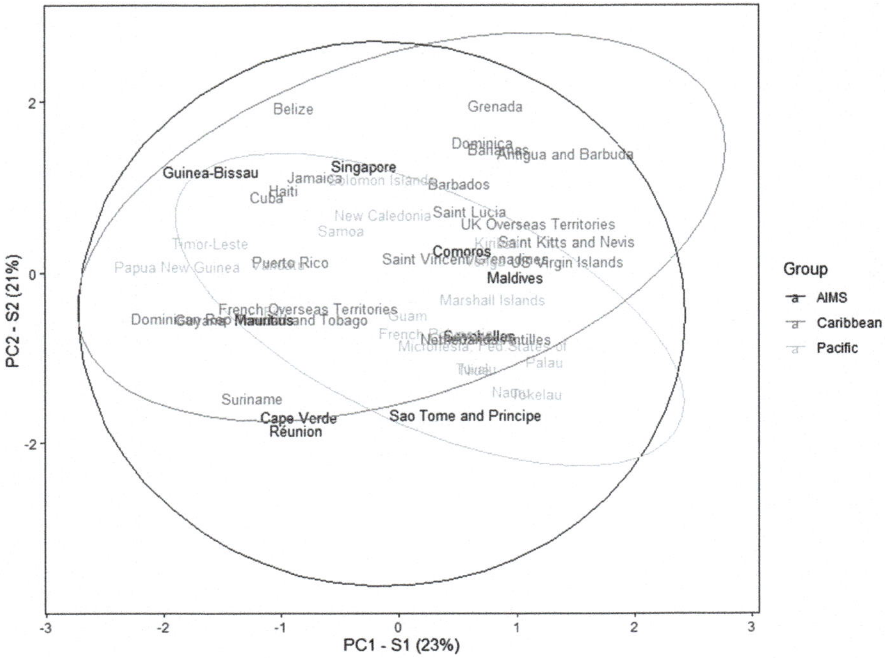

Fig. 4 Plot of the three SIDS groups along first two principal components for sensitivity. *(S1 = coastal vulnerability, S2 = coastal health)*

capacity. Pacific SIDS were also highly vulnerable in the final results, as a result of high scores in the components of sensitivity and adaptive capacity. AIMS SIDS had the lowest median ranking in the final vulnerability analysis, yet were highly vulnerable in sensitivity and adaptive capacity. This speaks to the different nature of the underlying vulnerability separating the country groups. Our results indicate that all three SIDS groups considered in the analysis are highly vulnerable in the sensitivity component (including coastal vulnerability, coastal health fisheries dependency and habitat health). These results support previous conclusions about SIDS with respect to coastal vulnerability and high fisheries sector dependency (Mimura et al., 2007; Nurse et al., 2014).

The sub-components indicate that Caribbean SIDS are particularly vulnerable to both sea level rise and sea surface temperature change and for sensitivity for coastal vulnerability and coastal health. AIMS SIDS are least vulnerable to sea surface temperature changes. In respect of sensitivity, the three country groups do not show a distinct pattern for the two PCAs which is in line with the findings for the overall component of sensitivity. In relation to adaptive capacity, the results show that Pacific and AIMS SIDS are particularly vulnerable in terms of socio-economic development as well as fisheries management. They should therefore focus on improving their capacity to diminish the impacts of these issues.

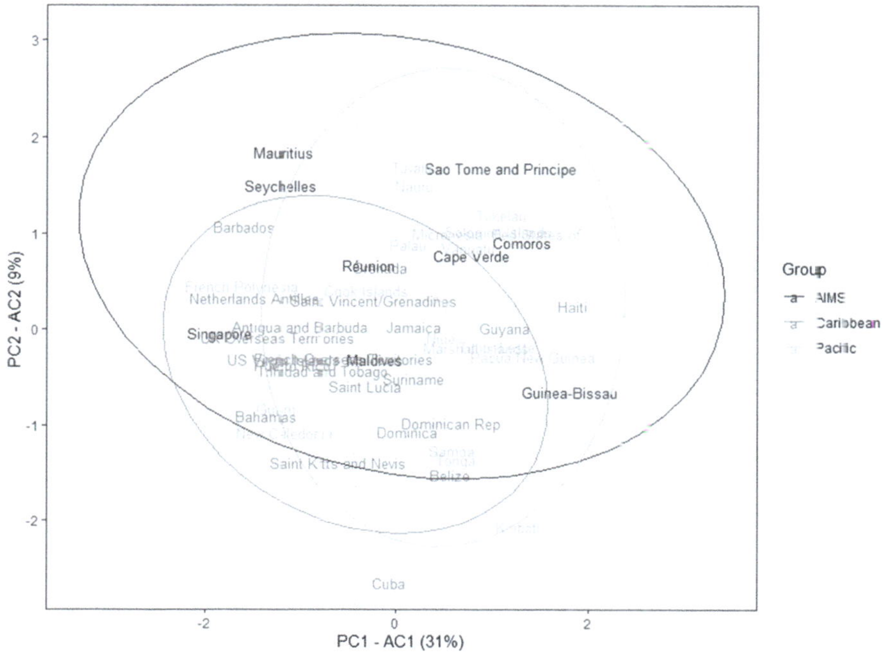

Fig. 5 Plot of the three SIDS groups along first two principal components for adaptive capacity. (AC1 = socio-economic vulnerability, AC2 = fisheries management and MPAs)

This analysis compares the vulnerability outcome between the three SIDS groups, yet we need to keep in mind that all scores are relative to those of other SIDS. Thus, low scores here may still represent high vulnerability in comparison to other country groups.

Our results indicate that the use of a single vulnerability value masks much of the information on the underlying (sub)components of vulnerability of the fisheries sector. Using final vulnerability scores for allocation of funds—for example, for climate change adaptation—is therefore too crude. A more detailed examination of the areas of vulnerability per SIDS group is needed in order to assess the true areas of vulnerability of each SIDS group. There remains a high level of variability within each of the three groups, and there is a lot of overlap of the groups in the biplots, which suggests that though there may be some regional geographic influences, there is also a lot of variability at country level—this also has implications for policy.

Since the precise impacts of climate change and the large differences in country groups are not fully known or understood, we need to take a precautionary approach and recognize that adaptation strategies might differ among countries in the different regions. In SIDS, fisheries governance will only be improved by employing a range of strategies, including the application of ecosystem-based approaches, wide stakeholder engagement, institutional building of fisheries administrations and stricter application of international fishery agreements to accommodate and support climate change related activities (Cochrane et al., 2009; Daw et al., 2009; Heenan et al., 2015). This study has shown the Caribbean to be highly vulnerable in terms of exposure and sensitivity. Adaptation measures already being undertaken in the Caribbean region include: development of mobile apps for tracking and early warning; investment in improved infrastructure and facilities; improving resilience of fisher folk and aquaculturists; and mainstreaming climate change into fishery policy and planning (Oxenford & Monnereau, 2018). These measures can support coping with their vulnerability. We also stress the importance of enhancing emergency preparedness and response, and development of insurance and social safety schemes in the fisheries sectors for all SIDS, in order to enhance their adaptive capacity.

Critical for future adaptation strategies and allocation of funding is the need for downscaled information on vulnerability of the fisheries sector at the regional, national and local level. Notably, results also indicate a large variation in scores within the three groups of SIDS. It is therefore crucial to conduct further analysis of the differences that exist within the three SIDS groups, to highlight more precisely the nature of the differences within the groups.

Acknowledgements This research was carried out with financial support from the Caribbean Community Climate Change Centre (CCCCC), Belize, and the Centre for Resource Management and Environmental Studies (CERMES), University of the West Indies, Cave Hill Campus, Barbados. The CCCCC has made the funds available to the University of the West Indies, CERMES for the project titled "Assessing the Vulnerability of Caribbean Fisheries to Climate Change". The duration of the project is from 1 September 2012 to 31 August 2014.

Appendix A: Indicators Used for Analysis

#	Component	Indicator	Source of data
Exposure			
1	Exposure	Sea Surface Temperature observed 1985–2005	Halpern et al. (2012)
2	Exposure	Sea level Rise projections 2050	Hinkel et al. (2014)
3	Exposure	Ocean acidification 1870–2000	Halpern et al. (2012)
4	Exposure	Thermal Stress (sea surface temperature prediction impacts on coral reefs) by 2050	Burke et al. (2011)
5	Exposure	UV radiation observed 1996–2004	Ocean Health Index (NASA) (2012)
Sensitivity			
6	Sensitivity	% of population living on land below 5 m above sea level	CIESIN (2010)
7	Sensitivity	% of population 10 km from the coast	CIESIN (2010)
8	Sensitivity	% of land below 5 m above sea level	CIESIN (2010)
9	Sensitivity	% of land 10 km from the coast	CIESIN (2010)

(continued)

#	Component	Indicator	Source of data
10	Sensitivity	Cities in coastal zone	McGranahan et al. (CIESIN) (2007)
11	Sensitivity	Population largest city (%)	World Development Indicators (2009)
12	Sensitivity	Biodiversity	Ocean Health index (2013)
13	Sensitivity	Habitat	Ocean Health Index (2013)
14	Sensitivity	Species	Ocean Health Index (2013)
15	Sensitivity	coral reef health	Ocean Health Index (2013)
16	Sensitivity	Coastal development threat	Burke et al. (2011)
17	Sensitivity	natural coastal protection status	Ocean Health index (2013)
18	Sensitivity	Exploitation status of fished stock	Ocean Health Index (2013)
19	Sensitivity	Fisheries employment	Monnereau et al. (2015)
20	Sensitivity	Coastal livelihoods and economies	Ocean Health Index (2013)
21	Sensitivity	Fisheries exports	FAO (2009)
22	Sensitivity	Fish catch	FAO (2010)
23	Sensitivity	Fish nutrition	FAO (2005–2009)
Adaptive Capacity			
24	Adaptive capacity	Healthy life expectancy	United Nations Healthy Life Expectancy (2007)
25	Adaptive capacity	Health: sanitation	World Bank (2011a, b)
26	Adaptive capacity	Health: infant mortality rate	World Bank (2011a, b)
27	Adaptive capacity	Education	CIA Factbook (2000–2010)
28	Adaptive capacity	Woldwide Governance	World Bank (2011a, b)
29	Adaptive capacity	Fisheries management capacity	Mora et al. (via OHI) (2009)
30	Adaptive capacity	Fisheries management capacity: MPAs	Environment Performance Index (2012)
31	Adaptive capacity	EEZ by coastline	Coastline is Hinrichsen, EEZ seaaroundus.org
32	Adaptive capacity	Resilience Marine livelihood	Ocean Health Index (2013)
33	Adaptive capacity	Resilience Wildfish caught	Ocean Health Index (2013)
34	Adaptive capacity	Tourism direct (real and norm)	World Bank (2013a, b)
35	Adaptive capacity	Tourism total (real and norm)	World Bank (2013a, b)
36	Adaptive capacity	liner shipping connectivity	World Bank (2009–2013)
37	Adaptive capacity	Agriculture % GDP	World Bank (2008–2010)

(continued)

#	Component	Indicator	Source of data
38	Adaptive capacity	Oil imports as % total oil consumption	U.S. Energy Information Administration (2009–2010)
39	Adaptive capacity	FDI%GDP	World Bank (2007–2011)
40	Adaptive capacity	OAD%GDP	World Bank (2009–2011)
41	Adaptive capacity	Remittances	World Bank (2008–2011)
42	Adaptive capacity	Gross Domestic Product per capita	World Bank (2011a, b)
43	Adaptive capacity	Nigh Light Development Index	NOAA (2012)
44	Adaptive capacity	Terms of trade	UNCTAD (2010–2011)
45	Adaptive capacity	Concentration of exports	UNCTAD (2013a, b)
46	Adaptive capacity	Diversification of exports	UNCTAD (2013a, b)

References

Allison, E. H. (2011). *Aquaculture, fisheries, poverty and food security*, WorldFish Working Paper 2011-65, Penang, Malaysia.

Allison, E. H., Perry, A. L., Badjeck, M.-C., Neil Adger, W., Brown, K., Conway, D., & Dulvy, N. K. (2009). Vulnerability of national economies to the impacts of climate change on fisheries. *Fish and Fisheries, 10*(2), 173–196. https://doi.org/10.1111/j.1467-2979.2008.00310.x

Bell, J. D., Johnson, J. E., & Hobday, A. J. (Eds.). (2011). *Vulnerability of tropical pacific fisheries and aquaculture to climate change*. Secretariat of the Pacific Community.

Betzold, C. (2015). Adapting to climate change in small Island developing states. *Climatic Change, 133*(3), 481–489.

Blasiak, R., Spijkers, J., Tokunaga, K., Pittman, J., Yagi, N., & Österblom, H. (2017). 'Climate change and marine fisheries: Least developed countries top global index of vulnerability. *PLoS ONE, 12*(6), e0179632. https://doi.org/10.1371/journal.pone.0179632

Boto, I., & Biasca, R. (2012). *Small Island Economies : from Vulnerabilities to Opportunties*. Brussels Rural Development Briefing #27, Brussels, Belgium. [online] Available from: https://brusselsbriefings.files.wordpress.com/2012/10/reader-br-27-small-islandeconomies-vulnerabilities-and-opportunities-eng.pdf. Accessed 1 Dec 2013

Brander, K. (2010). Impacts of climate change on fisheries. *Journal of Marine Systems, 79*(3-4), 389–402.

Brander, K. M. (2007). 'Global fish production and climate change. *Science, 104*(50), 44–46.

Briguglio, L. (1995). Small island developing states and their economic vulnerabilities. *World Development, 23*(9), 1615–1632. https://doi.org/10.1016/0305-750X(95)00065-K

Briguglio, L. (2003). *The vulnerability index and small Island developing states*. University of Malta.

Brooks, N., Adger, N., & Kelly, P. (2005). The determinants of vulnerability and adaptive capacity at the national level and the implications for adaptation. *Global Environmental Change, 15*(2), 151–163.

Burke, L., Reytar, K., Spalding, M., & Perry, A. (2011). *Reefs at risk revisited*. Washington, D.C. Available: http://pdf.wri.org/reefs_at_risk_revisited.pdf.

Cheung, W. L., Lam, V. W. Y., Sarmiento, J. L., Kearney, K., Watson, R., Zeller, D., & Pauly, D. (2010). Large-scale redistribution of maximum fisheries catch potential in the global ocean under climate change. *Global Change Biology, 16*(1), 24–35.

CIA Factbook. (2000–2010). https://www.cia.gov/library/publications/the-world-factbook/

CIESIN. (2010). http://sedac.ciesin.columbia.edu/data/collection/gpw-v3

Cinner, J., McClanahan, T., Graham, N., Daw, T., Maina, J., Stead, S., & Bodin, Ö. (2012). Vulnerability of coastal communities to key impacts of climate change on coral reef fisheries. *Global Environmental Change, 22*(1), 12–20.

Cinner, J. E., Huchery, C., Darling, E. S., Humphries, A. T., Graham, N. A., Hicks, C. C., & McClanahan, T. R. (2013). Evaluating social and ecological vulnerability of coral reef fisheries to climate change. *PloS One, 8*(9), e74321. https://doi.org/10.1371/journal.pone.0074321

Cochrane, K., De Young, C., Soto, D., & Bahri, T. (eds) (2009). *Climate change implications for fisheries and aquaculture: overview of current scientific knowledge*. FAO Fisheries and Aquaculture Technical Paper No. 530, Rome, pp. 212.

Crowards, T. (2002). Defining the category of small states. *Journal of International Development, 14*(2), 143–179. https://doi.org/10.1002/jid.860

Daw, T., Adger, W. N., & Brown, K. (2009). Climate change and capture fisheries: Potential impacts, adaptation and mitigation. In K. Cochrane, C. De Young, D. Soto, & T. Bahri (Eds.), *Climate change implications for fisheries and aquaculture: overview of current scientific knowledge. Fisheries and Aquaculture Technical paper 530* (pp. 107–153). Food and Agriculture Organization.

Environment Performance Index. (2012). http://epi.yale.edu/

FAO. (2005–2009). *Fish nutrition*. http://www.fao.org/fishery/statistics/en

FAO. (2009). *Fisheries exports*. http://www.fao.org/fishery/statistics/en

FAO. (2010). *Fish catch data*. http://www.fao.org/fishery/statistics/en

Guillaumont, P. (2009). An economic vulnerability index: Its design and use for international development policy. *Oxford Development Studies, 37*(3), 193–228. https://doi.org/10.1080/13600810903089901

Guillotreau, P., Campling, L., & Robinson, J. (2012). Vulnerability of small island fishery economies to climate and institutional changes. *Current Opinion in Environmental Sustainability, 4* (3), 287–291. https://doi.org/10.1016/j.cosust.2012.06.003

Halpern, B., Longo, C., Hardy, D., et al. (2012). An index to assess the health and benefits of the global ocean. *Nature, 488*, 615–620.

Heenan, A., Pomeroy, R., Bell, J., et al. (2015). A climate-informed, ecosystem approach to fisheries management. *Marine Policy, 57*, 182–192.

Hinkel, J. (2011). Indicators of vulnerability and adaptive capacity: Towards a clarification of the science–policy interface. *Global Environmental Change, 21*(1), 198–208.

Hinkel, J., Linckea, D., Vafeidis, A., et al. (2014). Coastal flood damage and adaptation costs under 21st century sea-level rise. *Proceedings of the National Academy of Sciences of the United States of America, 11*(9), 3292–3297.

Intergovernmental Panel on Climate Change (IPCC). (2001). *Climate Change 2001: Impacts, adaptation, and vulnerability, Contribution of Working Group II to the Third Assessment Report of the Intergovernmental Panel on Climate Change, Vol.* Cambridge University Press.

Khazai, B., Kunz-Plapp, T., Büscher, C., & Wegner, A. (2014). VuWiki: An ontology-based semantic wiki for vulnerability assessments. *International Journal of Disaster Risk Science, 5* (1), 55–73.

Klein, R. J. (2009). Identifying countries that are particularly vulnerable to the adverse effects of climate change: An academic or a political challenge? *Carbon & Climate Law Review, 3*(3), 284–291.

Lisowska, B. (2016). *Small island developing states: a case study of standards in defining supranational regions and groupings*, Discussion Paper No. 3, Development Initiatives. [online] Available from: http://juds.joinedupdata.org/wp-content/uploads/2016/04/DiscussionPaperNo3-1.pdf. Access Date Missing.

Mahon, R. (2002). *Adaptation of fisheries and fishing communities to the impacts of climate change in the CARICOM Region. Mainstreaming Adaptation to Climate Change (MACC) of the Caribbean Center for Climate Change (CCCC)* (p. 33). Organization of American States.

McGranahan, G., Balk, D., & Anderson, B. (2007). The rising tide: Assessing the risks of climate change and human settlements in low elevation coastal zones. *Environment and Urbanization, 19*(1), 17–37. https://doi.org/10.1177/0956247807076960

Mimura, N., Nurse, L., McLean, R. F., Agard, J., Briguglio, L. P., & Lefale, P. (2007). Small islands. In M. L. Parry, O. F. Canziani, J. P. Palutikof, P. J. van der Linden, & C. E. Hanson (Eds.), *Climate Change 2007: Impacts, adaptation and vulnerability. Contribution of Working Group II to the Fourth Assessment Report of the Intergovernmental Panel on Climate Change* (pp. 687–716). Cambridge University Press.

Monnereau, I., Mahon, R., McConney, P., & Nurse, L. (2015). *Vulnerability of the fisheries sector to climate change impacts in Small Island Developing States and the Wider Caribbean: early findings*. Barbados. CERMES Technical Report No. 56, 45 pp.

Monnereau, I., Mahon, R., McConney, P., Nurse, L., Turner, R., & Vallès, H. (2017). The impact of methodological choices on the outcome of national-level climate change vulnerability assessments: An example from the global fisheries sector. *Fish and Fisheries, 18*(4), 717–731.

Mora, C. (2013). The projected timing of climate departure from recent variability. *Nature, 502* (7470), 183–187. https://doi.org/10.1038/nature12540

Mora, C., Myers, R., Coll, M., Libralato, S., Pitcher, T., Sumaila, R., Zeller, D., Watson, R., Gaston, K., & Worm, B. (2009). Management effectiveness of the world's marine fisheries. *PLoS Biology, 7*(6), e1000131. https://doi.org/10.1371/journal.pbio.1000131

NOAA. (2012). *Night light development index*. https://ngdc.noaa.gov/eog/dmsp/download_nldi.html

Nurse, L. (2011). The implications of global climate change for fisheries management in the Caribbean. *Climate and Development, 3*(3), 228–241.

Nurse, L. A., McLean, R. F., Agard, J., Briguglio, L. P., Duvat-Magnan, V., & Pelesikoti, N. (2014). Small islands. In V. R. Barros, C. B. Field, D. J. Dokken, M. D. Mastrandrea, K. J. Mach, T. E. Bilir, M. Chatterjee, K. L. Ebi, Y. O. Estrada, R. C. Genova, B. Girma, E. S. Kissel, A. N. Levy, S. MacCracken, P. R. Mastrandrea, & L. L. White (Eds.), *Climate Change 2014: Impacts, Adaptation, and Vulnerability. Part B: Regional Aspects. Contribution of Working Group II to the Fifth Assessment Report of the Intergovernmental Panel on Climate Change* (pp. 1613–1654). Cambridge University Press.

Ocean Health Index. (2012). http://www.oceanhealthindex.org/

Ocean Health Index. (2013). http://www.oceanhealthindex.org/

Oculi, N., & Stephenson, S. (2018). Conceptualizing climate vulnerability: Understanding the negotiating strategies of Small Island Developing States. *Environmental Science and Policy, 85*, 72–80.

Oremus, K. L., Bone, J., Costello, C., Garcia Molinos, J., Lee, A., Magnin, T., & Salzman, J. (2020). Governance challenges for tropical nations losing fish species due to climate change. *Nature Sustainability, 3*, 277–280. https://doi.org/10.1038/s41893-020-0476-y

Oxenford, H. A., & Monnereau, I. (2018). Chapter 9: Climate change impacts, vulnerabilities and adaptations: Western Central Atlantic marine fisheries. In M. Barange, T. Bahri, M. Beveridge, K. Cochrane, S. Funge-Smith, & F. Poulain (Eds.), *Impacts of Climate Change on fisheries and aquaculture: Synthesis of current knowledge, adaptation and mitigation options. FAO Fisheries Technical Paper 627* (pp. 147–168). Food and Agriculture Organisation.

Polido, A., João, E., & Ramos, T. B. (2014). Sustainability approaches and strategic environmental assessment in small islands: An integrative review. *Ocean & Coastal Management, 96*, 138–148. https://doi.org/10.1016/j.ocecoaman.2014.05.005

Pörtner, H. O., Karl, D. M., Boyd, P. W., Cheung, W. W. L., Lluch-Cota, S. E., & Nojiri, Y. (2014). Ocean systems. In C. B. Field, V. R. Barros, D. J. Dokken, K. J. Mach, M. D. Mastrandrea, T. E. Bilir, M. Chatterjee, K. L. Ebi, Y. O. Estrada, R. C. Genova, B. Girma, E. S. Kissel, A. N. Levy, S. MacCracken, P. R. Mastrandrea, & L. L. White (Eds.), *Climate Change 2014: Impacts, adaptation, and vulnerability. Part A: Global and sectoral aspects. Contribution of Working Group II to the Fifth Assessment Report of the Intergovernmental Panel on Climate Change* (pp. 411–484). Cambridge University Press.

Smit, B., & Wandel, J. (2006). Adaptation, adaptive capacity and vulnerability. *Global Environmental Change, 16*(3), 282–292.

Thomas, A., Baptiste, A., Martyr-Koller, R., Pringle, P., & Rhiney, K. (2020). Climate change and small island developing states. *Annual Review of Environment and Resources, 45*(6), 1–6.27.

Tschakert, P., van Oort, B., St. Clair, A. L., & LaMadrid, A. (2013). Inequality and transformation analyses: A complementary lens for addressing vulnerability to climate change. *Climate and Development, 5*(4), 340–350.

UNCTAD. (2010–2011). *Terms of trade*. https://unctadstat.unctad.org/EN/

UNCTAD. (2013a). *Concentration of exports*. https://unctadstat.unctad.org/EN/

UNCTAD. (2013b). *Diversification of exports*. https://unctadstat.unctad.org/EN/

United Nations (UN). (1992). *United Nations framework convention on climate change*. FCCC/INFORMAL/84 GE.05-62220 (E) 200705. 24 pp. Retrieved August 14, 2013, from https://unfccc.int/resource/docs/convkp/conveng.pdf

United Nations Healthy Life Expectancy. (2007). https://www.un.org/esa/sustdev/natlinfo/indicators/methodology_sheets/health/life_expectancy.pdf

United Nations Office of the High Representative for the Least Developed Countries, Landlocked Developing Countries and Small Island Developing States (UN-OHRLLS), United Nations Educational, Scientific and Cultural Organization (UNESCO) & United Nations Division for Ocean Affairs and the Law of the Sea (UN-DOALOS). (2014). Report of the Expert Group Meeting on the Significance of Marine Science and Technology for SIDS: the Transfer of Marine Technology Transfer to SIDS to Support Sustainable Development, May 14–17, New York. [online] Available from: http://www.unohrlls.org/UserFiles/File/OHRLLS_EGM_report_recommendations_fin%20(1).pdf. Accessed 27 Jan 2021.

U.S. Energy Information Administration. (2009–2010). *Oil imports as % total oil consumption*. https://www.eia.gov/petroleum/imports/companylevel/

Von Tigerstrom, B. (2005). Small Island developing states and International Trade: Special challenges in the global partnership for development. *Melbourne Journal of International Law, 6*(2), 402–436.

World Bank. (2007–2011). *Foreign direct investment, net inflows (% of GDP)*. https://data.worldbank.org/indicator/BX.KLT.DINV.WD.GD.ZS

World Bank. (2008–2010). *Agriculture % GDP*. https://data.worldbank.org/indicator/NV.AGR.TOTL.ZS

World Bank. (2008–2011). *Migration and remittances data*. https://www.worldbank.org/en/topic/migrationremittancesdiasporaissues/brief/migration-remittances-data

World Bank. (2009–2013). *Liner shipping connectivity*. https://data.worldbank.org/indicator/IS.SHP.GCNW.XQ

World Bank. (2009–2011). *OAD/GDP*. https://data.worldbank.org/indicator/DT.ODA.ODAT.PC.ZS?view=chart

World Bank. (2009). *World Development Indicators 2009*. World Bank. https://openknowledge.worldbank.org/handle/10986/4367

World Bank. (2011a). *GDP per capita*. http://data.worldbank.org/indicator/NY.GDP.PCAP.CD

World Bank. (2011b). *Worlwide governance indicators*. https://databank.worldbank.org/source/worldwide-governance-indicators

World Bank. (2013a). *Tourism direct (real and norm)*. https://tcdata360.worldbank.org/indicators/tot.direct.gdp?country=BRA&indicator=24650&viz=line_chart&years=1995,2028

World Bank. (2013b). *Tourism total (real and norm)*. https://tcdata360.worldbank.org/indicators/tot.direct.gdp?country=BRA&indicator=24650&viz=line_chart&years=1995,2028

Part III
Places

Vulnerabilities to Climate Change and Enhancing Resilience in Caribbean Small Island Developing States: A Spatial Planning Framework

Michelle Mycoo

Abstract The inherent vulnerabilities of small island development states are exacerbated by weak implementation of spatial development policies and limited capacity to enforce against poor development practices. This chapter explores why the vulnerabilities of Caribbean small island development states to climate change are worsened by the undermining of spatial planning and presents a conceptual framework for planning reform that offers possibilities for enhancing resilience to climate change. It examines the spatial planning experience of Trinidad and Tobago and St. Lucia.

Keywords Spatial planning · Climate change vulnerability · Climate change resilience · Small Island developing states · Caribbean · Housing · Urban areas

1 Introduction

In recent years, climate change has become the focus of many governments in the Small Island Developing States (SIDS) of the Caribbean, yet limited attention by policymakers, decision-makers, including politicians, is given to the important role of spatial planning in tempering the vulnerabilities associated with climate change and its impact. In the English-speaking Caribbean, politicians and the public often ignore spatial zoning plans and site development standards. The political directorate frequently overturns the Planning Agency's decisions on major projects which inadvertently compromises ecosystems deemed critical to reducing vulnerability to climate change. It is often mooted in the Caribbean that spatial planning has lost its appeal as an approach to promoting orderly physical development and environmental management. However, Caribbean SIDS and in particular their coastal cities will be among the earliest and most affected by climate change in the coming decades

M. Mycoo (✉)
The Department of Geomatics Engineering and Land Management, The University of the West Indies, St. Augustine, Trinidad and Tobago
e-mail: Michelle.Mycoo@sta.uwi.edu

S. Moncada et al. (eds.), *Small Island Developing States*, The World of Small States 9, https://doi.org/10.1007/978-3-030-82774-8_12

(Mycoo & Donovan, 2017). The Habitat III Quito Declaration on a New Urban Agenda (2016) noted that "urban centres worldwide, especially in developing countries, often have characteristics that make them and their inhabitants especially vulnerable to the adverse impacts of climate change and other natural and human-made hazards…which particularly affect coastal areas, delta regions and small island developing states…" (UNGA, 2016, p. 1). Climate change presents a timely opportunity for spatial planning to reassert itself and to occupy centre stage in a concerted attempt to achieve climate change mitigation and adaptation along with creating safe, sustainable and resilient settlements in the Caribbean which is regarded as one of the world's most vulnerable region's to climate change.

Sea level rise (SLR) is perhaps one of the greatest challenges facing SIDS. By 2100, the cost of rising sea levels as a percentage of the gross domestic product (GDP) will be highest among SIDS and enormous relative to the size of their economies (Mycoo & Donovan, 2017). The average rate of SLR is estimated to be 1 millimetre in the Caribbean (UN-OHRLLS, 2008) and the annual projected change in sea level between two time periods nearly one hundred years apart (1986–2005 and 2081–2100) is 0.5 to 0.6 m in the Caribbean (Nurse et al., 2014). More recent estimates show a trend in the relative SLR for the coastal zones in the Caribbean East as 3.5 (±3.22) millimetres per year and the Caribbean Sea 2.0 (±3.14) millimetres (Rietbroek et al., 2016).

Many Caribbean states raised concerns at the Conference of Parties (COP) in 2015 that a 2 °C temperature rise was unacceptable as a safe threshold for the protection of small island states and that even a 1.5 °C increase would undermine the survival of communities (Tschakert, 2015). If the 1.5 °C warming level is surpassed, natural ecosystems will be further endangered and water and food security will be threatened by rising temperatures and drought. Climate model projections for the Caribbean of 1–4 °C showed drier conditions in the southern Caribbean with a strong tendency to drying in the traditional wet season from June to October as well as lengthening seasonal dry periods and increasing frequency of drought (Nurse et al., 2014).

In many Caribbean SIDS, vulnerability to climate change and natural hazards is made worse by anthropogenic drivers such as ill-informed land use practices. In the Caribbean, more than 50% of the population lives within 1.5 km of the shore and major infrastructure such as airports, roads and capital cities are spatially concentrated in the coastal zone (Mimura et al., 2007) where climate change is projected to have very significant impacts. In some cities, population resides in Low-Elevation Coastal Zones (LECZs) located less than ten metres above the sea level. Threats to public health and safety are occurring and economic impacts include GDP loss, inflation, increasing debt, revenue loss, and employment decline. These losses and damage will be high relative to the small size of Caribbean economies.

The capacity of human resources in Caribbean SIDS to respond to environmental degradation, economic vulnerability and climate change is limited and many these resources are stretched in critical fields including governance, spatial planning, economics as well as science and technology. A barometer of the brain drain is that they experience an average of 50% loss of skills or as much as 75% (UNEP,

2015). Stronger institutional capacity is paramount if spatial planning is to become more impactful in addressing climate change challenges.

It is especially timely to devise a more realistic planning model to help confront climate change and disaster related challenges which lie ahead. The devastating hurricanes which inflicted severe damage on some Caribbean SIDS in 2018 and the increasing frequency of inland and coastal flooding and droughts call for reflection among Caribbean policymakers and decision-makers on how to prepare for and adapt to climate change. The central role of spatial planning in strengthening adaptation to climate change and accompanying natural hazards is underscored by the United Nations (UN) Sustainable Development Goal (SDG) 11 which promotes the creation of safe, sustainable and resilient human settlements, the UN Habitat New Urban Agenda and the Conference of Paris (COP21) where Caribbean SIDS raised alarm about their survival if there was a temperature warming beyond 1.5 °C.

This chapter has a two-fold objective. One objective is to investigate the exacerbation of pre-existing vulnerabilities in selected Caribbean SIDS arising from weak spatial planning governance. The other objective is to present a conceptual framework for spatial planning reform which offers possibilities for enhancing resilience to climate change. It distils evidence of spatial planning irregularities from the English-speaking Caribbean case study countries of Trinidad and Tobago and St. Lucia.

The historical experience of English-speaking Caribbean SIDS in embarking on spatial planning is first examined. This is followed by an investigation into why spatial plans, policies and standards have not been implemented and the repercussions this has had on the case study countries' vulnerability to climate change. A conceptual framework, which articulates how resiliency to climate change may be enhanced through spatial planning, is then put forward. The final section of the chapter presents the main conclusion on mapping a way forward for spatial planning to influence climate change resilience among Caribbean SIDS.

1.1 History of Spatial Planning in the English-Speaking Caribbean

In the immediate post-independence era, considerable optimism prevailed among former British colonies, including Caribbean SIDS, that spatial planning was poised to play a transformative role in the development process of former British colonies. States anticipated the pursuit of physical development along with the simultaneous promotion of economic stimulation and sound environmental resource management (Mycoo, 2016). Among many ex-British colonies there was an expectation that spatial planning would achieve more orderly land development, deliver basic services efficiently, and promote well-planned, safe settlements. These newly independent countries inherited the British Town and Country Planning Act of 1947, which empowered the Minister to guide spatial development. However, McAuslan (1986) argued that political interference derailed the practice of planning in the Caribbean in the post-colonial era. By the turn of the twenty-first century, Rakodi (2001) posed the question "Forget planning, put politics first"?

In the early 1990s, Farvacque and McAuslan (1992) probed the main reasons for the inability of spatial planning to achieve the desired aspiration of developing countries to attain physical development which enhanced economic development while fulfilling social welfare concerns, the creation of thriving, liveable human settlements and safeguarding sensitive ecosystems. Their study found six common grievances among developing countries regarding planning deficiencies which were relevant to the English-speaking Caribbean SIDS: (1) most regulations were based on out dated and inappropriate planning legislation or urban planning codes that were inherited from colonial times; (2) master plans did not reflect the society's urgent needs because of the protracted plan preparation process which resulted in the plans becoming obsolete by the time they were completed; (3) the preparation and enforcement of spatial plans often based on obsolete planning ordinances that placed all planning powers and responsibilities with the central government; (4) institutional fragmentation; (5) lack of coordination with financial planning; and (6) delays in getting development approvals which in some countries took from 15 to many more steps, with an average time frame of two to seven years. Since this analysis of the state of planning in developing countries spatial planning systems in the Caribbean have evolved in a limited way.

The English-speaking Caribbean is one of the best examples of where spatial planning systems in the post-colonial era continue to be dysfunctional. Urban development in some countries has diverged from national spatial strategies and policies. In St. Lucia and Grenada the Prime Minister has exerted tremendous power over Development Control Authorities (Pugh, 2006). This feature of the legislation governing the authority of the Prime Minister has frustrated the planning process, the public and the planners (Pugh, 2006) since in many territories Ministers of Planning cannot be held liable for decisions relating to planning. In Barbados, the Minister responsible for Town and Country Planning, who is the Premier, may seek the advice of an Advisory Planning Committee in respect of development. The Committee cannot make decisions that are enforceable by law, but its advice is used by the Prime Minister to justify key development decisions and he/she is exempt from being liable for planning decisions (Pugh, 2006).

In many English-speaking Caribbean SIDS, weak regulation of land use planning and the highly politicized nature of the planning process are major drivers of ecosystem damage and increased vulnerability to natural hazards such as hurricanes and floods. Furthermore, the ambiguities in the existing plans allow politicians to make decisions on development projects that attract foreign investors at all costs in an era of deregulation and economic liberalisation (Potter & Pugh, 2002). Projects often proceed with the approval of the political directorate in the absence of environmental impact assessments or strategic environmental assessments. All too often planning decisions are made without public participation at the plan making or implementation phase, drawing resistance from individuals and environmental non-governmental organisations. Examples of this were luxury resort development at the Pitons in St. Lucia and Canouan in the Grenadines, which were both met with vehement public protests (Ishmael, 1990; Mycoo, 2005). Ishmael (1990) notes that in St. Lucia, the planning process is characterized by a lack of information, public consultation and genuine community participation in the planning process (Ishmael, 1990). A blanket of silence shrouds approval granted for projects, especially if these are controversial (Ishmael, 1990). Most of the public only learns about these projects when land is being cleared for construction, by which time it is too late to debate the pros and cons of the particular project (Ishmael, 1990).

In the case of Trinidad and Tobago, although a national physical development plan was produced by 1978, the Prime Minister favoured economic planning over spatial planning. Projects seemed more fashionable and more designed to play a greater pivotal role in socio-economic development. Examples of these projects were those promoted by the Urban Development Corporation of Trinidad and Tobago (UDeCOTT). These had the support of the Cabinet, and the Planning Agency had no opportunity to comment on these projects by the time they were approved. Moreover, city plans commissioned by UDeCoTT such as the Central Port of Spain Plan (2007) and the Scarborough Plan (2011) were not made available for public consultation. Furthermore, although the spatial development plans promoted the containment of urban sprawl, the political directorate ignored these plans. Urban sprawl was confirmed by the loss of approximately 40% of Trinidad's dense forest cover (UNEP, 2015), satellite imagery data which revealed that 50% of original wetland was lost because of urban land reclamation projects and the conversion of 40% of agricultural lands for state-sponsored housing. More recent events also reveal political interference in spatial planning. In October 2018, major flooding resulted from a combination of state tolerance toward squatting on steep hillsides which resulted in a denudation of vegetated slopes and the construction of state housing on the foothills, although the Planning Agency had cautioned that the site was unsuited for housing due to its high susceptibility to flooding.

Apart from shortcomings in development planning there are also challenges in development control. The development control process in Trinidad remains lethargic and rigid despite recommendations for reform made by Amos (1988) thirty years ago and more recent reviews by various committees. McHardy (2002) confirmed that Jamaica also suffered from administrative shortcomings. The subdivision approval was lengthy, averaging 171 days to process which resulted in high development costs and encouraged approximately 75% of development to occur outside the formal approval process (McHardy, 2002). Amos (1988) and Pugh and Potter (2003) contended that protracted decision-making encouraged political intervention as evidenced by politicians who made discretionary or ad hoc development decisions that did not adhere to formal processes.

A lack of confidence in the spatial planning system, non-compliance with development standards and spatial plans, out dated site development standards and building codes, centralised decision-making on planning matters, institutional fragmentation and the lack of institutional capacity for monitoring and enforcement do not augur well for Caribbean SIDS in the face of slow onset climate change phenomena and fast on-set natural hazards. Kelman (2014) asks whether climate change's rhetoric is detracting from dealing with wider development problems in SIDS. Others researchers ask if focusing upon climate change is deflecting attention "away from underlying political conditions of vulnerability towards the nature of the physical hazard itself?" (Kelman, 2014, p. 120; Pugh, 2017). It is argued in this chapter that politics has undermined the critical role of spatial planning and in so doing has heightened vulnerability and undermined resilience to climate change. The following section seeks to investigate how the role of spatial planning in Caribbean SIDS was diminished over time.

1.2 Spatial Planning Deficiencies and Climate Change Vulnerability in Case Study Countries

Politicians and the public have had a longstanding tension with planners in some English-speaking Caribbean countries. This section investigates the reasons spatial planning was ignored and how such action has increased vulnerability to climate change in the selected case studies.

1.2.1 Trinidad and Tobago: Spatial Development Planning

The National Physical Development Plan (NPDP) for Trinidad and Tobago was a statutory plan approved by Parliament in 1984, but the country's spatial development for the most part lacked conformity with the plan's proposals. There was consensus between Amos (1988) and professional planners who were employed in the public and/or private sectors between 1967 and 2014 that the plan was technically sound and therefore robust in its land use zoning proposals and spatial development strategies. However, its detailed analysis and proposals involved several years of preparation. For this reason, the plan was not reviewed every 5 years as stipulated in the planning legislation and so lacked relevance to the rapid socio-economic changes taking place in the 1980s and 1990s. Additionally, the Town and Country Planning Division and the Ministry of Finance overlooked resource requirements and mechanisms for plan implementation.

Empirical evidence confirms that weak plan implementation significantly increased the country's vulnerability to climate change impacts and compromised its resilience to these changes. A review of the National Physical Development Plan of 1984 undertaken in 2012 by consultants revealed that the proposed spatial strategy was unsuccessfully implemented. Four main findings pointed to failed plan implementation which intensified population vulnerability to climate change.

One major finding of the plan review exercise was the proliferation of squatter settlements throughout the country characterised by the landless poor settling in sensitive watersheds and coastal lands or on vulnerable sites already prone to flooding and landslides. Although the plan identified areas where settlement was permitted and where no settlement should occur in order to protect sensitive ecosystems, squatting encroached into these conservation zones. In the aftermath of the oil boom experienced by the country in the mid-1970s there was an influx of population from neighbouring Caribbean islands seeking employment opportunities. Inflated land values and housing prices resulted from the surge in the demand for land and housing from this migrant population. Neither economically disadvantaged local households nor in-coming poor immigrants could afford the purchase of state subsidised housing and land parcels in the main urban corridors. They therefore settled in marginal and ecologically sensitive areas which were invariably inadequately monitored state lands.

Another finding of the review of the national spatial development plan was that nearly half (46%) the lands deemed suitable for agricultural purposes was converted to housing settlements which in the present era of climate change has significant ramifications for the country's food security. A third issue was the high vulnerability of approved private sector housing as well as state subsidised housing settlements located in poorly drained coastal plains composed of impermeable clay. While state-built housing between 2000 and 2011 had some positive impact on reducing the housing shortage, the location of these projects was driven by the availability of state-owned land rather than through coordinated land use and transportation planning (Rajack & Frojmovic, 2016). This resulted in urban sprawl and negative environmental externalities. A fourth issue was the development of "dormitory" settlements as residents continued to work in the capital city of Port of Spain, while living in peri-urban and suburban areas of Greater Port of Spain. The primacy of the capital city was evident despite efforts to decentralise services through the creation of growth poles and new towns in the 1980s onwards. This combined with urban sprawl contributed to long commuting hours, traffic congestion, air pollution and eroded the opportunity to achieve low carbon urbanisation.

As elsewhere in the Caribbean, the development planning function was severely undermined by political expediencies, changes in government and economic vagaries in the post-independence era. Evidence of this included the non-review of spatial developments plans, the non-implementation of spatial strategies as articulated in the development plans, the substitution of plans with mega-projects, the growing trend toward ill-conceived development by the state, and unauthorised development by the private sector and Non-Governmental Organisations (NGOs) (Mycoo, 2015). Societal transformation occurred much faster than the instruments and practices of the orthodox planning system under which the planning agency operated. Additionally, politicians and the private sector held the view that planning should be rejected in favour of market forces which they perceived possessed the power to remedy planning related problems. Moreover, impatience permeated these stakeholders because of the long gestation of plans. Projects seemed more fashionable and set to play a greater pivotal role in socio-economic development. The country's planners began to question the lack of institutional capacity in policy and plan implementation.

A major constraint of the Planning Agency was staff shortages which resulted in delays in reviewing spatial plans. Vacancies in planning and technical posts were high because of an exodus of many professional planners to the private sector which offered far more attractive salaries. In 1989, 30% of the professional and technical posts were vacant compared to 2011 when it was almost double (57%).

In 2010, spatial development plans were prepared for fourteen regions in the country and a new national spatial development strategy was completed. Both these plans were undertaken by private consultants because of the lack of development planning capacity within the Planning Agency. In particular, the fourteen spatial development plans prepared almost a decade ago did not contain detailed climate change mitigation and adaptation strategies and policies. At that time, few planners were aware of the findings of the International Panel for Climate Change. As such, climate change was ignored as a major threat to sustainable development and was not fully embraced by planning professionals or the political directorate. Moreover, since then plan implementation has been slow and the few projects recommended in these plans to address climate change issues, foremost among them flooding, have been delayed because of the lack of finance and political will.

1.2.2 Trinidad and Tobago: Development Control

The level of non-compliance with development standards is very high according to available data and this translates into built development occurring outside the statutory planning framework which is designed to guide development to safe locations and to ensure public health and safety. "Between 1980 and 1990, 22% of new housing units were in full compliance with regulations and 65% was built "outside the formally sanctioned housing delivery system" (Pamuk & Dowall, 1998, pp. 292–295). This trend is compounded by the growth in informal urbanism without regard for planning regulations. An estimated 85,000 squatter households currently exists with over 396 squatter sites located on state lands. Approximately 55,000 households occupy public lands and an additional 30,000 households have encroached on private lands (Rajack & Frojmovic, 2016). Applying a conservative average household size of six persons in these settlements, it can be extrapolated that roughly 40% (510,000 persons) of the country's total population of 1.3 million persons is squatting.

The state was especially benevolent to squatting given its tolerance over the years where there were no forced evictions especially in the period of structural adjustment in the 1980s, which saw a tripling of squatting in only a decade (ECLAC 2000). State tolerance toward squatting continues to send a perverse incentive toward land grabbing in watersheds and coastal zones. Illegal land occupation in ecologically fragile zones is incentivised by the existing law of adverse possession and squatter regularisation policies (Toppin-Allahar, 2001). The Certificate of Comfort is another instrument used by the state to give squatters a sense of security on state lands. It prevents them from eviction despite their location in environmentally sensitive areas.

The inability of the formal housing sector to meet housing demand has significantly influenced the growth of informal settlements. Available evidence confirms that fewer affordable houses are being built for low-income households (Rajack &

Frojmovic, 2016). In 2014, the monthly median household income was about TT $10,500 (US$1600) and more than half of the demand for public housing came from low-income households earning less than TT$4000 (US$615) per month. These households cannot afford a mortgage and therefore resort to squatting or rent government housing. Interestingly, the 2011 census for Trinidad and Tobago revealed high home ownership, which may reflect that a very active informal sector is providing a large percentage of housing (McHardy & Donovan, 2016).

Inadequate governance such as delays in the planning approval process that exceeds the statutory eight-week period for town planning agencies to determine planning applications also contributes to illegal housing construction in both the formal and informal sectors (Mycoo, 2016). Furthermore, the process for obtaining construction permits is cumbersome, with many procedures and long processing times. It takes 16 procedures and 253 days to obtain all permits needed (IDB, 2017).

Non-compliance of these informal settlements with conventional planning standards means that planners miss the opportunity to ensure that built structures are safe for occupants and that they have access to basic water and sanitation services which are key to public health. Without planning guidance, squatters occupy sites which are vulnerable to sea level rise, flooding, storm surges and landslides associated with climate change. 'Uncontrolled urbanisation' is challenging for the Planning Agency which seeks to steer built development to optimal locations, at the right time and at the appropriate densities.

Weak monitoring and enforcement capacity of the Town and Country Planning Division have also resulted in the violation of site development standards by developers who build houses, commercial buildings and small factories. Site and building coverage and setback distances, which help minimise run-off and flooding, are often disregarded. Additionally, engineering standards designed to regulate land clearance, cutting of slopes or quarrying are violated, a consequence of which is flooding. On-site and off-site drainage standards are not adhered to if construction occurs outside the formal regulatory framework. Furthermore, informal land use activities do not have the benefit of environmental impact assessments which suggest ways of mitigating the impacts of climate change.

Monitoring and enforcement capacity are hampered by the lack of human resources, especially development control inspectors relative to the high level of non-compliance in the country's fourteen planning regions. Very limited use is made of data acquisition tools in development control. However, the big data revolution holds much potential to improve monitoring and enforcement capacity among professional planners and development control inspectors if training is provided in the use of earth observation data and drone technology.

A hillside development policy which restricted built development above the 91 metre contour line and on slopes in excess of 10 degrees was inconsistently implemented since the 1980s resulting in housing encroaching on steep areas which should have been left under natural vegetation. Although 60% of Trinidad's dense forest cover has survived urbanisation (UNEP, 2015), built development in steep sloping areas once under forest cover has triggered rapid surface water run-off. The need for housing has resulted in unprecedented land use changes, with the conversion of prime agricultural lands into housing areas. Urban sprawl has forced farmers to advance into unsuitable lands, even very steep hillsides, destroying vegetation cover and drastically changing hydrological systems. For example, a flooding episode in 2012 which occurred in Diego Martin a densely populated part of the metropolitan region was blamed by disaster officials on the denudation of hillsides by builders and "slash and burn" farming (Chadee & Sutherland, 2014). Recent studies confirmed small scale deforestation in watersheds significantly increases flood risks in Trinidad and progressive increase of flood damages starts below 85% forest cover (Brookhuis & Hein, 2016).

The island's coastal zone is more vulnerable to climate change as an estimated 50% of Trinidad's original wetland was destroyed by land reclamation. A significant percentage of wetland removal was undertaken to accommodate urban infrastructure, housing, hotels and port development. Much of the country's urban and industrial assets are vulnerable to sea level rise, storm surge and coastal flooding.

1.2.3 St Lucia: Spatial Development Planning

St. Lucia, like its other English-speaking Caribbean counterparts, inherited the British town-planning model and its legislation dating back to the Town and Country Planning Act of 1947, which gave the Minister the power to guide physical development. Additionally, in the decades following independence it continued to employ the master planning approach and oftentimes major developments occurred without reference to land use guidelines articulated in the physical development plans. The political directorate did not approve or explicitly endorse any of the plans prepared in the mid-1970s (Toppin-Allahar, 2001). Potter and Pugh (2002) found in St. Lucia this neo-liberal state was planning without plans. Furthermore, the ambiguities in the existing plans allowed politicians to make decisions that attracted foreign investors at all costs in an era of deregulation and economic liberalisation (Potter & Pugh, 2002).

The Town and Country Planning Ordinance of 1961 was replaced with the Physical Planning and Development Act of 2001 which gives power to a Development Control Authority composed of government representatives from various departments. However, decision-making power remains centralised and subject to the whims of the government. Political manipulation of the approval process is rife and economic expediencies may be to the detriment of environmental resources. Furthermore, St. Lucia is severely hampered by a shortage of trained, experienced town planners operating in the public sector. There is a gap in planning skills as most staff members working in planning unit are from allied professions such as architecture, engineering, surveying and geography.

1.2.4 St Lucia: Development Control

Over 50% of development has taken place without planning permission, occurs in areas prone to hazards and causes environmental degradation (King-Joseph & Lendor-Gabriel, 2013). Driven by poverty, squatters settle in highly sensitive watersheds and peri-urban areas, especially on steep hillsides and they often use poor construction practices and breach building codes (Prevatt et al., 2010). A rapid rate of legal or illegal expansion in the 1970s and 1980s led to significant loss of natural forests under state and private ownership (King-Joseph & Lendor-Gabriel, 2013). Although 80% of the country has slopes greater than 10 degrees, state-owned forest reserves cover only 12% of total land area, leaving the vast majority of steeply sloping lands unprotected from poor land use practices. Approximately 56% of natural forests are designated as forest reserves, which are located mainly in the central ridge of the island. An estimated 43% of forests are on private lands. Though forest reserves were designated in earlier spatial development plans for the protection of forest, soil, water and wildlife, squatters have converted 342 hectares of the reserve using deleterious agricultural practices and housing construction techniques.

Flooding in Castries is common during the wet season because unplanned human settlements have damaged watersheds and mangrove rendering them unable to protect the densely populated coastal plain. Deforestation impacts were made worse by Hurricane Tomas, which caused heavy siltation of the country's main water source, the Roseau Dam. The cost of repairing damage to the water supply, disposal and works subsector was approximately EC$124.47 million (US$46 million) which represented 32% of damage to the infrastructure sector and 14% of the total impact (UNECLAC, 2013).

Anthropogenic factors have also severely impacted on the island's coastal ecosystems. Changing human settlement patterns and direct impacts on shoreline processes present immediate erosion challenges in urban areas, including coastal zones where the population is concentrated. Furthermore, with the decline of the plantation and removal of preferential treatment for the banana industry, mass tourism has become St. Lucia's economic mainstay contributing 65% of the GDP (GOSL & World Bank, 2014).

Airport and marina construction associated with enhancing the tourism sector has resulted in land reclamation, mangrove depletion and increasing sedimentation in coral reefs and seagrass beds (Agard & Gobin, 2000). It is estimated that 50% of coastal wetlands have been lost falling from a total area of 320 to 193 hectares (GOSL, 2001). Many formerly unprotected mangrove sites were cleared to accommodate coastal development such as hotels and high-income housing in the Gros Islet region and offshore dredging to re-establish beaches removed by hurricanes (Agard & Gobin, 2000). Institutionally, most of the mangroves are legally designated protected areas. Nevertheless none of them are earmarked thus rendering management non-existent and further contributing to their decline.

Approximately 160 km^2 of coral reef exist around the island though an increase in coral disease, high sediment loading and the presence of macro algae have caused a rapid decline of coral cover. The reefs along the North West coast are among the only ones that have been accurately mapped. Data shows that approximately 82% of these reefs are either dead or in poor condition because of inadequate sewage treatment in areas where urban development occurred in coastal locations (GOSL, 2017). Much of the coral damage is attributed to increased levels of suspended particles generated by poor land use (GOSL, 2017). Additionally, high sediment loading from surface run off has led to the declining health of sea grass beds thereby placing additional stress on the coral reefs which act as buffer zones between the land and sea and reduce coastal erosion. With increasing storm surges associated with climate change the coral reefs cannot provide coastal lands with a first line of defence against stronger wave energy.

Although an approved robust policy is in place to address climate change, it has not received the wide cross-sectoral embrace necessary for meaningful advances in mitigation and adaptation efforts to effectively deal with the issues at the sector level, mainly because it is seen as the purview of the national climate change focal agency (GOSL, 2017). The government has strongly recommended that all national policies and plans should be reviewed to ensure that climate change mitigation and adaptation issues are factored into their frameworks and action plans.

Spatial development planning and development control in the two case studies revealed that a number of factors militate against the fulfilment of the main goals of spatial planning which are to safeguard public health and safety and promote natural resources management for on-going sustainable development. Past unsustainable anthropogenic activity has increased the vulnerability of the population, ecosystems and economies of these SIDS to climate change impacts. Improving resiliency to these impacts must be the next course of action. See Fig. 1 for summary of planning deficiencies.

2 A Conceptual Framework for Enhancing Climate Change Resilience in Caribbean SIDS

Resilience refers to a capacity to withstand shocks or stresses and to recover from the unexpected (Satterthwaithe & Dodman, 2013). Interdisciplinary researchers such as urban planners, geographers, ecologists, and environmental studies specialists embrace this definition of resilience (Olsson et al., 2015). Such researchers who study climate change adaptation and disaster risk reduction promote the social-ecological resilience paradigm. They state that resilience is the outcome of successful adaptation, which is a product of governments, enterprises, civil society organisations, households, and individuals with strong adaptive capacity (UN-Habitat, 2011).

Some researchers argue that "Resilience numbs the consciousness of policymakers and publics in SIDS into a bio-politics whose object is "excess adaptive capacity": no more, no less". Baldacchino (2018) and Pugh (2016) suggest

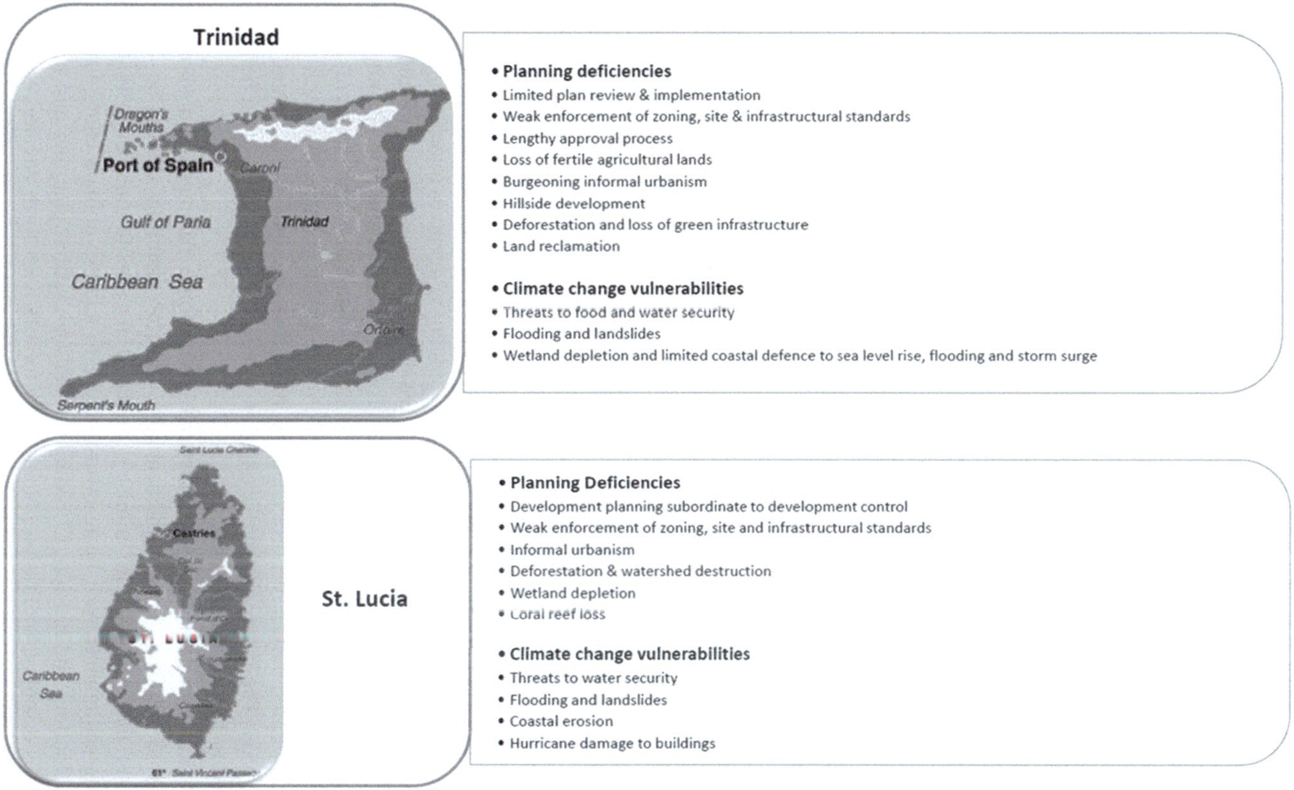

Fig. 1 Planning deficiencies in Trinidad and St. Lucia

that international interest in adaptation to climate change and resiliency building has diverted attention from chronic problems such as poverty, health, infrastructure provision, human resource capacity and international mobility. It becomes a struggle to service current endogenous concerns through exogenously determined projects driven by climate change (Baldacchino, 2018). However, "Resilience, as the flip side to vulnerability, is a welcome and optimistic take on SIDS and their ability to exploit opportunities" (Baldacchino, 2018). In practical terms resiliency is more than identifying and acting on specific climate change impacts. It usually requires a capacity to anticipate climate change and plan for adaptation in the interest of all citizens. Anticipation and planning are the forte of spatial planning. As was done earlier in the chapter, the underlying causes of vulnerability were first investigated. The proposed conceptual framework identifies components considered integral to enable spatial planning to enhance climate change resiliency in Caribbean SIDS.

2.1 Strengthening Spatial Plan Preparation and Implementation

The inordinate level of political interference in spatial planning has inadvertently increased the vulnerability of the population to climate change in the selected case studies. Bypassing the planning agency has been deemed necessary by the political directorate because of the lengthy delays in plan preparation and the determination of development applications which very often exceeds the statutory limit of eight weeks. Where this occurs in Caribbean SIDS, improving efficiency in plan review and preparation and the implementation of plans, policies and development guidelines, is central to minimising or eliminating political intervention which subverts the planning system to private and political party objectives.

Stronger effort must be made by the spatial planning agencies to achieve efficient plan review and plan making. This is necessary if Caribbean governments are to acquire greater confidence in the paramount role of spatial planning in attaining SDG 11 which focuses on the promotion of safer, sustainable and resilient human settlements especially in the context of climate change. Such initiatives should include increased human resource capacity, more rigorous data collection and robust data analysis. The latter will be discussed in more detail in a sub-section hereunder dedicated to new technologies.

Historical obstacles that have fostered institutional fragmentation should be removed to improve plan implementation. Spatial planning should be closely aligned to the infrastructure ministry to allow for institutional coordination in providing settlements with basic infrastructure to address climate change hazards such as flooding and erosion. Coordination between spatial planning and financial planning is essential; traditional planning has too often set forth development goals that have no bearing on their cost implications. Caribbean SIDS are replete with spatial plans which is the absence of financial resources failed and remained paper plans. Linking plan proposals to public sector budgets and private sector investments should be highly promoted.

Among the region's network of professional spatial planners there should be a systematic and science-driven debate on spatial policies which determine a more compact urban form through urban design and density standards, tames urban sprawl and promotes energy efficiency through transportation planning. The mainstreaming of low carbon urbanisation should be high on the spatial development debate.

2.2 *Implementing Development Standards for Climate Proofing*

Spatial planning agencies hold the key to ensuring that climate proofing is mainstreamed into policy-making and implementation. Planning authorities need to acknowledge that incorporation of risk reduction into the location and design of housing is of critical importance. As the Barbados Habitat III report indicates the ability of buildings to withstand high winds, flooding, and seismic activity is of paramount importance if the country is to maintain and expand its housing stock (SALISES, 2015). Housing quality should be improved in Caribbean SIDS, especially in older parts of the city where building stock has deteriorated and in areas where informal urban settlements have burgeoned.

The resilience of poor quality housing, often at risk from extreme weather, should be enhanced using structural retrofitting and interventions that reduce risks, such as flood-proofing houses or expanding drainage capacity to limit or remove flood risks. Housing and planning ministries should adopt new designs to increase the resilience of housing by: (1) ensuring that the floor levels of housing are above recorded flood levels; (2) improving standards for the foundations of housing to guarantee that structures can withstand dynamic water forces; and (3) altering roof designs to ensure adequate resistance in high winds (Mycoo & Donovan, 2017). Moreover, the future siting and design of housing should be informed by coastal hazard risk assessments and coastal setback requirements for new developments (Simpson et al., 2012) which form part of the retreat strategy for climate change adoption. Donor agencies have funded several projects in the Caribbean to build hurricane-resistant housing. The main lesson drawn from these projects is that strong enforcement capacity within planning agencies is necessary to ensure household compliance in adopting the building code for hurricane-resistant housing.

2.3 Adapting Water Supply Using Spatial Planning Standards

It is forecasted that the Caribbean will be the only insular region in the world to experience a decrease in water availability in the future. Droughts have already affected many Caribbean SIDS in the past decade, so much so that in many urban areas, households have had to purchase water. Stronger water management policies in SIDS should be implemented as a long term strategy to minimise water loss as suggested by participants in a survey conducted by Moncada and Bambrick (2019). More planning agencies in the region should impose regulations requiring the construction of cisterns in both residential and non-residential buildings to facilitate rainwater harvesting. Some countries in the region are promoting Domestic Rainwater Harvesting (DRWH), but finance appears to be a constraint for poor households (Peters, 2017). However, studies by Peters (2017) found households can self-finance DRWH systems if payments are based on 5% of household income and interest rates are less than 5% which suggests that DRWH can be scaled-using micro-financing. Additionally, as a planning condition of approval, commercial users such as tourism resorts with golf courses and landscaping should be required to recycle grey water as part of a broader water demand management approach. Moreover, spatial planners should play an important role in integrated water resources management. They should protect freshwater resources by ensuring that zoning standards are strictly enforced otherwise aquifers and rivers may be easily contaminated by land-based pollutants.

2.4 Promote Ecosystems-Based Approaches and Green Infrastructure

Many SIDS should mainstream Ecosystem-Based Approaches (EBAs) in their climate change adaptation plans since they have tremendous potential to minimise risks and vulnerabilities to natural hazards. EBAs, which form part of the accommodation strategy for climate change adaptation, have proven to be effective in coastal zones in Caribbean SIDS. For example, protecting and replanting mangroves is a cost-effective tool in safeguarding settlements from coastal hazards. EBA measures cost less to maintain than coastal engineering projects, are well suited to community involvement in implementation, are less harmful to the natural environment, promote environmental sustainability and generate livelihoods for communities. A study of SIDS carried by Moncada and Bambrick (2019) confirmed that coastal communities in Fiji recognised the importance of ecosystem services in providing food for daily consumption and storage in the aftermath of climate shocks.

Utilising green infrastructure may also help adaptation and build resilience to climate change. The term 'green infrastructure' refers to interventions to preserve the functionality of existing green landscapes (e.g., parks, forests, wetlands, and greenbelts) and to transform the built environment using phytoremediation and water management techniques and by introducing productive landscapes (Zhang et al., 2011). Many SIDS have embarked on watershed management, afforestation, and wetland conservation, including replanting mangroves, as part of their strategy to adapt to climate change. The Global Environment Facility of the United Nations Development Programme (UNDP-GEF), the United Nations Human Settlements Programme (UN-Habitat) and the World Bank have funded such projects in the capital cities of Jamaica, Cuba and Guyana respectively. Spatial planning should incorporate into plans and policies green infrastructure at the city level to influence the effectiveness of pervious surfaces used in storm water management, green/white/ blue roofs, wetlands used for flood protection, urban agriculture, and overall biomass production. Additionally, green infrastructure can be effective in mitigating the urban heat island effect found increasingly in Caribbean cities (Johnson, 2015) and can foster low carbon urbanisation.

At the site level, spatial planners should consistently invoke the use of the Tree Preservation Ordinance to mitigate flooding and landslides. In sub-division layouts they should insist that developers reserve green open spaces which may be gardens, parks, play lots and recreational grounds. At the national level, national parks and protected areas such as forests, rivers and wetlands should be conserved to provide ecosystem-provisioning services such as clean air, freshwater resources, flood reduction, biodiversity conservation and carbon sequestration.

2.5 *Managing Unauthorised Spatial Development*

Informal settlements have become widespread in Caribbean cities such as Kingston (Jamaica), Port of Spain (Trinidad), Georgetown (Guyana), Port-au-Prince (Haiti) and Santo Domingo (Dominican Republic). Approximately ten million people in The Bahamas, Barbados, Cuba, Dominican Republic, Guyana, Haiti, Jamaica, St. Lucia, Suriname and Trinidad and Tobago experience a housing deficit and building adequate housing for this population would cost about US$18 billion (Donovan & Turner-Jones, 2017). The Trinidad and St. Lucia case studies revealed that 75% and 50% of built development respectively were illegal and did not benefit from the guidance of zoning, site, building and engineering standards. Furthermore, almost 50% of Trinidad and Tobago's population occupy informal settlements. As previously mentioned, in Caribbean SIDS which have experienced illegal building activity, the human population and ecosystems are highly vulnerable to damage associated with climate change. Land tenure reform and innovative housing policies are instrumental in addressing squatting.

2.6 The Role of Market Mechanisms in Spatial Planning

A shift toward a more enabling approach with an emphasis on compliance is exhorted. Market mechanisms offer one approach to address low compliance with development regulations and weak enforcement capacity. These may include incentives or penalties designed to foster a bigger role of the market in encouraging efficiency by the planning agency and compliance among development applicants. Examples of incentives are subsidies, tax rebates and bonuses. Penalties consist of taxation and charges or fees. In most countries, the public generally complies when a reward system using economic incentives is used, but the reward must be large enough to make compliance worthwhile. Additionally, penalties must be very punitive to discourage non-compliance otherwise persons would contravene the law and pay the penalty.

Higher levels of compliance with environmental regulations, which is critical in the context of climate change challenges, may be achieved through post-construction monitoring of large and small developments that have significant impacts on sensitive ecosystems. The environmental management authority will require additional human resources to conduct post-construction monitoring and evaluation.

Informal development occurs on predominantly state lands in many Caribbean SIDS, but in many former colonies these lands were designated to protect forested areas including sensitive watersheds. However, state resources to monitor these vast lands are limited. The private sector, NGOs and community-based organizations should be paid to conduct surveillance on these lands, which not only play an important role in carbon sequestration, but also protect freshwater resources and mitigate landslides and downstream flooding. Performance incentives in the form of bonuses may be granted to employees of these agencies who make the highest number of reports on informal land occupation. It is critical, however, that the state guarantees the security and strict confidentiality of such persons who are mobilised to protect them from victimisation by persons engaged in illegal land development.

The imposition of fees to process applications should be used to reduce the backlog of development applications created by speculation among developers whose main objective is to "test the waters" of the planning agency and its reaction to a project idea.

Applying targets and offering incentives should also reinforce improved performance of planning professionals in the public sector. Performance should be assessed on the basis of the speed with which applications are determined and the quality of decisions.

2.7 Education, Empowerment and Behavioural Change

As Leal Filho and Hemstock (2019, p. v) note "The role of education is not only about helping people to understand the impacts of global warming. Rather, education on matters related to climate change is also about catalysing changes in attitudes and behaviour, which may encourage individuals to take a more active role in both climate change mitigation and adaptation efforts." Politicians, businesses, households, NGOs and faith-based organisations are important stakeholders who should be actively engaged in knowledge transfer and uptake which result in behavioural change. Education on the anthropogenic drivers of climate change and the adoption of a new outlook on finite land supply and fragile island ecosystems may help bring about a conversion among these actors to comply with the proposals contained in spatial plans and site development standards. In the information age, several tools exist to enhance stakeholder empowerment and improved decision-making regarding climate change. Education and communications experts should utilise such tools for knowledge transformation and behavioural change.

A multidisciplinary approach to climate change education is advocated because of the complexity and inter-linkages associated with climate change (Briguglio & Moncada, 2019). No single discipline has all the answers to climate change challenges facing SIDS. The success of a multidisciplinary approach to teaching climate change will depend, however, on how well teaching is organised and coordinated (Briguglio & Moncada, 2019).

Good governance should facilitate mediation of policies and decisions across different actors, spheres of influence, sources of information, and resources to co-produce knowledge and support learning and action over time. Furthermore, Beckford (2018) strongly argues that local and traditional knowledge in the Caribbean is an invaluable resource and cultural capital that can play a major role in addressing climate change concerns. Beckford (2018) suggests the establishment of a Caribbean Local and Traditional Knowledge Network as a mechanism to harness local traditional knowledge and integrate it in regional climate change agendas.

2.8 New Technologies

Caribbean SIDS have been hamstrung in their efforts to monitor and enforce spatial planning policies and development standards. However, the advent of new technologies holds much promise where human resource constraints exist. Additionally, they provide planners with the tools for more innovative and robust, evidence-based decision-making. Perhaps the biggest advantage is that the big data revolution may help planners persuade and educate politicians and policymakers on the impacts of climate change and facilitate their visualisation of climate change scenarios, which may enhance the mainstreaming of climate change mitigation and adaptation strategies into spatial plans and projects.

The big data revolution allows spatial planners to harvest data quickly and at minimal or no cost. In the Caribbean, particularly post-2017 hurricane devastation, the use of big data analytics is now often seen as key to dealing with increased and intensified hurricanes (Whyte, 2017). Remotely sensed imagery can be provided at no cost almost daily from earth observation data. It can identify roads and other infrastructure that are flooded to better direct rescue efforts. Quick, easy and centralised access to high-quality imagery helps to create maps that are useful to disaster management authorities. Landsat data, which is now freely available on the Internet, and cloud mapping also hold much potential to improve spatial planning Caribbean SIDS that have struggled with a paucity of data. Digital photography showing infringements such as squatting and building operations can enhance the surveillance and monitoring capacity of spatial planners. The availability of earth observation data provides planners with an opportunity to track land use changes over time and empowers them to conduct vulnerability assessments and prescribe 'no-development' zones in response to climate change and disaster risks challenges. The potential for mining data from new sources, for instance mobile phones and drones, is also high. A drone is a low-cost and easy to use technology which can be effectively applied for data capture, spatial analysis, synthesis and visualisation of climate change impacts and vulnerabilities. Such data can be shared with planning agencies.

Geotagged web content is promising for urban and environmental policy formulation. Data contribution can occur with mobile devices in situ, or uploaded later via a web browser. The advantage of this data source is that the high volume of contributors increases the reliability of the dataset (Goldberg et al., 2017). Many digital cameras including those available on current generations of smart phones have the ability to automatically geotag photographs with coordinates acquired via built GPS chips. This technology is valuable in assisting scientists map ecosystem services and in informing policy and guiding decision-making.

A geographic information system is a tool that many of the region's planners are trained to use in assessing the impact of development proposals on population and ecosystem vulnerability to climate change impacts. It has been increasingly used to produce flood and landslide susceptibility maps throughout the Caribbean which assist planners in land use zoning and applying building standards. The insurance industry is increasingly using such maps to calculate premiums for households and businesses.

New technologies which can create smart maps, 3-dimensional maps and facilitate participatory mapping are particularly useful for spatial planners in conducting public consultations on climate change impacts, adaptation and resiliency building.

2.9 Corporate Social Responsibility

With growing unsustainable land use and building practices among 'big business' throughout the English-speaking Caribbean, which impact on population and

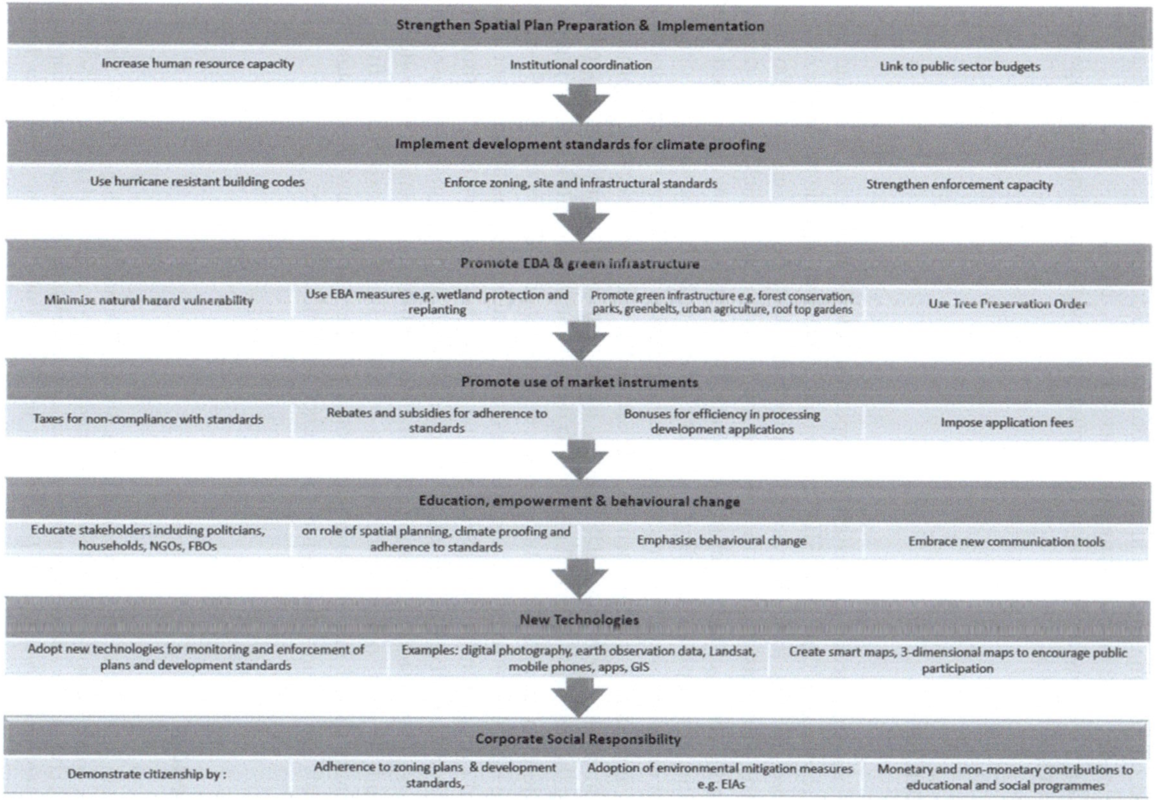

Fig. 2 Conceptual framework for climate change resiliency in Caribbean SIDS

environmental vulnerability to climate change, greater corporate social responsibility is needed. Companies can demonstrate citizenship by conforming to site development and building standards, embracing environmental mitigation measures and contributing to educational and social programmes. New reflections have also emerged on greening citizenship as the basis for sustainable development and on the role of the state in promoting this process (Scerri, 2013).

See Fig. 2 below illustrating a conceptual framework for climate change resiliency in Caribbean SIDS.

3 Conclusion

Spatial planning needs to reassert its important role in managing land and other natural resources if Caribbean SIDS are to be made less vulnerable to climate change and more resilient to its impacts. For many decades, politicians have favoured engineering projects because they deliver more visible and quick results deemed important given electoral cycles every 5 years. However, these have produced short-term gains in addressing flooding and other natural hazards. The time has come to focus on more sustainable efforts which spatial planning can deliver largely due to its holistic approach to development as opposed to the traditional sectoral approach to development in Caribbean SIDS. Today, more than ever, an integrated approach utilising the expertise of spatial planners to remedy the vulnerabilities of the region's population, economies and natural resources is imperative. Caribbean SIDS need to increase their efforts to meet these climate change challenges frontally and not allow electoral cycles or traditional developmental approaches to dictate their destiny. The Fijian *Talanoa* concept which refers to an inclusive, transparent dialogue based on a process of sharing stories, building empathy and reaching decisions for the collective good and, as such, relies on the pooling of ideas, skills and experience from all participants should inspire Caribbean spatial planners and renew their confidence and resolve to enhance the region's resilience to climate change.

References

Agard, J. B. R., & Gobin, J. F. (2000). The Lesser Antilles, Trinidad and Tobago. In C. R. Sheppard (Ed.), *Seas at the millennium: An environmental evaluation* (pp. 627–641). Elsevier Science.

Amos, F. (1988). *Role and administration of town and country planning. Report prepared for the Government of Trinidad and Tobago*. Port of Spain: United Nations Centre for Human Settlements

Baldacchino, G. (2018). Seizing history: Development and non-climate change in Small Island developing states. *International Journal of Climate Change Strategies and Management, 10*(2), 217–228. https://doi.org/10.1108/IJCCSM-02-2017-0037

Beckford, C. (2018). Climate change resiliency in Caribbean SIDS: Building greater synergies between science and local and traditional knowledge. *Journal of Environmental Studies and Sciences, 8*(1), 42–50.
Briguglio, L., & Moncada, S. (2019). The benefits and downsides of multidisciplinary education relating to climate change. In W. Leal Filho & S. L. Hemstock (Eds.), *Climate change and the role of education* (pp. 169–187). Springer. https://doi.org/10.1007/978-3-030-32898-6
Brookhuis, B., & Hein, L. (2016). The value of the flood control service of tropical forests: A case study for Trinidad. *Forest Policy and Economics, 62*, 118–124.
Chadee, D., & Sutherland, J. M. (2014). Introduction to flooding and climate change. In D. D. Chadee, J. M. Sutherland, & J. B. Agard (Eds.), *Flooding and climate change: Sectoral impacts and adaptation strategies for the caribbean region* (pp. 1–10). Nova Publishing.
Donovan, M. G., & Turner-Jones, T. (2017). Caribbean housing is expensive and scarce. Here's how to change that. *Americas Quarterly*, 20 April. [online]. Retrieved January 27, 2021, from www.americasquarterly.org/content/exit-caribbean-housing-trap
Farvacque, C., & McAuslan, P. (1992). *Reforming urban land policies and institutions in developing countries*. World Bank Publications.
Goldberg, G., D'Iorio, M., & McClintock, W. (2017). Applied marine management with volunteered geographic information. In D. Bartlett & L. Celliers (Eds.), *Geoinformatics for marine and coastal management* (pp. 269–298). Taylor and Francis.
Government of Saint Lucia (GOSL). (2001). *Integrating the management of watersheds and coastal areas in St. Lucia*. Government of St. Lucia.
Government of Saint Lucia (GOSL). (2017). *Third National Communication on Climate Change for St. Lucia*. Castries.
Government of Saint Lucia (GOSL) & World Bank. (2014). *Joint rapid damage and needs assessment: St. Lucia flood event of December 24-25, 2013*. Castries: Government of St. Lucia.
Inter-American Development Bank (IDB). (2017). *Comparative Project Evaluation of IDB Support to Low Income Housing Programs in Four Caribbean Countries*. [online]. Retrieved November 5, 2018, from https://publications.iadb.org/publications/english/document/Comparative-Project-Evaluation-of-IDB-Support-to-Low-income-Housing-Programs-in-Four-Caribbean-Countries.pdf
Ishmael, L. M. (1990). *Soufriere and the Pitons: The price of development*. Proceedings of the Urban Management Conference: Understanding the Economic and Environmental Linkages, USAID, Bridgetown, Barbados, 5–7 September 1990, pp. 112–125
Johnson, C. (2015). Green networks: A solution to the urban heat island effect. *Transactions on Ecology and The Environment, 194*, 205–214. https://doi.org/10.2495/SC150191
Kelman, I. (2014). No change from climate change: Vulnerability and small island developing states. *The Geographical Journal, 180*(2), 120–129. https://doi.org/10.1111/geoj.12019
King-Joseph, A., & Lendor-Gabriel, G. (2013). *Saint Lucia land policy issues paper*. Report prepared for the social and sustainable development division of the Organisation of Eastern Caribbean States (OECS). OECS, Morne Fortune, Castries, St. Lucia.
Leal Filho, W., & Hemstock, S. L. (2019). Climate change management: Preface. In W. Leal Filho & S. L. Hemstock (Eds.), *Climate change and the role of education* (pp. v–vi). Springer. https://doi.org/10.1007/978-3-030-32898-6
McAuslan, J. P. (1986). *Planning law in the Commonwealth Caribbean in the development of relevant principles and practice for town and country planning in the Caribbean*. Proceedings of the Second Convocation of the Caribbean Conference for Town and Country Planning, Port-of-Spain, Trinidad.
McHardy, P. (2002). *Urban and regional planning in Jamaica*. Minerva Press.
McHardy, P., & Donovan, M. G. (2016). *The state of social housing in six Caribbean*. Inter-American Development Bank
Mimura, N., Nurse, L., McLean, R. F., Agard, J., Briguglio, L. P., Lefale, P., Payet, R., & Sem, G. (2007). Small islands. In M. L. Parry, O. F. Canziani, J. P. Palutikof, P. J. van der Linden, & C. E. Hanson (Eds.), *Climate change 2007: Impacts, adaptation and vulnerability. Contribution*

of Working Group II to the fourth assessment report of the intergovernmental panel on climate change (pp. 687–716). Cambridge University Press.

Moncada, S., & Bambrick, H. (2019). Extreme weather events in Small Island Developing States: Barriers to climate change adaptation among coastal communities in a remote island of Fiji. In C. Klöck & M. Fink (Eds.), *Dealing with climate change on small islands: Towards effective and sustainable adaptation?* (pp. 217–247). Göttingen University Press. https://doi.org/10.17875/gup2019-1218

Mycoo, M. A. (2005). Minimising foreign control of land: A case study of St. Lucia. *Land Use Policy, 22*, 345–357.

Mycoo, M. A. (2016). Reforming spatial planning in Anglophone Caribbean countries. *Planning Theory and Practice, 18*(1), 1–20. https://doi.org/10.1080/14649357.2016.1241423

Mycoo, MA & Donovan, M.G. (2017), A Blue Urban Agenda: Adapting to climate change in coastal cities of Caribbean and Pacific Small Island Developing States, , .

Nurse, L. A., McLean, R. F., Agard, J., Briguglio, L. P., Duvat-Magnan, V., Pelesikoti, N., Tompkins, E., & Webb, A. (2014). Small islands. In V. R. Barros, C. B. Field, D. J. Dokken, M. D. Mastrandrea, K. J. Mach, T. E. Bilir, M. Chatterjee, K. L. Ebi, Y. O. Estrada, R. C. Genova, B. Girma, E. S. Kissel, A. N. Levy, S. MacCracken, P. R. Mastrandrea, & L. L. White (Eds.), *Climate Change 2014: Impacts, adaptation, and vulnerability. Part B: Regional aspects. Contribution of Working Group II to the Fifth Assessment Report of the Intergovernmental Panel on Climate Change* (pp. 1613–1654). Cambridge University Press.

Olsson, L., Jerneck, A., Thoren, H., Persson, J., & O'Byrne, D. (2015). Why resilience is unappealing to social science: Theoretical and empirical investigations of the scientific use of resilience. *Science Advances, 1*(4), 99–104.

Pamuk, E., & Dowall, D. D. (1998). The price of land for housing in Trinidad: Role of regulatory constraints and implications for affordability. *Urban Studies, 35*(2), 289–299.

Peters, E. J. (2017). Financing domestic rainwater harvesting in the Caribbean. *Journal of Sustainable Development, 10*(5), 107–115. https://doi.org/10.5539/jsd.v10n5p107

Potter, R. B., & Pugh, J. (2002). Planning without plans and the neo-liberal state: The case of St. Lucia, West Indies. *Third World Planning Review, 23*, 323–340.

Prevatt, D., Dupigny-Giroux, L., & Masters, F. (2010). Engineering perspectives on reducing hurricane damage to housing in CARICOM Caribbean Islands. *Natural Hazards Review, 11*, 140–150. https://doi.org/10.1061/(ASCE)NH.1527-6996.0000017

Pugh, J. (2006). Physical development planning in the anglophone Caribbean: The re-articulation of formal state power. In J. Pugh & J. H. Momsen (Eds.), *Environmental planning in the Caribbean* (pp. 7–20). Ashgate Publishing.

Pugh, J. (2016). The relational turn in island geographies: Bringing together Island, sea and ship relations and the case of the Landship. *Social and Cultural Geography, 17*(8), 1040–1059.

Pugh, J. (2017). Postcolonial development, (non)sovereignty and affect: Living on in the wake of Caribbean political independence. *Antipode, 49*(4), 867–882. https://doi.org/10.1111/anti.12305

Pugh, J., & Potter, R. B. (2003). *Participatory planning in the Caribbean. Lessons from practice*. Ashgate Publications.

Rajack, R., & Frojmovic, M. (2016). *Housing and land markets in Trinidad and Tobago: Inputs to the IDB's country challenges document for Trinidad and Tobago*. Washington, DC.

Rakodi, C. (2001). Forget planning, put politics first? Priorities for urban management in developing countries. *Journal of Applied Earth Observation and Geoinformation, 3*, 209–223.

Rietbroek, R., Brunnabend, S., Kusche, J., Schroter, J., & Dahle, C. (2016). Revisiting the contemporary sea-level budget on global and regional scales. *Proceedings of the National Academy of Sciences of the United States of America, 113*(6), 1504–1509.

Satterthwaithe, D., & Dodman, D. (2013). Editorial: Towards resilience and transformation for cities within a finite planet. *Environment and Urbanisation, 25*(2), 291–298.

Scerri, A. (2013). Green citizenship and the political critique of injustice. *Citizenship Studies, 17*(3), 293–307.

Simpson, M. C., Mercer Clarke, C. S. L., Clarke, J. D., Scott, D., & Clarke, A. J. (2012). *Coastal setbacks in Latin America and the Caribbean: A study of emerging issues and trends that inform guidelines for coastal planning and development*, (Technical Note No. IDB - TN – 476). Inter-American Development Bank.

Sir Arthur Lewis Institute for Social and Economic Studies and Centre for Environmental Resources Management and Environmental Studies (SALISES). (2015). *Barbados Habitat III Report*. SALISES.

Toppin-Allahar, C. (2001). A comparative analysis of environmental assessment law and planning practice in the Commonwealth Caribbean. *Caribbean Law Review, 11*(1).

Tschakert, P. (2015). 1.5°C or 2°C: A conduit's view from the science-policy interface at COP20 in Lima, Peru. *Climate Change Responses, 2*(1). https://doi.org/10.1186/s40665-015-0010-z

ECLAC (UN Economic Commission for Latin America and the Caribbean). (2000). *Poverty and social integration in the Caribbean*. United Nations.

United Nations Economic Commission For Latin America And Caribbean (UNECLAC). (2013). *Caribbean Synthesis Report, Third International Conference on Small Island Developing States*. UNECLAC.

United Nations Environment Programme (UNEP). (2015). *Success stories in mainstreaming ecosystem services into macro-economic policy and land use planning: Evidence from Chile, Trinidad and Tobago, South Africa and Viet Nam.* Nairobi: UNEP.

United Nations General Assembly (UNGA). (2016). *New Urban Agenda*, Resolution adopted by the General Assembly on 23 December 2016, A/RES/71/256. [online]. Retrieved March 26, 2018, from http://www.un.org/en/development/desa/population/migration/generalassembly/docs/globalcompact/A_RES_71_256.pdf

United Nations Human Settlements Programme (UN-Habitat). (2011). *Addressing Urban Issues in National Climate Change Policies* (Cities and Climate Change Initiative Policy Note. 3)

United Nations Office of the High Representative for the Least Developed Countries and Small Island Developing States (UN-OHRLLS). (2008). *The impact of climate change on the development prospects of the least developed countries and small Island developing states*. United Nations.

Whyte, A. V. (2017). Preparing for the next Hurricane Irma, or Harvey How analytics and smart cities will help. *Huffington Post*. 7 November. [online]. Retrieved January 27, 2021, from https://www.huffingtonpost.com/entry/preparing-for-the-next-hurricane-irma-orharvey-how_us_5a01cb19e4b085d72ae06d2b

Zhang, J., Sui, Y. H., & Geng, X. B. (2011). Landscape design of urban green space adaptive to global climate change: A review. *Advanced Materials Research, 243*, 6842–6845.

Human Mobility and Disasters in Pacific and Caribbean Small Island Developing States

Lilian Yamamoto and Miguel Esteban

Abstract Human mobility in the context of disasters has been steering the international community to develop initiatives and consensus with regards to the protection of the affected people. Small Island Developing States (SIDS) have been heavily involved in this discussion. This article will examine how SIDS governments in the Pacific and Caribbean have been implementing measures to deal with human displacement in their norms and policies related to disaster. To do so, the authors will examine human mobility in the context of disasters in the regional agendas of Pacific and Caribbean SIDS, and will then look at disaster displacement policies. While the authors do not believe that islands will disappear in the future, looking at human mobility and building of skills as adaptation strategies can be considered 'no-regret' strategies, which can help improve the long-term economies and resilience of populations at risk. For this to happen, it would undoubtedly be of benefit that surrounding countries establish policies that permit those displaced by episodic disasters to remain in their soil.

Keywords Climate change · Pacific · Caribbean · Human mobility · Displacement · Small Islands Developing States · Islands · Migration · Natural hazards · Sea level rise · Disaster risk

1 Introduction

This chapter intends to analyse to what extent Caribbean and Pacific Small Island Development States (SIDS) integrate human mobility—in the context of natural hazards—into climate change adaptation, which will be done through a literature review of papers on this topic, including a non-exhaustive review of migration.

L. Yamamoto (✉)
South American Network for Environmental Migrations (RESAMA), Sao Paulo, Brazil

M. Esteban
Research Institute of Sustainable Future Society, Waseda University, Tokyo, Japan

S. Moncada et al. (eds.), *Small Island Developing States*, The World of Small States 9, https://doi.org/10.1007/978-3-030-82774-8_13

national adaptation plans, and other regional instruments. The scope covers different kinds of mobility: voluntary and involuntary, economic and non-economic, temporary and permanent, and domestic, regional and international. In order to understand how different types of SIDS are addressing this issue, a comparative assessment between Pacific and Caribbean islands policies regarding human mobility in the context of disasters will be carried out. To do so, the authors will start by outlining what types of natural hazards may cause displacement, followed by a review of the global and regional agenda on disaster displacement. Finally, some examples of human mobility policies will be provided for both of the geographical areas targeted in the present study.

The concept of human mobility includes not only forced displacement, but also voluntary migration. This term is very useful, since for the case of environmental degradation it is difficult to distinguish between migration and displacement (IDMC & NRC, 2017). Natural hazards can have severe impacts on the communities of SIDS, as they can displace people from their homes. A possible way to mitigate the potential consequences of these hazards and increase resilience would be to relocate the populations that live in areas prone to suffer from them. In this context, disaster risk reduction plays an essential role to prevent authorities from being overwhelmed with potentially having to deal with large-scale future displacements.

One of the most prominent examples of potential future human mobility in the context of climate change and sea level rise regards the case of the so-called 'disappearing atoll islands', such as Tuvalu, Kiribati and Maldives. The perils facing these islands have been widely documented by the world's media, and their eventual disappearance has in many cases been described as inevitable (Methmann & Oels, 2015), to an extent that even normalises their loss (Barnett, 2017). Both these narratives—the inevitability and the normalisation—are not supported by detailed evidence and deserve to be challenged. Even though this discussion has drawn a large amount of attention from global fora, academia and mass media, the issue of human mobility in SIDS has not yet found a consolidated strategy from the government of SIDS. Instead, the onerous predictions of the potential consequences of climate change have alarmed governments and academia, and helped highlight that there is a lack of a global instrument to protect those that could suffer from climate change displacement.

Essentially, Pacific SIDS that are mainly or exclusively composed of atolls are low-lying and depend on coral reefs and foraminifera to compensate for Sea Level Rise (SLR). If the worst predictions regarding the effects of climate change come true, many or all their inhabitants could be deprived of land under their feet and be forced to relocate to other islands or countries. However, it is worth emphasising that at this stage it is doubtful that atoll-based SIDS will indeed disappear, given the considerable resilience coral islands display both from a morphological and human adaptation point of view (Yamamoto & Esteban, 2014; Esteban et al., 2019; Narayan et al., 2020).

Up to now, it is clear that SLR and climate change have had little effect on these islands, and McLean and Kench (2015) found (in a study which analysed the physical changes in over 200 islands on 12 atolls in the central and western Pacific during the past decades) that there is little evidence on whether any islands have reduced in size. Jamero et al. (2016), looking at small coral islands in the Philippines that experienced instantaneous relative sea level rise (due to subduction during an earthquake), found that the local communities were mostly being able to adapt to the changing conditions. Nevertheless, there are some limits to how far adaptation strategies can go (Crichton & Esteban, 2018). Although these islands might not disappear, the impoverishment that could be brought about by the death of the coral reefs and the cost of adapting, together with a potential increase in tropical cyclone intensity (as highlighted in the Intergovernmental Panel on Climate Change 5th Assessment Report, IPCC AR5, 2014) could eventually drive many people to leave their homes.

Conversely, Caribbean SIDS have not been the target of many studies on human mobility in the context of climate change and natural hazards, as they are typically not made up of atolls and low-lying, and therefore, the eventual complete loss of their territory is not a concern. In a scenario in which sea levels rise by 1 metre (m), only the Bahamas is likely to be significantly impacted, with between 5 and 7% of its total population likely to be affected (Dasgupta et al., 2007). This comes from a fundamental geological-morphological difference, as the Bahamas is made of low-lying limestone islands, while Tuvalu, Kiribati and the Marshall Islands are atolls (Connell, 2013). Among the 44 member states of Association of Small Island States (AOSIS, which also has five observer members), only three have a maximum elevation above sea level of under 10 m, and they are all located in the Indian and Pacific Ocean, i.e., Maldives (altitude max. 2 m), Tuvalu (5 m) and Marshall Islands (6 m). Nauru (65 m) and Kiribati (81 m) have altitudes lower than 100 m. Bahamas is the only Caribbean archipelago which is considered to be low-lying (Piguet, 2012), although it has a high point which is 64 m above sea level, though it is extremely unlikely that it sea level rise would make it disappear. Hence, when it comes to the discussions on the threatened Island States, Maldives, Tuvalu, Marshall Islands, Nauru and Kiribati are the most cited cases.

However, sea level rise and recurring storms can still severely affect Caribbean SIDS, affecting coastal areas and significantly eroding beaches. This can have severe economic impacts given that many of the islands rely on tourism as one of their main sources of income. In the course of the last decade, St. Vincent, for example, lost about 18 m to 30 m of beach (World Bank, 2014), with many other beaches in the Caribbean affected (EU, 2017). Several sea ports, such as Castries in St Lucia, St Georges in Grenada and Kingstown in St. Vincent are threatened by flooding due to sea level rise and coastal erosion (World Bank, 2014, the problems faced by Caribbean ports were also the subject of the United Nations Convention on Trade and Development (UNCTAD) workshops organised in St Lucia and Jamaica in May 2017.

Bahamas counted 2800 people as being displaced in 2015 by hurricane Joaquin (IDMC & NRC, 2016). In 2016, hurricane Matthew prompted small-scale evacuations in St Vincent and the Grenadines, St Lucia, Barbados, the Turks and Caicos Islands and Jamaica. The Bahamas were also hit as and around 5000 people evacuated to safer places. In Cuba, the hurricane forced the evacuation of over one million people. However, it was Haiti which suffered the worst impacts, and around 500,000 were displaced (IDMC & NRC, 2017). However, it is worth noting that it is not only the hazards that can be exacerbated by climate change which can lead to movements of people, and earthquakes such as the one hit Haiti in 2010 can lead to massive displacement, and the country was also badly affected by hurricane Matthew in October 2016 (Jones & Holpuch, 2016). These sudden-onset processes may cause the destruction of houses and businesses, damage critical infrastructure such as water supplies, transportation networks and electrical utilities, resulting in high numbers of casualties and out-migration from the affected area (McLeman, 2014).

The relevance of climate-induced displacement has been recognised by the IPCC AR5 (2014) in its chapter 12, entitled "Human Security", which indicated that extreme weather events provide the most direct pathway from climate change to migration. Nevertheless, there is scarce empirical evidence that climate change has so far been the main cause of migration in SIDS, taking into consideration that other drivers also cause people to migrate. Studies from the Pacific, for example, have demonstrated that culture, lifestyle, and a connection to place are more significant drivers of migration (Barnett & Webber, 2010), and even when islands are flooded (and not necessarily due to climate change) their inhabitants will not necessarily migrate or migrate far (Ballu et al., 2011; Jamero et al., 2017).

Nevertheless, potential for sudden-onset processes to bring about massive displacement is clear, particularly as a result of high magnitude floods, tsunamis and tropical cyclones (amongst other types of hazards). In 2015, Cyclone Pam displaced around a quarter of Vanuatu's population, and Tuvalu and other SIDS in the region were also affected, resulting in 55% of Tuvalu's population being displaced. The same cyclone displaced 2500 people in Kiribati. Also, the IPCC AR5 (2014) summarises a wealth of studies showing that atmospheric-induced events could increase in intensity due to global warming, leading to changes in the rates of precipitation amongst different regions, alterations in wind patterns, and potentially an intensification in tropical cyclones and their associated storm surges (see also Nakamura et al., 2020). More of such experiences could give greater strength to push factors, convincing many amongst the population in areas at risk that it would be better to seek better opportunities elsewhere. Hence, human mobility includes a wide variety of population movements, which can be voluntary or forced, assisted or spontaneous, long or short-term, and long-or short-distance (Guadagno, 2016).

2 Natural Hazards

SIDS have a number of characteristics that increase the challenge presented by natural hazards and climate change: small size, remoteness, small but growing populations, high inequalities which divide urban elites from the rural poor, high youth unemployment, and high public service and institutional costs (Everest-Phillips, 2014). As a consequence, the coastal areas of these islands may suffer from beach erosion, coral bleaching or the salinization of drinking water, which in turn may lead to losses in the tourism industry (Kaenzig & Piguet, 2014), usually one of the main economic sectors in these types of islands. The impact of sea level rise will depend on the topography of the islands: if the coastline is steep and low lying areas are narrow less land will be lost, though the loss of beaches can be severely detrimental to the tourism industry. If the low-lying areas are more extensive, then significant damage might be caused to agricultural land, leading to the salinization of the water table and damage to infrastructure (Simpson et al., 2009; Jamero et al., 2016).

In the next section, the authors will describe the topographic characteristics of Pacific and Caribbean SIDS.

2.1 Pacific SIDS

Coastal erosion, more often than not, is caused by interference of human activities on geomorphological processes (Shibayama, 2008). Several studies mention a scenario in which some of the SIDS in the Pacific could be 'sinking' due to SLR (Rayfuse, 2010; McAnaney, 2012). Based on such studies, a number of policy makers, academics and governments have been developing strategies to deal with this possibility (even though, as mentioned earlier, this possibility is rather remote and really represents a simplification of the geological and morphological nature of these islands, as summarised by Yamamoto & Esteban, 2014) The IPCC AR5 (2014) points out that many SIDS have undergone deep changes in their socioeconomic and environmental conditions, which may have masked on-going processes, making it is difficult to determine to what extent coastal erosion is caused by climate change.

The consequences of SLR on SIDS is an important topic of study, since human communities and infrastructure in these islands are usually located in coastal zones with limited possibility of them being relocated further inland, especially for the case of atoll islands (Woodroffe, 2008). Nevertheless, these islands have survived past fluctuations in water levels, and there is considerable geomorphological response capability in the systems (Webb & Kench, 2010; McLean & Kench, 2015). Consequently, the authors underline the need to develop a more realistic and sound scientific approach to the narrative that *sea-level rise will result in island inundation, erosion, destruction and ultimately island abandonment and out-migration of people* (as also noted in Jamero et al., 2016). It is also worth noting that even when islands are flooded due, for example, to subsidence, islanders can continue to inhabit coral islands through a range of adaptation strategies (Jamero et al., 2017).

Regardless of this narrative concerning whether islands will disappear or not, extreme climatic events such as tropical and extratropical cyclones and changing rainfall patterns, ocean chemistry, drought and storm events are likely to intensify in the future due to climate change (IPCC AR5, 2014). Rainfall events that occur once every 20 years on average are expected to occur four times per 20-year period, on average, by 2055, and seven times per 20-year period, on average, by 2090 under the high-emissions scenario in the IPCC AR5 (2014). Tropical cyclones, coastal erosion, salinity intrusion and drought, are expected to become even more damaging events in places like Tuvalu (Government of Tuvalu, 2007). The warming of the sea could lead to increased coral mortality, which together with ocean acidification might decrease islanders' ability to obtain food (Carr et al., 2014; Yamamoto & Esteban, 2014; Crichton & Esteban, 2018). Due to a combination of tidal ranges and inter-annual variability in sea levels, caused by the possible intensification of phenomenon such as El Nino, most Pacific islands will be subject to more intense and frequent flooding by 2100.

2.2 Caribbean SIDS

Caribbean SIDS range from developed tourist or agriculture centred economies to less economically developed states, which include countries such as Haiti with its chronic poverty and underdevelopment. Caribbean islands are frequently affected by tropical storms and hurricanes, which often result in large numbers of casualties and material losses (Kaenzig & Piguet, 2014) due to the storm surges, floods and the strong winds associated with these events. The Bahamas is located in the hurricane belt, and several of these storms were particularly destructive in recent times: Hurricane Andrew (1992), Hurricane Floyd (1999) (Government of The Bahamas, 2014) and Hurricane Matthew (2016). Climate projections for Cuba, based on different climate models, indicated a warmer, dryer and extreme climate in which the air temperature could rise by up to 4 °C, with a decrease in annual rainfall that, depending on the scenario, would range between 15 and 63%. Studies on SLR and the impact of migration in the region is sparse (Deprez, 2010). Regarding SLR, it is expected that the group of keys and the Zapata swamp will disappear and that coastlines may recede by up to a maximum of 7 km (Government of Cuba, 2015).

Climate change can also bring about more intense and long-lasting droughts. Recently, Antigua and Barbuda was suffering from the impacts of a chronic drought that lasted for 24 months (Government of Antigua and Barbuda, 2015). The country has experienced coastal erosion along the Northwest coast, in addition to storm surges and other extreme events (Government of Antigua and Barbuda, 2015). Haiti has the highest vulnerability rating when it comes to hurricanes, with 96% of its population living at risk. It is extremely vulnerable to wind damage, flooding, landslides, torrential debris flows, and coastal surges, not to mention that the country is located in a seismically active zone (World Bank & GFDRR, n.d.). It has also suffered in the recent past from civil unrest, long-term civil conflict and vulnerability patterns. In addition, very few alternate sources of income may also drive people to find a new home (IDMC, 2013).

Jamaica is, along with Haiti, located in one of the most seismically active regions in Central America. Climate change models have predicted that Jamaica will face a warming and drying trend, and hence will suffer more frequent heat waves and droughts, increased intensity rainfall and rising sea levels, as predicted for the rest of the Caribbean and consistent with the projected global median (World Bank &

GFDRR, n.d.). All these effects are likely to have a considerable impact on human socio-economic systems, such as ports (Monioudi et al., 2018).

3 Human Mobility and Disaster Displacement in the Global Agenda

Displacement in the context of natural hazards has been dealt with by various international organisations such as the United Nations High Commissioner for Refugees (UNHCR), the International Organization for Migration (IOM), the International Federation of Red Cross and Red Crescent Societies (IFRC), the United Nations (UN) Office for Coordination of Humanitarian Affairs (OCHA), and the World Health Organization (WHO). The UN system centralises its response to humanitarian emergencies with the Inter-Agency Standing Committee (IASC), which plays a central role in such processes (Klepp, 2017).

The relationship between human mobility and hazards acquired a new status within the Sendai Framework of Disaster Risk Reduction 2015–2030 (SFDRR), since several paragraphs mention human mobility.[1] The previous instrument on

[1]The SFDRR includes migrants in the design and implementation of DRR: Paragraph 7: "governments should engage with relevant stakeholders, including [. . .] migrants [. . .] in the design and implementation of policies, plans and standards" (UNISDR, 2015a, p. 10). Paragraph 27(h): "empower local authorities, as appropriate, through regulatory and financial means to work and coordinate with [. . .] migrants in disaster risk management at local level" (UNISDR, 2015a, p. 18). Paragraph 36(a)(vi): "Migrants contribute to the resilience of communities and societies and their knowledge, skills and capacities can be useful in the design and implementation of disaster risk reduction" (UNISDR, 2015a, p. 23); it adapts land use policy to demographic changes: Paragraph 6 "[. . .] More dedicated action needs to be focused on tackling underlying disaster risk drivers, such as [. . .] demographic change [. . .]" (UNISDR, 2015a, p. 10). Paragraph 30(f): "Promote the mainstreaming of disaster risk assessments into land-use policy development and implementation [. . .] and the use of guidelines and follow-up tools informed by anticipated demographic and environmental changes" (UNISDR, 2015a, p. 19); management of relocation: Paragraph 27(k) "Formulate public policies, where applicable, aimed at addressing the issues of prevention or relocation, where possible, of human settlements in disaster risk zones, subject to national law and legal systems" (UNISDR, 2015a, p. 18); management of evacuation: Paragraph 33(h) "Promote regular disaster preparedness, response and recovery exercises, including evacuation drills, training and the establishment of area-based support systems, with a view to ensuring rapid and effective response to disasters and related displacement, including access to safe shelter, essential food and non-food relief supplies, as appropriate to local needs" (UNISDR, 2015a, p. 21). Paragraph 33(m): "Strengthen the capacity of local authorities to evacuate persons living in disaster-prone areas" (UNISDR, 2015a, p. 22). Paragraph 28(d): "Promote transboundary cooperation to enable policy and planning for the implementation of ecosystem-based approaches with regard to shared resources, such as within river basins and along coastlines, to build resilience and reduce disaster risk, including epidemic and displacement risk" (UNISDR, 2015a, p. 18). Paragraph 33(j): "[. . .] Integrate post-disaster reconstruction into the economic and social sustainable development of affected areas. This should also apply to temporary settlements for persons displaced by disaster" (UNISDR, 2015a, p. 22).

DRR, the Hyogo Framework for Disaster Risk Reduction (HFDRR), only mentioned to "(. . .) ensure (. . .) that programmes for displaced persons do not increase risk and vulnerability to hazards" (§ 19, ii, i). Hence, SFDRR represents an advance by demonstrating recognition that climate change could aggravate the levels of displacement. With regards to the specific case of SIDS, SFDRR states that "[t]he effects of disasters, some of which have increased in intensity and have been exacerbated by climate change, impede their progress towards sustainable development" (Paragraph 42, UNISDR, 2015, p. 22).

Given the variety of studies which predict an increase in the intensity of certain hazards in SIDS, an approach which focuses on building the resilience of local communities is to be expected. Higher resilience requires DRR policies which are effective. However, climate change adaptation, DRR related laws, policies and strategies generally do not address disaster displacement, migration and planned relocation (Protection Agenda, 2015). DRR aims to minimise the factors which determine economic, socio and environmental impacts of hazards by providing a specific theoretical framework on these relationships. It tries to understand how to reduce vulnerability and risk and prevent these processes from resulting in the worst scenario, and in case that the worst scenario occurs, to recover from shocks and stresses (Guadagno, 2017). Human mobility can be seen in many ways as a process that can help build resilience, by relieving pressure from densely populated island communities, and providing funds for local adaptation strategies to take place, through the remittances of those who relocate to other countries. However, the reason why human mobility is not more readily recognised as an adaptation strategy may be due to the fact that governments and international agencies see displacement as a disruptive process, which should require restrictive measures (Tacoli, 2009).

Unfortunately, hazard zone management and implementation has generally been neglected in many countries, and there is high concentration of people and economic activities in hazard-prone areas such as coastlines and river deltas. This means that there is an increasing number of people exposed to natural hazards (IDMC, 2013), though by improving disaster management techniques more people are expected to survive disasters, even if their homes are destroyed.

Other conferences that have tackled the subject of human mobility also indicated the need for governments to address disaster displacement. The UN's 2030 Agenda for Sustainable Development mentions that global challenges include the "forced displacement of people". The Agenda for the Protection of Cross-Border Displaced Persons in the Context of Disasters and Climate Change (2015) makes recommendations for the prevention and management of these displacements, identifies practices and instruments for the protection of displaced people and mechanisms of cooperation between States in the same region. During the World Humanitarian Summit (2016) the Platform for Disaster Displacement was launched, which will work to implement the recommendations of the Protection Agenda. The World Humanitarian Summit has also pointed out disaster displacement as an emerging humanitarian challenge. The Paris Agreement (2015) included for the first time within the UNFCCC Framework a compulsory norm on climate change displacement. Furthermore, the New York Declaration for

Refugees and Migrants (UNGA, 2016) included a mention of those displaced by the adverse effects of climate change, natural hazards and other environmental factors.

The process of reduction of disaster displacement has three stages:

(1) A Pre-displacement phase, which aims to reduce the risk of long-term displacement, and which is connected to displacement prevention by reducing vulnerability and improvements in the resilience of the population. When prevention is not possible, preparations should be carried out.
(2) Displacement is when the actual displacement takes place, which requires emergency humanitarian responses and a recovery plan.
(3) Post-displacement phase takes place when the immediate threat of a given hazard ceases and recovery activities begin. It includes durable solutions such as local integration, resettlement and return (Government of Bangladesh, 2015).

When talking about human mobility, it is important to distinguish between climate-induced displacement and migration. Climate- induced displacement is a forced displacement which is caused by extreme climatic events, such as hurricanes cyclones, floods and drought. While this climate-induced displacement assumes that a given person is unable to remain in his community, migration includes a voluntary element to human mobility. Thus, a migrant is able to remain in his or her place of origin, but prefers to leave in order to seek better life conditions. The situation in his place of origin may not be calamitous, even though his livelihood may be suffering from socioeconomic problems or slow onset processes, such as a drought that might slowly force the population of an area to move.

For the case of SIDS, studies have suggested that bilateral agreements and the use of soft law are probably the more realistic approaches to protect those displaced (Williams, 2008; Yamamoto & Esteban, 2014, 2017). There have also been suggestions to use regional conventions, and there is already one example, the African Convention for the Protection and Assistance of Internally Displaced Persons in Africa (AU, 2009), of a regional convention which includes climate displacement in its scope of protection. However, most of the displacement will remain internal and the Guiding Principles on Internal Displacement (1998), which is a non-legally binding instrument, should guide the protection of the human rights of internally displaced persons.

With the exception of the 2018 Vanuatu National Policy on Climate Change and Disaster-Induced Displacement (Government of Vanuatu, 2018), in comparison to climate-induced displacement and migration, planned relocation has been receiving much less attention, although it requires special treatment from policy-makers. Natural hazards only become disasters if there is no careful preventive planning and action. People may need to relocate if they live in areas prone to sudden-onset natural hazards, which could increase, diminish, or be unchanged in severity as a result of climate change (coastal areas, landslide and floods prone areas). However, even those who live in areas threatened by the slow-onset processes of climate change (such as the salinization of water resulting from sea level rise or droughts) might also be affected in the long run (Nansen Initiative, 2014b).

Temporary relocations are very common, and are usually referred to as evacuations. Evacuations often fall within the domestic legal framework of civil protection and national disaster management. These laws refer to the process by which evacuations may be ordered, and have been developed by national disaster management organisations (Nansen Initiative, 2014b). Permanent relocations often happen in the context of building infrastructure or problems related with development, such as the construction of dams, with those relocated facing the risk of losing their land, employment, shelter, access to common resources, economic marginalisation, increase morbidity and mortality, food insecurity and negative cultural and psychological impacts (de Sherbinin et al., 2011). Since they are usually under-budgeted, they are rarely successful and should be avoided as far as possible (IDMC & NRC, 2017).

Human mobility as an adaptation strategy can be implemented through resettlement and relocation, remittances from migrants to contribute to increasing the resilience of island communities, and adopting adequate land use policies. Migration, despite being a resourceful adaptation measure, if not properly planned can result in maladaptation. Thus, it is often seen by many policy-makers as a last resort adaptation strategy for islands, especially given that many island populations are unwilling to relocate, even when facing constant flooding (Jamero et al., 2016). Thus, educational incentives for the younger generations might be a better way to achieve reductions in population density (Jamero et al., 2018).

4 Human Mobility in the Context of Disasters in the Regional Agendas of Pacific and Caribbean SIDS

Recognising that a global instrument to protect environmental migration is not feasible in the short term, the international community has been advocating regional governance on this topic. Essentially, it is argued that regional governance would be more efficient (Mayer, 2013) in the context of necessity-based approach, which would require the international community to create protection for environmental migrants. This is the basis of the Protection Agenda (2015) which also "supports an approach that focuses on the integration of effective practices by States and (sub-) regional organizations into their own normative frameworks in accordance with their specific situations and challenges." Sciaccaluga (2015) suggests that a necessity-based approach, would take into account two elements, in addition to the regional context: degree of duress of displacement and importance of environmental factors (Sciaccaluga, 2015). In addition to the regional approach, the suggestion of using soft law and bilateral agreements are considered as more pragmatic solutions. While soft law would be a first step to have the protection of environmental migrants recognised, bilateral agreements could also address displacement and would deal more specifically with the needs of both parties.

The 2012 Alliance of Small Island States Leaders' Declaration raised an alarm on the

> urgent need to consider and address the security implications of climate change, including violation of territorial integrity, more frequent and severe climate-related disasters, threats to water and food security, increased natural resource scarcity, and forced displacement. (AOSIS, 2012)

The 2015 Suva Declaration on Climate Change states that "climate change is already resulting in forced displacement of island populations and the loss of land and territorial integrity". In addition, human mobility issues are visible in regional declarations, including the Niue Declaration on Climate Change (2008), Moana Declaration Outcome instrument of Pacific Conference of Churches (2009) and Moana Declaration Outcome Statement of Pacific Parliamentarians (2013). Victor David from the University of Caledonia also presented a draft of a regional convention for Pacific SIDS in the 3rd World Conference for Disaster Risk Reduction in 2015 and in the COP21 in Paris in the same year (personal correspondence by Victor David, 2017).

The need to improve disaster preparedness, population evacuation and emergency relief for people in vulnerable situations were also (paragraph 52, c) emphasised in the outcome document of the Third International Conference on Small Island Developing States, held in Samoa in 2014. At the Regional Meeting on Climate Change and Migration the Pacific in December 2016, government officials emphasised the need to create a regional framework to protect those displaced by climate (Corendea, 2017).

Latin America has advanced when it comes to regional instruments to protect migrants. The Cartagena Declaration uses an expanded definition of refugees and includes "persons who have fled their country because their lives, security or freedom have been threatened by generalized violence, foreign aggression, internal conflicts, massive violation of human rights or other circumstances which have seriously disturbed public order". Although the Cartagena definition of refugees does not include environmental migrants, specialists argue that States can choose to apply the Cartagena definition to this category of people (ACNUR, 2013). In the occasion of Cartagena +30 process, the 2014 Brazil Plan of Action foresees the establishment of a regional consultative mechanism in the Caribbean and management of mixed migration (UNHCR, 2015). The Brazil Declaration and Plan of Action recognises "the challenges posed by climate change and natural disasters, as well as by the displacement of persons across borders that these phenomena may cause in the region" and the need to "give more attention to this matter" The Association of Caribbean States (ACS) would be an interesting organisation to treat the topic.[2] One of the areas of concern is natural hazards, given the vulnerability

[2]The Convention Establishing the Association of Caribbean States (ACS) was signed on 24 July 1994 in Cartagena de Indias, Colombia, with the aim of promoting consultation, cooperation and concerted action among all the countries of the Caribbean, comprising 25 Member States and three

of all countries and territories of the Greater Caribbean to the physical and economic consequences of them (ACS, n.d.).

5 Human Mobility and Disaster Displacement Policies

Natural hazards can displace people in Pacific and Caribbean SIDS, which require States and regional organisations to address these questions in their environmental, migration and climate change policies and laws. This section will review the national policies of SIDS in the Caribbean and Pacific with regards to disaster displacement.

5.1 Caribbean SIDS

Some countries in the region have enacted norms and regulations that incorporate human mobility. That is the case of Dominican Republic, which provides the possibility of resettlement after an assessment at risk communities. The Institute of Housing will be responsible for the overall coordination and municipal committees shall report the presence of high-risk communities (IFRC & UNDP, 2014). Barbados established the Vulnerable Persons Committee, which aims to coordinate data on vulnerable persons (such as children, elderly or those with disabilities) in order to provide assistance to them. The Committee also reviews the Evacuation of Vulnerable Persons Plan, and visits vulnerable areas and shelters (Government of Barbados, 2014). Grenada has shelters that are made accessible to vulnerable communities, although they are vulnerable to earthquake, volcanic, technological and biological hazards. There was also a relocation of individuals from Moon Shadow Park, which were located in a flood prone area (Government of Grenada, 2014).

The Caribbean region has also been affected by large-scale displacement due to non-climatic hazards, such as in the case of the volcanic eruption starting in 1995 on Montserrat, a United Kingdom Overseas Territory. The population living close to the Soufriere Hills volcano in the south of the island was gradually evacuated to the north, which was not developed at the time. As the volcanic activity intensified, the UK government supported relocation of sectors of the population to the UK (Monteil & Simmons, 2017). Thus, finding land to relocate their inhabitants was not a problem, although it disrupted the economy and communities.

A more recent transboundary disaster displacement example was the Mw (moment magnitude) 7.0 earthquake which struck Haiti in 2010, killing more than 160,000 people and displacing nearly 1.5 million (IDMC & NRC, 2016). Most of these appeared to have been internally displaced, though quite a large number also

Associate Members. Eight other non-independent Caribbean territories are eligible for associate membership.

left the country. In response to this disaster, Caribbean countries have showed a tendency of being flexible to "regularise" the status of those displaced by disasters, even when their migration laws lack specific provisions to fit their situation (Cantor, 2015). The Political Council of ALBA[3] held an emergency meeting and adopted a "Plan of solidary contribution of the countries of ALBA to the efforts of reconstruction of Haiti", which included a migration amnesty which would regulate the migratory status of Haitian citizens residents in countries that are members of ALBA (Serna, 2015). As a result, Dominica allowed Haitians to apply for visa and to extend their stay for an additional six months. Antigua and Barbuda granted a visa waiver for Haitian migrants who were living in the territory to bring close relatives from Haiti to stay with them, as long as they demonstrated enough economic capacity to receive them (Cantor, 2015).

Other countries such as The Bahamas suspended the forced returns of Haitians, with those who were detained being released and granted temporary legal status. The Dominican Republic also announced it would stop deporting Haitians who were staying in the country illegally. The Jamaican government stated that Haitians arriving on Jamaican shores would not be forcedly returned (UNHRC, 2012). In Cuba, special measures to receive Haitian citizens were not necessary since disaster-displaced people were already included in the domestic definition of refugees:

> ...those aliens and persons lacking citizenship whose entry to the national territory is authorized due to leaving their country owing to social or warlike calamity, due to cataclysm or other phenomena of nature and who will remain temporarily in Cuba, until normal conditions are re-established in their country of origin. (1978 Regulations, in Government of Cuba, 2012)

While Caribbean SIDS demonstrated solidarity to receive people who were displaced across borders, they have not developed long-term strategies to deal with the potential future internal displacement of their own inhabitants. The Bahamas, for example, mentions the relocation of communities from the shoreline in its Intended Nationally Determined Contribution (INDC). Adaptation strategies focus on building and strengthening coastal defences, as well as the relocation of communities from the shoreline. However, there is a lack of concrete strategies, actions and policy frameworks on human mobility (Haley-Benjamin, 2016). One of the main causes of displacement in urban areas is the risk of landslides, which can affect poor housing in certain areas. Keeping in mind this problem, Cuba has mentioned a strategy to enhance the resilience of human settlements in coastal areas that are threatened by sea level rise and hurricanes (Oxfam, 2011).

The Guna indigenous community from the island of Gardi Sugdub in Panama represents a case of slow-onset disaster-driven migration. The community is

---

[3]The Bolivarian Alliance for the Peoples of our America (ALBA) is a regional organisation which aims to promote social, political and economic integration. It was founded in 2004 through the Cuba-Venezuela agreement as an alternative to the Free Trade Area of the Americas (FTAA), proposed by the US to be an extension of NAFTA. Its members are socialist and social democratic governments, including Antigua and Barbuda, Bolivia, Cuba, Dominica, Ecuador, Nicaragua, Saint Lucia, Saint Vincent and the Grenadines and Venezuela.

planning their relocation due to the combined result of a lack of space caused by rising sea levels and increasing population. However, the relocation has not been implemented so far, as they are waiting for the relocation facilities and housing to be built by the government. The Yandub Muladup community is also planning their relocation within a land that they own, but feasibility studies to assess the best place to relocate were not made (Displacement Solutions, 2016).

5.2 *Pacific SIDS*

Although some authors fear the possible disappearance of the territory of atoll island states, a slow-onset phenomenon, the policies among SIDS in the Pacific region are not always aligned with this narrative. While the Kiribati government, for example, has the most developed migration as a strategy at the international level (McNamara et al., 2016), the tragic narrative of loss of territory is perceived by Tuvaluans as an international discourse that is a tool to fail to curb greenhouse gas emissions (Farbotko & Lazrus, 2012). Tuvalu, in fact, is often portrayed as an example of future "climate refugees" (Farbotko & Lazrus, 2012), although few Tuvaluans would wish for that (Kelman, 2015).

The Republic of Kiribati is a small island nation consisting of 32 atolls and one raised coral island with a total land area of 800 sq. km spread over an oceanic Exclusive Economic Zone (EEZ) of 3.5 million sq. km (Government of Kiribati, 2012). The high density of population impacts the country's sustainability, since it increases demands on food, energy, water, social services, and infrastructure (World Bank, 2016). Kiribati agriculture is constrained by limited arable soils, isolation (which prevents it from accessing global or regional markets) and environmental vulnerabilities aggravated by population growth (Carden, 2003). The Government of Kiribati has adopted the program "migration with dignity" which provides the opportunity to I-Kiribati to develop skills to make them more attractive to receiving countries (Government of Kiribati, n.d.). The training is held in Kiribati, New Zealand and Australia, and is funded by scholarships and public funding (Duong, 2015). The country already has experience with relocation, and in the 1980s Kiribati resettled people internally due to population pressure in the Line Islands (ADB, 2002). The government has been developing projects to manage water supply, build coastal protection, at the same time as planning the relocation of its entire population (Wyett, 2014).

There have been several internal relocations that have been carried out in Fiji, such as in Vunidogoloa, Narikoso village and Denumanu Village, Vuya. These villages are subsiding, and the communities are to be relocated in the next five to ten years (Gharbaoui & Blocher, 2017). Fiji's relocation guidelines are still being drafted, and are expected to underline the role of communities in starting technical and financial assistance on relocation (Gharbaoui & Blocher, 2017). The relocation is going to be done in close consultation with both affected and host communities (The Nansen Initiative, 2014a). Hence, by assessing if the community is vulnerable and has a high chance of having to relocate, the government is effectively planning potential relocations in advance.

On 29 September 2009 a Mw 8.1 earthquake struck the south of Samoa, which was followed by a tsunami which killed 143 people and caused damage and economic losses estimated at US$124 million. Some 859 households (about 5000 people), spread over 51 villages, were affected. The government's emergency response to the 2009 tsunami was guided by its National Disaster Management Plan (Koskinen-Lewis et al., 2016).

Communities affected by the tsunami had the option to relocate inland to other land areas owned by them (as part of the same village). Many voluntarily moved to higher grounds (Mikami & Shibayama, 2015), living first in makeshift shelters. The Government (in close consultation with village representatives) worked to provide much needed services, including water, electricity, housing and road networks to the new relocation areas. Numerous development partners also contributed to the effort. By the end of 2010, the housing program had provided 502 new homes and rehabilitated 360. The government provided support based on a cyclone-resistant, Samoan-styled home, including sanitation facilities and water tanks. Most new schools and facilities were also completed by then (Koskinen-Lewis et al., 2016).

6 Conclusions

Pacific and Caribbean SIDS share similar socioeconomic conditions, and a future increase in the intensity in natural hazards could lead to greater numbers of their citizens being temporarily or permanently displaced. While Pacific SIDS have been actively arguing the case of climate-induced displacement before international fora and have claimed that their territories are sinking, these claims are not supported by the scientific literature, though the same reasoning cannot be applied to Caribbean SIDS. Caribbean SIDS are generally not low-lying and their statehood is thus not deemed to be threatened due to sea level rise. However, the potential future increase in the intensity of climatic extreme events may affect the livelihood of the populations in this region and lead to large-scale displacement, at least temporarily in the aftermath of major disasters.

The authors hold the view that the predictions of the complete disappearance of Pacific SIDS are unlikely to take place, though the possibility that this could happen has prompted the Kiribati government to seek solutions to this potential future challenge. Their policy solutions involve human mobility as an adaptation strategy, which involves preparing for the worst-case scenario by building the skills of their citizens. In a sense, this could be viewed as a 'no-regrets' strategy, as even if the worst predictions do not come to pass, investment in education and skills will likely result in socio-economic improvements in a given population. In the case of many other islands relocation is considered as a last resort strategy, since in several cases it has worked poorly, leading to cultural and economic losses in the communities involved.

Disaster displacement research in the Caribbean SIDS has been focused on the case of the Haitian earthquake in 2010, which tested to what extent the region was prepared to receive those displaced. The Caribbean SIDS, with the exception of Cuba, do not have specific norms which encompasses the protection of disaster displaced people, but several SIDS have taken measures to allow Haitian citizens to remain in their soil. Nevertheless, these temporary visas that were granted due to natural hazards should be provided for in domestic laws to ensure that the rights of the displaced people are respected in the receiving country.

Acknowledgements A part of the presented research was performed as activities of the Research Institute of Sustainable Future Society, Waseda Research Institutefor Science and Engineering, Waseda University.

References

ACNUR. (2013). *Reunión de expertos Interpretación de la definición ampliada de refugiado contenida en la Declaración de Cartagena sobre Refugiados de 1984 Montevideo, Uruguay 15 y 16 de octubre de 2013*. Resumen de las conclusiones sobre la interpretación de la definición ampliada de refugiado de la Declaración de Cartagena de 1984. [online] Retrieved May 23, 2017, from http://www.acnur.org/t3/fileadmin/Documentos/BDL/2014/9651.pdf

African Union (AU). (2009). *African Union Convention for the Protection and Assistance of Internally Displaced Persons in Africa (Kampala Convention)*. [online] Retrieved May 23, 2017, from https://www.au.int/web/en/treaties/african-union-convention-protection-and-assistance-internally-displaced-persons-africa

Alliance of Small Island States (AOSIS). (2012). Alliance of Small Island States Leaders' Declaration. [online] Retrieved May 30, 2017, from http://aosis.org/wp-content/uploads/2012/10/2012-AOSIS-Leaders-Declaration.pdf

Asian Development Bank (ADB). (2002). *Kiribati - Monetization in an Atoll Society - managing economic and social change*. [online] Retrieved May 28, 2017, from https://www.adb.org/sites/default/files/publication/28818/kiribati.pdf

Association of Caribbean States (ACS). (n.d.). *About the ACS*. [online] Retrieved May 22, 2017, from http://www.acs-aec.org/index.php?q=about-the-acs

Ballu, V., Bouin, M. N., Siméoni, P., Crawford, W. C., Calmant, S., Boré, J. M., Kanas, T., & Pelletier, B. (2011). Comparing the role of absolute sea-level rise and vertical tectonic motions in coastal flooding, Torres Islands (Vanuatu). *PNAS, Proceedings of the National Academy of Sciences, 108*(32), 13019–13022.

Barnett, J. (2017). The dilemmas of normalising losses from climate change: Towards hope for Pacific atoll countries. *Asia Pacific Viewpoint, 58*(1), 3–13. https://doi.org/10.1111/apv.12153

Barnett, J., & Webber, M. (2010). *Accommodating migration to promote adaptation to climate change*. Policy Research Working Paper No. 5270. Background Paper prepared for the Secretariat of the Swedish Commission on Climate Change and Development and the World Bank World Development Report 2010: Development and Climate Change, *The International Bank for Reconstruction and Development/The World Bank, Development Economics Vice Presidency,* Washington, DC, USA, p. 62.

Cantor, D. (2015). *Law, policy and practice concerning the humanitarian protection of aliens on a temporary basis in the context of disasters*. Background Paper, Nansen Initiative. [online] Retrieved May 28, 2017, from http://disasterdisplacement.org/wp-content/uploads/2015/07/150715_FINAL_BACKGROUND_PAPER_LATIN_AMERICA_screen.pdf

Carden, Y. H. (2003). Solid waste-level rise on atoll nation states: A less publicised environmental issue in the Republic of Kiribati. *Australasian Journal of Environmental Management, 10*, 35–45. https://doi.org/10.1080/14486563.2003.10648571

Carr, M., Rubenstein, M., Graff, A., & Villarreal, D. (2014). Sea level rise in a changing climate-What do we know? In M. Gerrard & G. Warnier (Eds.), *Threatened Island Nations-legal implications of rising seas and a changing climate* (pp. 15–54). Cambridge University Press.

Connell, J. (2013). *Islands at risk? Environments, economies and contemporary change*. Edward Elgar.

Corendea, C. (2017). *Pacific countries advance regional policy towards climate migration*. United Nations University. [online] Retrieved May 20, 2017, from http://ehs.unu.edu/blog/articles/pacific-countries-advance-regional-policy-towards-climate-migration.html

Crichton, R., & Esteban, M. (2018). Limits to coastal adaptation in Samoa: Insights and experiences. In F. W. Leal & J. Nalau (Eds.), *Limits to climate change adaptation. Climate change management*. Springer. https://doi.org/10.1007/978-3-319-64599-5_16

Dasgupta, S., Laplante, B., Meisner, Wheeler, D., & Yan, D. J. (2007). *The impact of sea level rise on developing countries: A comparative analysis*. World Bank Policy Research Working Paper No. 4136. [online] Retrieved May 10, 2017, from https://ssrn.com/abstract=962790

de Sherbinin, A., Castro, M., Gemenne, F., Cernea, M. M., Adamo, S., Fearnside, P. M., Krieger, G., Lahmani, S., Oliver-Smith, A., Pankhurst, A., Scudder, T., Singer, B., Tan, Y., Wannier, G., Boncour, P., Ehrhart, C., Hugo, G., Pandey, B., & Shi, G. (2011). Preparing for resettlement associated with climate change. *Science, 334*(6055), 456–457.

Deprez, A. (2010). *Climate migration in Latin America: A future of "flood refugees" to the North?*, Council of Hemispheric Affairs. [online] Retrieved May 9, 2017, from https://www.climate-diplomacy.org/news/climate-migration-latin-america-future-%E2%80%98flood-refugees%E2%80%99-north

Displacement Solutions. (2016). An overview on the relocation of Guna indigenous communities in Gunayala, Panama-Mission Report. [online] Retrieved May 8, 2017, from http://displacementsolutions.org/wp-content/uploads/2017/01/Gunayala-Planned-Relocation-Jan-2017.pdf

Duong, S. (2015). *Rising islands- enhancing adaptive capacities in Kiribati through migration with dignity*. Master Thesis, Stockholm, Stockholm University. [online] Retrieved May 3, 2017, from http://www.diva-portal.org/smash/record.jsf?pid=diva2%3A826447&dswid=-1928

Esteban, M., Jamero, L., Nurse, L., Yamamoto, L., Takagi, H., Nguyen, T. D., Mikami, T., Kench, P., Onuki, M., Nellas, A., Crichton, R., Valenzuela, V. P., Chadwick, C., Avelino, J. E., Tan, N., & Shibayama, T. (2019). Adaptation to sea level rise on low coral islands: Lessons from recent events. *Ocean and Coastal Management, 168*, 35–40.

European Union (EU). (2017). Global Surface Water Explorer. [online] Retrieved May 31, 2017, from https://global-surface-water.appspot.com/

Everest-Phillips, M. (2014). *Small, so simple? Complexity in small island developing states*. UNDP Global Centre for Public Service Excellence. [online] Retrieved May 25, 2017, from http://www.undp.org/content/dam/undp/library/capacity-development/English/Singapore%20Centre/GPCSE_Complexity%20in%20Small%20Island.pdf

Farbotko, C., & Lazrus, H. (2012). The first climate refugees? Contesting global narratives of climate change in Tuvalu. *Global Environmental Change, 22*(2), 382–390. https://doi.org/10.1016/j.gloenvcha.2011.11.014

Gharbaoui, D., & Blocher, J. (2017). The reason land matters: Relocation as adaptation to climate change in Fiji islands. In A. Milan, B. Schraven, K. Warner, & N. Cascone (Eds.), *Migration, risk management and climate change: Evidence and policy responses. Volume 6 of the series global migration issues* (pp. 149–173). Springer.

Government of Antigua and Barbuda. (2015). Third National Communication. *Government of Antigua and Barbuda, Department of Environment, Ministry of Health and Environment*. [online] Retrieved May 29, 2017, from http://unfccc.int/resource/docs/natc/antnc3.pdf

Government of Bangladesh. (2015). National strategy on the management of disaster and climate induced internal displacement - comprehensive disaster management programme. *Government of Bangladesh, Ministry of Disaster Management and Relief (ModDMR).* [online] Retrieved May 28, 2017, from http://www.preventionweb.net/files/46732_nsmdciidfinalversion21sept2015withc.pdf

Government of Barbados. (2014). Country document for disaster risk reduction. *Government of Barbados, Department of Emergency Management.* [online] Retrieved May 28, 2017, from http://dipecholac.net/docs/files/784-documento-pais-barbados-web.pdf

Government of Cuba. (2012). Edición actualizada del Decreto No. 26, Reglamento de la Ley de Migración (Cuba), de 19 de Julio de 1978. Gaceta Oficial, No. 44, 1373–1387, Article 80, 16 October, Government of Cuba, Ministry for Justice (2012).

Government of Cuba. (2015). República de Cuba - Segunda Comunicación Nacional de la Convención Marco de las Naciones Unidas sobre Cambio Climático. [online] Retrieved May 10, 2017, from http://unfccc.int/resource/docs/natc/cubnc2.pdf

Government of Grenada. (2014). Country Document for Disaster Risk Reduction. *Government of Grenada, National Disaster Management Agency.* [online] Retrieved May 14, 2017, from http://dipecholac.net/docs/files/871-documento-pais-grenada-web.pdf

Government of Kiribati. (2012). Kiribati Integrated Environment Policy. [online] Retrieved April 7, 2017, from http://www.mfed.gov.ki/sites/default/files/KIEP%20FINAL.pdf

Government of Kiribati. (n.d.). *Relocation.* [online] Retrieved April 10, 2017, from http://www.climate.gov.ki/category/action/relocation/

Government of The Bahamas. (2014). The Second National Communication Report of the Commonwealth of the Bahamas under the UN Framework Convention on Climate Change. [online] Retrieved May 27, 2017, from http://unfccc.int/resource/docs/natc/bhsnc2.pdf

Government of Tuvalu. (2007). Tuvalu's National Adaptation Programme of Action: Under the auspices of the United Nations Framework Convention on Climate Change. Government of Tuvalu, Ministry of Natural Resources, Environment, Agriculture and Lands, Department of Environment. [online] Retrieved December 20, 2020, from https://reliefweb.int/sites/reliefweb.int/files/resources/TUV_National_Adaptation_Programme.pdf

Government of Vanuatu. (2018). National Policy on Climate Change and Disaster-Induced Displacement. [online] Retrieved January 28, 2020, from https://www.iom.int/sites/default/files/press_release/file/iom-vanuatu-policy-climate-change-disaster-induced-displacement-2018.pdf

Guadagno, L. (2016). Human mobility in the Sendai framework for disaster risk reduction. *International Journal of Disaster Risk Science, 7*(1), 30–40. https://doi.org/10.1007/s13753-016-0077-6

Guadagno, L. (2017). Human mobility in a socio-environmental context: Complex effects on environmental risk. In K. Sudmeier-Rieux, M. Fernández, I. Penna, M. Jaboyedoff, & J. Gaillard (Eds.), *Identifying emerging issues in disaster risk reduction, migration, climate change and sustainable development-shaping debate and policies* (pp. 13–31). Springer.

Haley-Benjamin, V. (2016). It is time to help the world's most vulnerable people. *The Environmental Forum, 6*, 54.

Internal Displacement Monitoring Centre (IDMC). (2013). Technical paper: The risk of disaster-induced displacement-Central America and Caribbean. [online] Retrieved April 25, 2017, from https://reliefweb.int/sites/reliefweb.int/files/resources/201312-am-risk-of-disaster-induced-displacement-en.pdf

Internal Displacement Monitoring Centre (IDMC) & Norwegian Refugee Council (NRC). (2016). Global Report on Internal Displacement. [online] Retrieved April 25, 2017, from https://www.internal-displacement.org/globalreport2016/#:~:text=There%20were%2019.2%20million%20new,of%2025.4%20million%20each%20year

Internal Displacement Monitoring Centre (IDMC) & Norwegian Refugee Council (NRC). (2017). Positioned for action. Displacement in the Sendai Framework for Disaster Risk Reduction. [online] Retrieved December 20, 2020, from https://www.internal-displacement.org/sites/default/files/publications/documents/20170216-idmc-briefing-paper-drr.pdf

International Federation of Red Cross and Red Crescent Societies (IFRC) & United Nations Development Programme (UNDP). (2014). Effective law and regulation for disaster risk reduction: A multi-country report. [online] Retrieved April 25, 2017, from http://www.undp.org/content/undp/en/home/librarypage/crisis-prevention-and-recovery/effective-law%2D%2D-regulation-for-disaster-risk-reduction.html

IPCC AR5. (2014). Small Islands. [online] Retrieved April 15, 2017, from https://www.ipcc.ch/pdf/assessment-report/ar5/wg2/WGIIAR5-Chap29_FINAL.pdf

Jamero, L., Esteban, M., & Onuki, M. (2016). Potential in-situ adaptation strategies for climate-related sea-level rise: Insights from a small island in the Philippines experiencing earthquake-induced land subsidence. *J-SustaiN - International Journal of Sustainable Future for Human Security, 4*(2), 44–53.

Jamero, L., Onuki, M., Esteban, M., Billones-Sensano, X. K., Tan, N., Nellas, A., Takagi, H., Thao, N. D., & Valenzuela, V. P. (2017). Small island communities in the Philippines prefer local measures to relocation in response to sea-level rise. *Nature Climate Change, 7*(8), 581–586.

Jamero, L., Onuki, M., Esteban, M., & Tan, N. (2018). Community-based adaptation in low-lying islands in the Philippines: Challenges and lessons learned. *Regional Environmental Change, 18* (8), 2249–2260. https://doi.org/10.1007/s10113-018-1332-8

Jones, S., & Holpuch, A. (2016). Haiti death toll rises to 842 after Hurricane Matthew, reports say. *The Guardian*, 7 October. [online] Retrieved May 15, 2017, from https://www.theguardian.com/world/2016/oct/06/hurricane-matthew-haiti-rescuers-battle-reach-remote-areas

Kaenzig, R., & Piguet, E. (2014). Migration and climate change in America and the Caribbean. In E. Piguet & F. Laczko (Eds.), *People on the move in a changing climate, global migration issues 2* (pp. 155–176). Springer.

Kelman, I. (2015). Difficult decisions: Migration from small island developing stats under climate change. *Earth's Future, 3*(4), 133–142.

Klepp, S. (2017). *Climate change and migration*. Oxford Research Encyclopedia of Climate Science. [online] Retrieved April 5, 2017, from http://climatescience.oxfordre.com/view/10.1093/acrefore/9780190228620.001.0001/acrefore-9780190228620-e-42?rskey=qWfBmV&result=3

Koskinen-Lewis, P., de Carvalho, A., Dias, C. M., Fernandes, C., Diogo, O., Taulealo, L., Evalu, F., & Simi, N. (2016). *Managing population retreat from at-risk areas*. SISRI Knowledge Note No. 3. Small Island States Resilience Initiative. Washington DC, The World Bank and Global Facility for Disaster Reduction and Recovery (GFDRR).

Mayer, B. (2013). Environmental migration: Prospects for a regional governance in the Asia-Pacific region. *Asia Pacific Journal of Environmental Law, 16*, 77–103. https://doi.org/10.1017/S204425131200029X

McAnaney, S. C. (2012). Sinking islands? Formulating a realistic solution to climate change displacement. *New York University Law Review, 87*(4), 1172–1209.

McLean, R., & Kench, P. (2015). Destruction or persistence of coral islands in the face of 20th and 21st century sea-level rise? *Wires Climate Change, 6*, 445–463. https://doi.org/10.1002/wcc.350

McLeman, R. (2014). *Climate and human migration-past experiences, future challenges*. Cambridge University Press.

McNamara, K. R., Bronen, R., Nishara, F., & Klepp, S. (2016). The complex decision-making of climate-induced relocation: Adaptation and loss and damage. *Climate Policy, 18*(1), 111–117. https://doi.org/10.1080/14693062.2016.1248886

Methmann, C., & Oels, A. (2015). From 'fearing' to 'empowering' climate refugees: Governing climate-induced migration in the name of resilience. *Security Dialogue, 46*(1), 51–68.

Mikami, T., & Shibayama, T. (2015). Tsunami disasters in remote islands: 2009 Samoan and 2010 Mentawai Island Tsunamis. In M. Esteban, H. Takagi, & T. Shibayama (Eds.), *Handbook of coastal disaster mitigation for engineers and planners* (pp. 75–91). Butterworth-Heinemann (Elsevier).

Monioudi, I. N., Asariotis, R., Becker, A., Dowding-Gooden, D., Esteban, M., Mentaschi, L., Nikolaou, A., Nurse, L., Phillips, W., Smith, D. A. Y., Satoh, M., Snow-Bhat, C., Trotz, U. O., Velegrakis, A., Vousdoukas, M., & Bob Witkomp, B. (2018). Climate change impacts on critical international transportation assets of Caribbean Small Island Developing States (SIDS): The case of Jamaica and Saint Lucia. *Regional Environmental Change, 18*, 2211–2225.

Monteil, C., & Simmons, P. (2017). Immigration in Montserrat after the volcanic disaster: A tool and a challenge for the recovery of the island. In L. Guadagno, M. Fuhrer, & J. Twigg (Eds.), *Migrants in disaster risk reduction - Practices for inclusion* (pp. 104–109). International Organization for Migration and Council of Euope.

Nakamura, R., Shibayama, T., Esteban, M., Iwamoto, T., & Nishizaki, S. (2020). Simulations of future typhoons and storm surges around Tokyo Bay using AR5 RCP 8.5 scenario in multi global climate models. *Coastal Engineering Journal, 62*(1), 101–207.

Nansen Initiative. (2014a). Impacts of climate change and disasters on human mobility in the Pacific: Challenges and opportunities. Outcome Report, Pacific Regional Civil Society Meeting, Lami Bay, Fiji, 18–20 August.

Nansen Initiative. (2014b). Planned relocations, disasters and climate change: Consolidating good practices and preparing for the future. Background Document, San Remo Consultation, 12–14 March.

Narayan, S., Esteban, M., Albert, S., Jamer, M. L., Crichton, R., Hech, N., Goby, G., & Jupiter, S. (2020). Local adaptation responses to coastal hazards in small island coastal communities: Insights from 4 Pacific island nations. *Environmental Science and Policy, 104*, 199–207.

Oxfam. (2011). The climate changes, threatens and demands adaptation: A look at the Cuban experience of protection against climate change. [online] Retrieved April 14, 2017, from http://www.uhasselt.be/images/internationaal/North-South/2017/Oxfam%20report%20-%20A%20look%20at%20the%20Cuban%20experience%20of%20protection%20against%20climate%20change.pdf

Piguet, E. (2012). Des apatrides du climat? *Annales de Geographie, 683*, 86–100.

Rayfuse, R. (2010). *International law and disappearing states: Utilizing maritime entitlements to overcome the statehood dilemma*, University of New South Wales Faculty of Law Research Series, Paper 52. [online] Retrieved April 25, 2020, from http://papers.ssrn.com/sol3/papers.cfm?abstract_id=1704835

Sciaccaluga, G. (2015). *Climate change-related disasters and human displacement: Towards an effective management system*, Disaster Law Working Paper Series, Paper No.4, IFRC. [online] Retrieved April 25, 2017, from http://www.ifrc.org/PageFiles/125745/Working%20Paper%20No%204%20(Sciaccaluga).pdf

Serna, N. R. (2015). *Human mobility in the context of natural hazard-related disasters in South America*, Background Paper, the Nansen Initiative. [online] Retrieved April 15, 2017, from https://www.nanseninitiative.org/wp-content/uploads/2015/12/14122015_FINAL_BACKGROUND_PAPER_SOUTH_AMERICA_screen.pdf

Shibayama, T. (2008). *Coastal processes: Concepts in coastal engineering and their application to multifarious environment*. World Scientific.

Simpson, M. C., Scott, D., New, M., Sim, R., Smith, D.,Harrison, M., Eakin, C. M., Warrick, R., Strong, A. E., Kouwenhoven, P., Harrison, S., Wilson, M., Nelson, G. C., Donner, S., Kay, R., Geldhill, D. K., Liu, G., Morgan, J. A., Kleypas, J. A., Mumby, P. J., Christensen, T. R. L., Baskett, M. L., Skirving, W. J., Elrick, C., Taylor, M., Bell, J., Rutty, M., Burnett, J. B., Overmas, M., Robertson, R., & Stager, H. (2009). *An overview of modeling climate change impacts in the Caribbean region with contribution from the Pacific Islands*. United Nations Development Programme (UNDP), Barbados, West Indies. [online] Retrieved April 28, 2017, from https://coralreefwatch.noaa.gov/satellite/publications/UNDP_Final_Report.pdf

Tacoli, C. (2009). Crisis or adaptation? Migration and climate change in a context of high mobility. *Environment and Urbanization, 21*(2), 513–525.

United Nations Conference on Trade and Development (UNCTAD). (2017). Climate Change Impacts and Adaptation for Coastal Transport Infrastructure in the Caribbean (24 May–1 June

2017). [online] Retrieved May 31, 2017, from http://unctad.org/en/Pages/DTL/TTL/Legal/Climate-Change-Impacts-on-SIDS.aspx

United Nations High Commissioner for Refugees (UNHCR). (2015). *Overview of UNHCR's operations in the Americas*. Executive Committee of the High Commissioner's Programme, Standing Committee 62nd meeting, 10 February. [online] Retrieved May 22, 2017, from http://www.unhcr.org/54fd6f769.pdf

United Nations General Assembly (UNGA). (2016). *New York Declaration for Refugees and Migrants*. UN Doc RES/71/1 (3 October). Retrieved May 31, 2017, from https://undocs.org/a/res/71/1

United Nations Human Rights Council. (2012). Report of the Independent Expert on the situation of human rights in Haiti, Michel Forst. Addendum, Forced returns of Haitians from third states, UN Doc A/HRC/20/35/Add.1 (4 June). Geneva, United Nations Human Rights Council, pp. 6–7.

United Nations International Strategy for Disaster Reduction (UNISDR). (2015). Sendai Framework for Disaster Risk Reduction. [online] Retrieved May 31, 2017, from http://www.unisdr.org/we/coordinate/sendai-framework

Webb, A., & Kench, P. (2010). The dynamic response of reef islands to sea-level rise: Evidence from multi-decadal analysis of island change in the Central Pacific. *Global Planet Change, 72* (3), 234–246. https://doi.org/10.1016/j.gloplacha.2010.05.003

Williams, A. (2008). Turning the tide: Recognizing climate change refugees in international law. *Law and Policy, 30*(4), 503–529.

Woodroffe, C. D. (2008). Reef-island topography and the vulnerability of atolls to sea level rise. *Global and Planetary Change, 62*(1–2), 77–96.

World Bank. (2014). *Can you imagine a Caribbean minus its beaches? It's not science fiction, it's climate change*, 5 September. [online] Retrieved April 23, 2017, from http://www.worldbank.org/en/news/feature/2014/09/05/can-you-imagine-a-caribbean-minus-its-beaches-climate-change-sids

World Bank. (2016). *Databank on Kiribati*. [online] Retrieved April 22, 2017, from http://data.worldbank.org/country/kiribati

World Bank & the Global Facility for Disaster Reduction and Recovery (GFDRR). (n.d.). *Disaster Risk Management in Latin America and the Caribbean Region*, GFDRR Country Notes. [online] Retrieved April 20, 2017, from http://dipecholac.net/docs/files/521-drm-lac-countryprograms.pdf

Wyett, K. (2014). Escaping a rise tide: Sea level rise and migration in Kiribati. *Asia & Pacific Policy Studies, 1*(1), 171–185. https://doi.org/10.1002/app5.7

Yamamoto, L., & Esteban, M. (2014). *Atoll island states and international law-climate change displacement and sovereignty*. Springer.

Yamamoto, L., & Esteban, M. (2017). Migration as an adaptation strategy for atoll island states. *Journal of International Migration, 55*(2), 144–158. https://doi.org/10.1111/imig.12318

Climate Change and the Case of Grenada's Blue Growth Plan: Using the SDGs to Propose a Policy Planning Framework for SIDS's Sustainable Development

John N. Telesford

Abstract Grenada, a Small Island Developing State (SIDS) in the Caribbean, proposed a blue growth plan that is inclined to investments in tourism infrastructure along the coast. This raises concerns about the vulnerability of SIDS to the pressures of climate change. But this tourism centric approach is expected, as the sector contributes significantly to economic growth and development. However, with the identified vulnerability of SIDS, some proposed projects in blue growth plans should be critically assessed. The vulnerability/resilience debate is interrogated as a precursor for the sustainable development of SIDS. The chapter then proposes that the Sustainable Development Goals (SDGs) can provide a framework for analysing the implementation of blue growth projects. Using Grenada as a case study, the blue growth investment plan is critiqued from the perspective of building resilience to climate change pressures and sustainable development. The proposed framework is then illustrated using a sample of SDGs and blue growth projects. Specifically, synergies and trade-offs between the SDGs and targets are analysed. The chapter concludes with a discussion on how the framework may enhance policy planning that leads towards sustainable development and resilience building of SIDS.

Keywords Blue economy · Climate resilience · Coastal and marine resources · Economic growth · Grenada · Sustainable Development Goals (SDGs) · Small Island Developing States (SIDS) · Tourism · United Nations · Sea level rise

---

J. N. Telesford (✉)
T. A. Marryshow Community College (TAMCC), Tanteen, St. George, Grenada

Institute of Island Studies, University of Prince Edward Island, Charlottetown, PEI, Canada
e-mail: j.n.telesford@rgu.ac.uk

S. Moncada et al. (eds.), *Small Island Developing States*, The World of Small States 9, https://doi.org/10.1007/978-3-030-82774-8_14

1 Introduction

This chapter proposes the conceptualisation of a policy planning framework for development using the Sustainable Development Goals (SDGs), and demonstrates its application for assessing blue growth plans in a Small Island Development State (SIDS). Due to the vulnerability of islands to climate change pressures, blue growth plans may have negative impacts on their sustainable development. Moreover, building island resilience can be hindered with the implementation of such plans. Therefore, as islands seek to transition their economies to 'blue economies', careful consideration should be given to ensuring island resilience and sustainable development.

The blue economy provides an excellent prospect for the future economic development of islands. In this regard, the wealth contained along the shoreline and in the volume of blue-ocean, known as the Exclusive Economic Zone (EEZ), can be tapped into for development. In many cases, the EEZs of islands are about fifty (50) times more expansive than their land masses. Therefore, SIDS have an intricate link to the seas and the oceans that surround them. According to the UNEP (2012, p. 20):

> [W]ith most SIDS populations and economic activities located in the narrow coastal belt, one of the closest connections between humanity and the blue economy is found among those who harvest the seas for a living.

Coastal tourism is a key feature of island economies, as they depend on the shoreline for tourist attractions, such as beaches and tourist accommodations, e.g. hotels. UNEP (2012, p. 20) suggests that: "[T]ourism, another dominant economic activity in SIDS, is the largest source of foreign exchange for more than half of all countries". However, due to "Human development pressures on one hand, coupled with sea level rise and associated impacts such as erosion on the other...", the phenomenon- coastal squeeze is created (UNEP, 2014, p. 25). With the intensification of severe storms and the expected surge in sea level rise (SLR), coastal squeeze could be exacerbated (UNEP, 2014).

The significance of the blue economy and blue growth strategies to the Caribbean region was further highlighted by Patil et al. (2016, p. 25): "[P]erhaps nowhere else is the ocean economy relevant than in the Caribbean Sea, where many states and territories are defined by the Ocean". This relevance of the blue economy may have sparked the development of a Blue Growth Plan by the small Caribbean island of Grenada. The Blue Growth Coastal Master Plan, which contains nine (9) incentive zones for infrastructure development, proposes 55% of the total investment for the Grand Anse Bay Tourism Development, located in the south of the island. This raises the alarms for the future resilience of the coastal corridor. However, the tourism centric plan is understood, as tourism is a significant contributor to Caribbean SIDS's economies. The World Travel and Tourism Council (WTTC, 2018a) reported that in 2017, travel and tourism contributed 15.2% to the gross domestic product (GDP) and accounted for 13.8% of all jobs in the Caribbean.

The development and eventual implementation of blue growth plans and strategies therefore, require critical and deliberate assessment of the projects, especially tourism-based projects. In this regard, the concept of the blue economy is instructive, that is addressing "... the improvement of human well-being and social equity, while significantly reducing environmental risks and ecological scarcities" (UNCTAD, 2014, p. 2). Blue economy planners and policy makers must take cognisance of the potential risks that some tourism focused projects can pose to coastal ecologies, thus stymying island sustainability and resilience (see e.g. Berry, 2015). From this perspective, the sustainable development agenda 2030 and the goals that guide it, can provide a practical approach to assessing blue growth projects' contribution to sustainable development. This chapter attempts to demonstrate how this can be accomplished.

The chapter consists of the following sections. Section 2 interrogates the literature on resilience and vulnerability, and suggests the need for resilience building. Section 3 discusses the Blue Economy concept. In the fourth section the case study island and nature of the blue growth plan are discussed. Section 5 proffers the planning framework and demonstrates it application, while Sect. 6 discusses the policy implications of the planning framework for the sustainable development of SIDS in the face of climate change. The final section concludes the chapter.

2 Resilience and Vulnerability: Two-Sides of the Same Coin?

Due to the nature of many SIDS, they are mainly characterised and described by vulnerabilities. In this regard, Haskins (2012) notes:

> In the case of small island [developing] states, vulnerability arises from the fact that the environment and economies of these states are to a large extent shaped by forces outside their control.

It is widely accepted therefore, that most SIDS are vulnerable to external shocks brought on by global economic crisis and climate change. More specifically, and in the context of this chapter, the vulnerability of SIDS to climate related hazards are in question and to the extent that, coastal physical infrastructure development will hinder building the future resilience of SIDS. Therefore, the vulnerability/resilience nexus is important to the future sustainable development of many SIDS; and forging ahead with building resilience in SIDS, must be considered in the context of vulnerability.

In this regard, Kelman (n.d., p. 8) recommends: "Irrespective of the choice of definition or background to interpret resilience, connections must be made with vulnerability". This recommendation appears to be associated with some SIDS. For example, the results of one study conducted in the Seychelles, concluded that the two concepts were "... not mutually exclusive, but complementary and even symbiotic" (Philpot et al., 2015, p. 32). In this regard, five (5) connections between the concepts and their conceptualization were detected: conflict between the concepts; and compromise, complementarity, symbiosis and transformation (see Philpot et al., 2015, pp. 43–44). Island vulnerability and resilience may be perceived and/or conceived as concepts on either side of the same coin.

Using evidence from their research, Philpot et al. (2015), provided some explanations and examples of their findings. However, for purposes of this chapter, the suggestions of vulnerability and resilience likened to, different sides of the same coin, suggests a symbiotic relationship that SIDS policy makers and planners should be cognisant of. Philpot, Gray and Stead wrote:

> Symbiosis means that there is a dynamic interaction and mutual dependence between conceptions of vulnerability and resilience. Far from being in conflict, they need each other: a conception of vulnerability implies a conception of resilience and a conception of resilience presupposes a conception of vulnerability. . .Indeed there is a synergistic relationship between them. . . Vulnerability is a necessary (if not a sufficient) precondition of resilience (Philpot et al., 2015, p. 43).

It is a suggested requirement therefore, that SIDS policy makers attempt to address this complex relationship by seeking to build resilience. In other words, SIDS policy makers should strive for "transformation", which means that ". . . a SIDS decisively shakes off the threat of vulnerability, transforming it into a secure, ecological, economic, social and political state. . ." (Philpot et al., 2015, p. 44). However, vulnerability is inherent to many SIDS, and this ". . . is largely due to intrinsic characteristics . . . including: small size, remoteness, vulnerability to external shocks; . . . and exposure to global environmental challenges" (UNDESA, 2010). In this regard, Briguglio (2010, p. 1) noted that: ". . . the risk of being harmed by external shocks is the outcome of inherent vulnerability features and policy induced resilience building". Moreover, Haskins (2012) reiterated: "While vulnerability is inherent to small islands, strategies can be put in place to reduce the risk of being harmed by external shocks by increasing their resilience". Building the resilience of SIDS therefore, is a policy-oriented process.

However, resilience building in the face of climate change threats such as sea level rise (SLR) must be considered in the broader context of development and disaster risk reduction. From this perspective, Kelman et al. (2016) argue: considering vulnerability and resilience within the narrow context of climate change as one environmental stressor, can frustrate other pressing development issues such as poverty. Kelman et al. wrote:

> . . . vulnerability and resilience have been analysed theoretically and empirically to demonstrate 'multiple exposures' at 'multiple scales'. Vulnerability to hazards such as earthquakes and vulnerability to hazard drivers such as climate change, alongside poverty, inequality, injustice, . . . resource distribution, and lack of access to education and health opportunities-among other challenges- are linked to each other and tend to overlap to cause the most adverse impacts for those who have the fewest choices, resources, and abilities for addressing the challenges (Kelman et al., 2016, pp. S134–S135).

Using this as the point of departure, the Sustainable Development Agenda 2030, with the SDGs, suggest a framework, for analysing projects that are proposed along vulnerable communities such as the coastal regions of SIDS. The SDGs focus on the pillars of people, planet, prosperity, peace and the intent is to integrate these into a whole. In this regard, a SDGs framework should encompass the 'multiple exposures' necessary for assessing vulnerabilities and resilience building. In the next section, the concept of the Blue Economy, as one of the most recent thrust for the development of SIDS in such sensitive areas is interrogated. The Blue Economy therefore, can impact resilience building and by extension the sustainable development of SIDS. The chapter proposes that the SDGs can be used as a policy planning framework within which blue economy projects can be assessed for their (lack of) contribution to the sustainable development and resilience building of SIDS.

3 The Concept of the Blue Economy

There is a new thrust from many SIDS to take a more focused approach to economic growth and development based on the Blue Economy. The UNEP (2015) highlighted several Blue Economy case studies at the regional, national and community levels. The two case studies at the national level were SIDS. It is apparent therefore, that SIDS have taken a keen interest in transitioning to blue economies. But what is the Blue Economy? The Oceans have been exploited for decades by many nations and this has fuelled economic growth. However, this exploitation has changed ocean ecosystems "... in ways that have threatened to reduce their benefits to future generations" (Patil et al., 2016, p. 8). In this regard

> ... this threat has given rise to a new concept or framework for action to reduce the risks: the transition of the current ocean economy to a 'blue economy' with 'blue growth'. This vision foresees even higher numbers of people making their living from the ocean, but in ways that protect its resources and ecological integrity (Patil et al., 2016, p. 8).

The Blue Economy concept therefore, seeks to tap into the ocean for economic growth, while ensuring social equity and the maintenance of healthy ocean ecosystems. The Blue Economy should contribute to the sustainable development of the SIDS that are transitioning to such an economy. According to the Roberts and Ali (2016, p. 24), for small states "the key sectors" in the blue economy "... include fishing, marine transport and tourism".

The Seychelles Blue Economy strategy focuses on such sustainable development as they strive for:

> increased investment in existing ocean-based sectors (particularly tourism and fisheries); the development of new economic sectors based on existing marine resources (for example marine based aquaculture); greater protection for Seychelles' ocean spaces and resources...; new research, innovation and generation of knowledge... (Purvis, 2015, p. 15).

Berry (2015, p. 20) further writes:

> With the inclusion of socio-ecological frameworks, the Blue Economy aims for the sustainable use of coastal and marine resources, while at the same time acknowledging the importance of socio-economic necessity for current and future generations.

However, Spalding (2016) warns, that the traditional activities/enterprises associated with the Ocean Economy may not achieve the sustainable development gains intended by the Blue Economy.

Therefore, strategies and projects, especially coastal tourism-based projects, must be carefully conceptualised and implemented if the intent of the Blue Economy concept is to be achieved and maintained over a long period of time. From the perspective of coastal development projects, the resilience of coastal ecosystems should be maintained and/or enhanced. Berry concludes:

> The Blue Economy is a concept that aims at increasing human well-being and social equity through the long-term sustainable use of coastal and marine resources. For this to become a reality, the Blue Economy needs to consider the full gamut of both coastal and marine ecosystems, along with their structure, function and interconnectivity (Berry, 2015, p. 24).

In other words, coastal and marine ecosystems are vulnerable to harm from SLR and as such resilience building should be a focus of Blue Economy plans, if a sustainable development outcome is to be achieved.

Additionally, there are socio-economic concerns relating to the Blue Economy projects along vulnerable coastal regions. Baldacchino and Kelman (2014, p. 9) proffer "losers" and "winners" of climate change impacts especially in low lying islands and coastal regions of islands. From this perspective, the inevitable losers are those who are "poor and powerless" and who may be inhabitants of areas most vulnerable to climate impacts and unattractive to tourism investors. On the other hand, there are "rich and powerful" losers, who consist of investors in tourism infrastructure in coastal regions. The claim to economic growth from the Blue Economy is often wrapped up in a promise of jobs, especially from new tourism investments. However, due to the vulnerability to SLR, such investments can be threatened or may even disappear and with it the jobs that were provided. According to Grenada's Second National Communication (SNC) to the UNFCCC: "As beaches degrade, price structures for coastal accommodation may decline, affecting profitability and potentially leading to closures, unemployment and further decline in overall destination range (Scott et al., 2012)" (GOG, 2017a, p. 229). Although the 'rich and powerful' investors loose, they can easily recover investments through insurance and move on. However, the 'poor and powerless' workers may be back on the breadline, thus hindering sustainable development from a socio-economic perspective.

4 Grenada's Blue Growth Plan: Towards Sustainable Development?

Grenada is a small, mountainous, volcanic island in the southern Caribbean. Measuring about 344 sq. km and with a perimeter of only 120 km., the space for physical infrastructure development is limited. Moreover, as the island is predominantly mountainous, infrastructure development is mainly coastal, including the vast space utilised for tourism development along the southern shores. Grenada's economy is highly dependent on services, especially focused on the tourism industry. According to the CIA World Factbook (2018), services comprise about 75% of the total GDP of Grenada's economy. The WTTC (2018b) reveals that the total contribution of travel and tourism to the GDP of Grenada was 23.3% in 2017 and that was projected to grow by 2.6% in 2018. Therefore, the need to expand and diversify the economy, including the tourism industry, is enshrined in Grenada's proposed Blue Growth Master Plan. According to the government of Grenada (GOG) (2017b) in its Investment Prospectus for Blue Growth 2017: boutique tourism, marine research, eco-tourism, are future investment sectors for developing Grenada's economy.

The expansion of the development prospects for Grenada, as with all other small island developing states is required, and the turn to the vast economic exclusive zone (EEZ), including the near-shore assets can be justified. Grenada's EEZ is about 26,000 sq. km making it about 75 times larger than the island. There are prospects for further infrastructure development, but these must be carefully planned and located.

Grenada's blue growth plan has provided the opportunity for the government of Grenada to expand its economic growth possibilities, while attempting to maintain the sustainability of the island. Thus, "Grenada's blue growth vision is: to optimise the coastal, marine, and ocean resources to become a world leader and international prototype for Blue Growth and Sustainability" (GOG, 2016, p. 1).

The plan further considers the environmental impacts that may be associated with the planned infrastructure investments. For example, Integrated Coastal Zone Management (ICZM) should guide the implementation of the plan. In this regard, consideration must be given to the ICZM policy which

> ...provides a vision for the future use, development and protection of the nation's coastal zone by setting out policies to guide the relationships among resource users, community facilities and activities and physical development and infrastructure (GOG, 2016, p. 3).

The principles of sustainability; the precautionary principle; good governance and equity, will dictate the future management of the coastal zone and the implementation of the projects in the plan (GOG, 2016, p. 4). This upfront focus on coastal zone management is required, however, the nature of and focus on the infrastructure and physical developments along the coast raises concerns about the sustainable development of the island as a whole. This concern is addressed in detail in subsequent Sects. 5 and 6.

The proposed investments are clustered into nine (9) development incentive zones: Petite Martinique (marine service, including marina and marine housing communities); Carriacou (boutique tourism, but also including marinas,); Northern Grenada (Ocean research, including marinas); Victoria (Eco-tourism); Gouyave (fisheries and agriculture, including fishing fleets, marinas); St. George's/Grand Anse (global tourism, resorts, casinos, medical tourism); Southern Grenada (science and technology); Southeast Grenada (retirement residence, including boutique tourism) and Grenville (shipping and industry relocation of main port) (GOG, 2016, p. 19). This chapter focuses on the large island of Grenada and a 'screen-short' of the development zones is provided in Fig. 1.

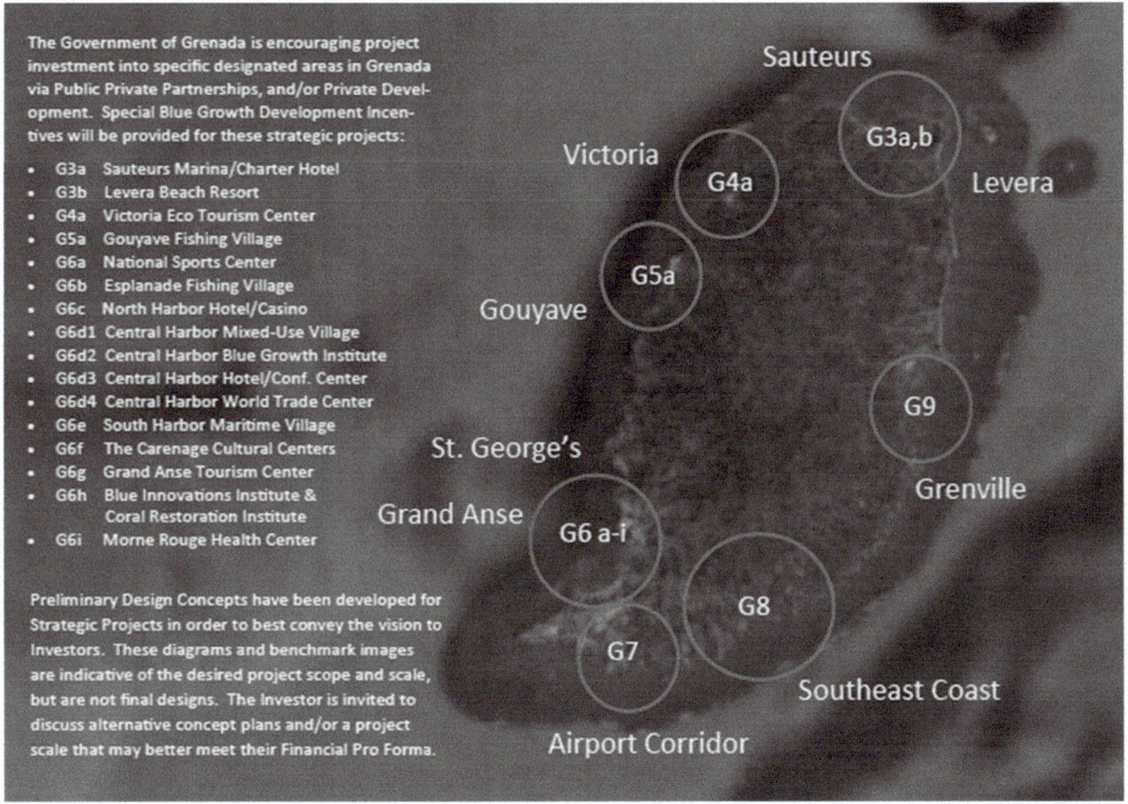

Fig. 1 Development zone on the island of Grenada. *Source: Screenshot from the Blue Growth Master Plan 2016 by the Author, GOG* (2016)

The proposed investment outlay for the development in the zones is estimated to be US$2.7 to $3.3 billion. Such investments may be required to pursue the further development of Grenada's economy. However, the nature of some of the proposed projects can be questioned in the context of achieving island sustainability. For example, the Northern zone (G3) project located adjacent to a National Park is also a popular location for the spawning of the endangered Leather Back Turtle. Similarly, some of the projects earmarked for the Grand Anse Bay Tourism zone (G6 a-l), stretching from the town of St. George to the famous Grand Anse Beach Area is focused mainly on the development of more tourism infrastructure that attracts the majority (approximately 55%) of the estimated investment funds. These include for example, a hotel with casino, the maritime area with proposed hotels and mega yacht docking and other hotel projects, which are located in sensitive ecological zones and may be vulnerable to threats of SLR. According to Grenada's SNC: "Higher sea levels will accelerate beach erosion due to wave action... For example, Grand Anse, a popular beach for tourists, is projected to lose up to 77% beach area under a 2 m SLR (Simpson et al., 2012)" (GOG, 2017a, p. 229).

In that same area, extensive material stocks, represented by buildings such as hotels, the airport and other critical infrastructure, have accumulated along the south-western coast, see Fig. 2. This stock is highly vulnerable to climate related threats such as SLR and intense storms. In 2004, Grenada experienced one of the most devastating storms. According to the World Bank Report (World Bank, 2004, p. 10): "Close to 70 per cent of tourism infrastructure... was made unusable", after the passage of the storm. Therefore, it may be of some concern that more infrastructure development is proposed for that area in the face of future SLR and more frequent and severe storms.

Additionally, a number of the projects suggest an economic growth focus, invoking the contested idea of blue growth (e.g. Ertör & Hadjimicahel, 2020). This neo-liberal growth perspective further implies that the economy as one pillar of sustainable development is now shrouded in a 'blue cloak', much to the possible detriment of the other two pillars—environment and society. With this thrust, 'mass tourism' appears to be the way forward. However, many proponents (see for example Roberts & Ali, 2016; Spalding, 2016) of the Blue Economy concept warn against mass tourism and the possible threats to coastal regions as a result of poor infrastructure development. In this regard, Roberts and Ali (2016, p. 27) note: "Rising sea levels is the biggest threat to coastal areas in the long term, but uncontrolled mass tourism is the most pressing". Moreover, Spalding (2016, p. 7) writes:

> Simply stated, there is sustainable and responsible tourism; and there is unsustainable and disruptive tourism. ... Tourism infrastructure can seriously damage coastal areas if not planned with socially and environmentally recognised best practices.

In other words, unsustainable tourism development, especially along coastal regions can hinder the achievement of island sustainability. Sustainable tourism should be the preferred approach for the sustainable development of the country.

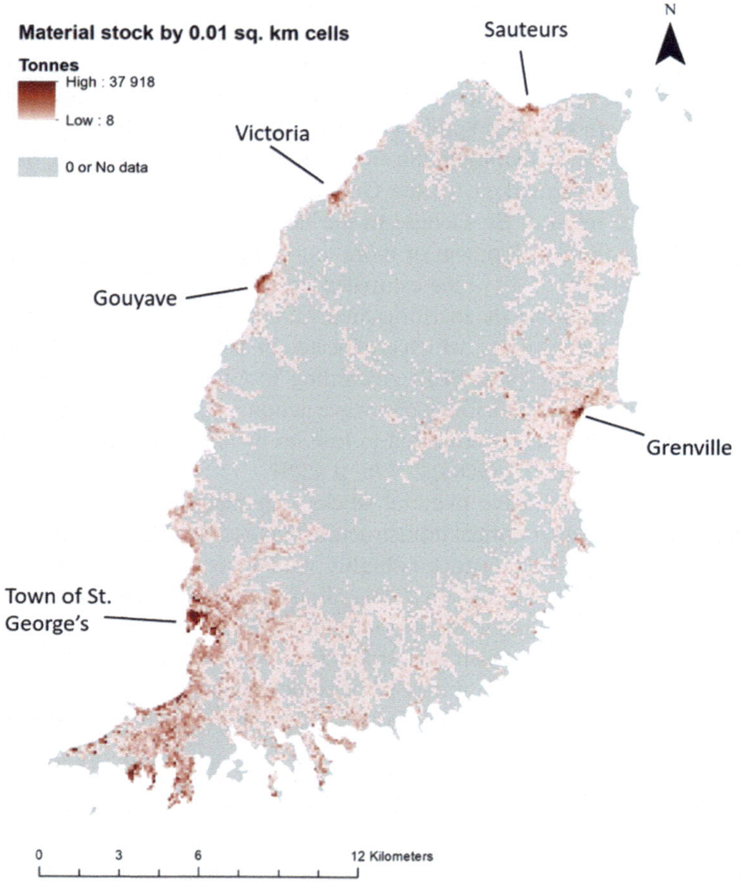

Fig. 2 Accumulation of material stocks (concrete, aggregate, timber and steel), along the south-western tip. *Source: Symmes et al.* (2020)

The key question therefore is: with some of the projects in the current blue growth plan, is island sustainability 'still' achievable? Although the blue growth plan is intended to be implemented within an ICZM plan, a holistic approach to the future development of Grenada will be critical to the success of the blue growth strategy. It is proposed therefore, that the SDGs and specifically SDG 14, its targets and interactions with other SDGs and their targets can be used for gauging progress towards the sustainable development and resilience building of islands.

5 The SDGs as a Planning Framework

The Sustainable Development Agenda 2030 was agreed to at the United Nations Summit in 2015. 17 Sustainable Development Goals and 169 targets guide the agenda (UNGA, 2015). The agenda and by extension the SDGs, focus on people, planet, prosperity, peace and partnerships. In this regard, the SDGs should not be viewed as a 'tick-box' for development, but rather as a set of interacting goals and targets that can be used as a policy planning tool and for analysing progress towards sustainable development at global and national scales (Xu et al., 2020). According to Nilsson et al. (2016, p. 320):

> Implicit in the SDG logic is that the goals depend on each other - but no one has specified exactly how. International negotiations gloss over trade-offs. Still balancing interests and priorities is what policy-makers do- and the need will surface when the goals are implemented.

Therefore, the interdependent nature of the SDGs, allows them to be used as a policy planning framework for SIDS sustainable development. In this regard, Prajal et al. (2017) note: "In order to make policy happen, the dependencies among the goals in terms of potential interactions need to be evaluated, across... and within... the SDGs". From this perspective, synergies and trade-offs (see Eisenmenger et al., 2020) between SDGs and their targets can be assessed and used to inform policy direction towards sustainable development. According to the International Council for Science (ICSU) (2017, p. 07): "Understanding possible trade-offs as well as synergistic relationships between the different SDGs is crucial for achieving long-lasting sustainable development". Le Blanc reiterates:

> Lack of integration across sectors in terms of strategies, policies, and implementation has long been perceived as one of the main pitfalls of previous approaches to sustainable development. Insufficient understanding and accounting for trade-offs and synergies across sectors have resulted in incoherent policies. (Le Blanc, 2015, p. 1)

From this perspective, problem-shifting and displacement can occur, when major development projects, such as the blue growth plan, that should place a country on to a path of sustainable development are proposed and implemented (Baumgartner & Kohonen, 2010). Therefore, Blue Economy projects should be carefully assessed for their overall contribution to sustainable development. The World Wildlife Fund (WWF) proposed a number of principles and sub-sets of principles that can help to define and implement a "sustainable Blue Economy". One relevant principle suggested by the WWF (n.d.) is that: "To create a sustainable Blue Economy, public and private actors must... Set clear, measurable, and internationally consistent goals and targets for a Sustainable Blue Economy". In the context of SIDS, striving to build resilience and to achieve sustainability through Blue Economy activities, this principle is very critical. According to the WWF:

> Governments, economic sectors, individual businesses and other sectors must all set relevant and measureable goals and targets for a Sustainable Blue Economy to provide planning, management and activities with a clear direction. Goals and targets for different economic,

> social and ecological areas- as well as related policies and activities- must be made as integrated and coherent as possible to avoid conflicts and contradictions. (WWF, n.d.)

The SDGs, which are universally accepted, including by all SIDS, provide an excellent platform for implementing this sub-principle. The goal in focus is SDG 14: *Conserve and sustainably use the oceans, seas and marine resources for sustainable development.* In the context of islands and the new paradigm of blue growth strategies, SDG target 14-7: *By 2030, increase the economic benefits to Small Island Developing States and least developed countries from the sustainable use of marine resources, including through sustainable management of fisheries, aquaculture and tourism*, is the locus of concern. In-keeping with the need to ensure the interactions of SDG target 14-7 and all the SDGs, the World Bank (WB) and UNDESA (2017) noted:

> Target 14-7 is related to all other SDGs and is instrumental for reaching many of them. At the same time, success in achieving other SDGs often underlies the ability to reach target 14-7 and the other SDG 14 targets.

Although the interconnectedness of the SDG 14 targets is demonstrated, (see World Bank & UNDESA, 2017), it lacks an analytical approach for identifying synergies and trade-offs, that will inform policy towards the sustainable development of SIDS.

Systematically analysing trade-offs and synergies between SDGs and their targets has been the focus of recent research (e.g. Nilsson et al., 2016; ICSU, 2017; Prajal et al., 2017; Weitz et al., 2018). The application of this approach is novel and instructive for SIDS policy makers, as it provides them with a framework which may be used to systematically think about the many interactions amongst the SDGs and to craft 'integrated policies'. In many public sectors and especially in SIDS's governments, policy is led from the responsible ministries, but if applied, this framework may bring policy makers on the same page.

Moreover, the private sector, which drives investments can also be brought in sync with the policy direction for development. Other stakeholders, such as community groups, will also be more meaningfully engaged. In this regard, the framework can be used to critically analyse project ideas and develop integrated policies geared towards building island resilience and sustainable development.

The proposed analytical framework uses a scale to assess the interactions between the SDGs and their targets along a continuum shown in Table 1. The highest level of interaction is rated as '+3; indivisible', while at the other end '−3: cancelling' denotes most negative interaction. In this regard, a synergy/trade-off continuum can be envisioned. For example, the ICSU (2017) suggests: "Reduction of air pollution (SDG 12.4) is indivisible from improved health and reducing non-communicable diseases"; on the other hand, "[d]eveloping infrastructure (SDG 9-1) could be cancelling the reduction of degradation of natural habitats in terrestrial ecosystems".

The application of the framework is illustrated using SDG 14 and targets and their interactions with other SDGs. Nilsson et al. (2016) suggest that SDG interactions should start with the SDG of focus and interactions with all other SDGs should be

Table 1 Indivisible/cancelling continuum of interactions between SDGs and their targets, ranging from '+3', indivisible, to '−3', cancelling. *Positive scores suggest a synergistic relationship, while negative scores may be a trade-off. Sources: Adapted from Nilsson et al.* (2016)*, ICSU* (2017*), and Weitz et al.* (2018*)*

	Goals score	Description
Indivisible	+3	The strongest form of positive interaction in which one objective is inextricably linked to the achievement of the other
Reinforcing	+2	One objective directly creates contributions that lead to the achievement of another objective
Enabling	+1	The pursuit of one objective enables the achievement of another objective
Consistent	0	A neutral relationship where one objective does not significantly interact with another or where interactions are deemed to be neither positive nor negative
Constraining	−1	A mild form of negative interaction when the pursuit of one objective sets a condition or a constraint on the achievement of another
Counteracting	−2	The pursuit of one objective counteracts another objective
Cancelling	−3	The most negative interaction where progress in one goal makes it impossible to reach another goal and possibly leads to deteriorating state of the second. A choice has to be made between the two

analysed. In this regard, SDG 14, focusing on the blue economy, is assessed with SDG 1, considered to be one of the most important socio-economic SDGs. The intent is illustrative and therefore, one example is chosen considering the restrictions on the length of the chapter.

5.1 An Illustration of Applying the Framework to the Blue Growth Plan

The SDGs in question are therefore: SDG 14: *Conserve and sustainably use the oceans, seas and marine resources for sustainable development*; and SDG 1: *End poverty in all its forms everywhere.* The key interactions between example targets will be demonstrated and examples of how some of the proposed blue economy projects contribute to (un)sustainability is unfolded. After describing trade-off and synergies therefore, the implications for policy-makers attempting to apply blue economy strategies for sustainable development are illuminated.

The scores shown in each of the interactions below are adopted from the ICSU (2017) proposal. This was done as the intent of the demonstration is not to score possible interactions and argue for or against a score, but rather to uncover the possible synergies and identify possible trade-offs that may be hidden in the Blue Growth projects. The examples provided are merely the opinion of the Author and as such will be limited in scope. However, the illustration establishes a foundation upon which further deliberation on blue growth projects on islands can occur, and may guide the selection of coherent policies that will move SIDS towards sustainable development and resilience building. This exercise therefore is not intended to be exhaustive.

5.1.1 SDG14 + SDG 1

The interactions between SDGs 14 and 1 and some example targets are shown in Table 2 below From the onset the ICSU (2017) suggests that: "SDG 14 is a critical enabler of poverty alleviation, and environmentally sustainable economic growth and social well-being (blue growth), particularly in SIDS...". They further propose however, that,

> Strengthening the resilience of oceans and coasts, for example through conservation and protection of coastal wetlands, will help reduce shock exposure and enhance resilience of poor coastal populations to extreme climate-related events. However, **blue growth** (Author's emphasis) policy interventions aimed at achieving economic growth to lift people out of poverty might impair ocean health and promote exploitation of marine resources. (ICSU, 2017)

Table 2 Examples of possible trade-offs and synergies between examples of SDG 14 and SDG 1 targets. *Source: adapted from ICSU (2017)*

Targets	Key interactions	Score
14.2: ...sustainably manage and protect marine and coastal ecosystems to avoid significant adverse impacts, including by strengthening their resilience, and take action for their restoration in order to achieve healthy and productive oceans + 1.1: ...eradicate extreme poverty for all people everywhere, currently measured as people living on less than $1.25/day 1.2: ... reduce at least by half the proportion of men, women and children of all ages living in poverty in all its dimensions according to national definitions	Healthy and productive oceans benefit small-scale fisheries, improve tourism and increase potential for blue carbon markets. In other words, meeting target 14.2 has the potential to improve the livelihood of persons in poverty, thus hopefully eradicating poverty.	+2
14.7: ... increase the economic benefits to small island developing states... from the sustainable use of marine resources, including through sustainable management of fisheries, aquaculture and tourism + 1.1, 1.2	Sustainable tourism, fisheries, ... can also create jobs and income; improve the livelihoods of both men and women, thus reducing poverty. This results in an 'indivisible' interaction among the targets	+3
14.7 + 1.1, 1.2	Increased economic activity to reduce poverty, creates more pressure on coastal and marine resources and more environmental harm, which may constrain meeting target 14.7	−1

The first two interactions are '+2' and '+3', indicating reinforcing and indivisible relationships, respectively, between the targets shown in Table 2. Indeed, the many coastal development projects, proposed by the plan especially in the North-zone G3 and South- zones G 6, 7 and 8, may have tremendous potential to crate many jobs that will lift persons out of poverty. Moreover, in Gouyave- zone 5a, the development of facilities to support the fisheries sector has similar potential. However, there appears to be a constraining factor on meeting target 14-7, especially in zones G6, 7 and 8, as represented by the '−1' or constraint shown in Table 2. Due to the scale of the projects and their close proximity to sensitive coastal ecosystems, more pressure on coastal and marine resources over time can be created. For example, in both zones G3 and G8, there are critical mangrove ecosystems that support the fisheries for Grenada, and which provide natural resilience to threats from climate impacts. If these mangrove ecosystems are compromised, the fisheries can be diminished and with it the jobs and livelihoods that were created by the sector may disappear.

Similarly, the scale of the infrastructure development along zones G6, 7 and 8, as was previously recognized, may be placing more infrastructure in harm's way, while also constraining the ability to achieve target 14-7. In this regard, Fig. 2 shows a massive build-up of construction material stock in the south-western part of the island. Most of this stock provides services for the accommodation of tourists. With more construction earmarked for that location, and the vulnerability to SLR, the future of these facilities is questioned. One study (Symmes et al., 2020) predicts that about 26–33% of material stocks in the tourist accommodation sector can be lost under 1m and 3m SLR scenarios, respectively. This is instructive for policy makers, for the economic benefits and services that these stocks will provide are threatened in the long term as the resilience to SLR along the coast may be diminished by the stock. In this regard, the natural capacity of the coast to adapt to SLR (see for example Berry, 2015) can be compromised by the increased stock accumulation. By reducing resilience to climate shocks, the infrastructure that have led to improved livelihoods, may be lost and so too the jobs that have pulled persons out of poverty.

The scale of some of the projects in the blue economy plan further suggests a 'mass tourism' approach, for example the North Harbor Hotel/Casino in zone G6. It was previously argued that 'mass tourism' as a blue growth strategy may not achieve the object of sustainable development or even build the resilience of the coastal region of the island. According to Spalding (2016, p. 7): "Tourism infrastructure can seriously damage coastal areas if not planned with socially and environmentally recognized best practices". Therefore, island policy makers and planners must carefully consider the scale of the projects, in the context of the socio-economic and environmental problems that may manifest in the future.

Therefore, the 'indivisible' interaction among the three targets demonstrated in Table 2 specifically calls for sustainable tourism and fisheries management. The Victoria Eco-tourism project, zone G4a, epitomises this synergistic interaction. Sustainable tourism projects such as eco-tourism may be one approach that can be considered in more detail under the blue growth plan. For example, it was already noted that turtles use the beach located in close proximity to the boutique hotel proposed in zone G3. One community group already uses the opportunity to protect the nesting of the turtles and provides tours for tourists to view the nesting. Therefore, it will be prudent that any such development in that area should take cognisance of the socio-economic and environmental benefits that can accrue from a sustainable tourism approach. If this is done, policymakers and private developers will be ensuring that the tightly coupled interaction among the SDG targets demonstrated in Table 2 is maintained over time.

Other projects within the blue growth plan such as the blue institutes proposed for zone G6, have tremendous potential to improve socio-economic benefits, while enhancing the terrestrial environment. In this regard, educational tourism has been one of the largest contributors to the GDP of Grenada. Therefore, these institutes have the potential to employ more persons, while restoring coral reefs as is the intent of the Blue Innovation Institute proposed by the plan. If planned and implemented effectively, these institutes have the potential to ensure that the indivisible coupling between the example targets shown in Table 2 are achieved and sustained into the future.

Finally, fishing is the focus of two projects in zones G1a and G5. These projects appear to enhance the capacity of the fisher folks on the island by providing improved facilities, such as storage and processing facilities. However, the sustainability of the fishing industry also includes sustainable fishing mechanisms and plans that will ensure the fisheries stocks are maintained over time. Therefore, policy makers should ensure that sustainable fisheries' plans are developed, implemented and enforced if SDG 14 and its targets are to be met. It must be kept in mind, that the sustainability of the fishing sector should not only be measured by the number of boats and capacity for storage, but that the maintenance of the fisheries stock is equally important, if not the most important measure of the sustainability of the sector. This example highlights a possible hidden trade-off, as if unsustainable fisheries is promoted through increased fishing fleet and other amenities, then the SDG 14 and its targets may not be met and maintained.

6 Using the SDGs Framework: Implications for Policy

6.1 Integrated Policy, Goal-Setting and Planning Towards Sustainable Development

This illustrative framework suggests some critical implications for policy makers. Firstly, the framework provides a platform for 'goals-based' policy setting, and in this regard, supports integrated policy planning and actions. Sachs (2015), coined the phrase 'goals-based development' to highlight the importance of the SDGs and the role they can play in focusing policy makers on sustainable development. Le Blanc (2015) suggests that the integration of strategies and policies are critical for sustainable development; while the WWF (n.d.) notes that goal setting is important for the sustainable Blue Economy. Moreover, this approach also sets a stage for bringing other stakeholders to the policy table.

In this regard, the sustainable development framework can be used to set clear, measurable goals and targets for a Blue Economy, which policy makers, blue growth planners, private investors and communities can all use for planning activities and projects for a sustainable blue economy. In other words, for a particular project, say eco-tourism in zone G4, the private investor, government and the community where the project is to be developed can all be exposed to and deliberate on the pros and cons of the project as it relates to a sustainable blue economy, building resilience and the sustainable development of the island. All stakeholders therefore, can simultaneously interrogate questions relating to the project, such as is inclusive economic development possible? Is full and productive employment and decent work, which will bring people out of poverty, proposed and can it be sustained? Is the project going to enhance environmental and socio-economic resilience?

On the targets level, a more granular approach can be achieved. Specific targets can be used to establish a baseline from which future assessment of the projects' contribution to sustainable development and resilience can be measured. For example, in Table 2, the interactions between 14-7 that addresses increasing economic activities in SIDS and 1.1- … *eradicate extreme poverty for all people, currently measured as people living on less than $1.25 a day*, and 1.2- … *reduce at least by half the proportion of men, women and children of all ages in poverty in all its dimensions according to national definitions*, can be used at the community level to measure how the project is contributing to poverty reduction. In other words, all stakeholders will have a baseline to gage the contribution of the specific project to the selected goals. Moreover, this interaction was seen as 'indivisible' thus making these targets, extremely important for sustainable development and building climate and socio-economic resilience.

Therefore, the proposed framework can show how polices such as the proposed ICZM policy, climate change policy, investment policy, tourism policy and poverty reduction policy can all be integrated into a whole for moving towards sustainable development and building resilience. The SDGs provide a foundation upon which such policies can be developed. With its 'indivisible' intent and the ability to identify trade-offs, an integrated policy direction can be achieved.

The SDGs platform can also bring the policy makers in different departments and ministries on to the same page. In this regard, the framework can catalyse discourse and partnerships among ministries and other agencies responsible for policy direction and planning. It was previously discussed that fragmented policy planning for islands may result from the fact that policies are developed in disparate ministries/departments. In the case of this blue growth plan, the core policy direction appears to be one focused on ICZM. So maybe ministries responsible for fisheries and/or environment will drive such a policy. However, the blue growth plan has implications for other socio-economic issues, such as poverty reduction, inclusive job creation and other community development issues. With the proposed framework all key policy makers can be forced to talk, as they seek together to bring the island on to path of sustainable development.

6.2 Policy, Resilience Building and Sustainable Development

Due to the inherent vulnerability of SIDS, policy makers must continue to focus attention on building island resilience. From this perspective, resilience building becomes an important foundation for the sustainable development of islands. Policy geared towards resilience building therefore, must take a holistic approach, and this becomes more critical as blue growth strategies are implemented. Blue economy plans have tremendous potential to improve the sustainable development of SIDS. In this regard, the framework in this chapter suggests one approach for crafting such integrated policies.

The framework illustrated how the intent of the SDGs, which is to represent an 'indivisible whole' of sustainable development, can be integrated. However, in implementing the SDGs, trade-offs are possible. Drawing on the illustration in the previous section, the synergy/trade-off analysis can assist policy makers to pay careful attention to simultaneously meeting the obligations to reduce poverty, while supporting coastal and marine ecosystems' conservation. According to Kelman et al. (2016) (see Sect. 2): considering vulnerability and resilience within the narrow context of climate change as one environmental stressor, can stymie other pressing development issues such as poverty. In this regard, building socio-economic resilience, through sustained job creation is as critically important as maintaining coastal environmental integrity, or resilience to climate change. The synergistic relationship described in Table 2 (+3) therefore, will assist with building both socio-economic and environmental resilience. However, the trade-off identified (−1), shows both socio-economic and environmental resilience can be minimised if the projects in the blue growth plan are not carefully implemented. For example, the tourism project in zone G3, if implemented in a sustainable manner can maintain the natural resilience of the area, with its mangrove ecosystems, while providing more jobs to community members. However, the constraints can be invoked, if the project reduces the natural resilience of the area, which can provide just a temporary respite for community members whose improved livelihoods can be eroded over time.

Another example illustrates. There is a possible future threat to SLR predicted for the proposed tourism development in zones G6 and G7. SDG 14-1 speaks to sustainable management and protection of marine and coastal ecosystems to avoid significant adverse impacts, including by strengthening their resilience and taking action for their restoration in order to achieve healthy and productive oceans; which reinforces SDG 13, *strengthen resilience and adaptive capacity to climate related and natural disasters*. However, the scale and location of some projects, such as the hotel/casino in zone 6, which will have the potential to improve socio-economic resilience, through job creation, may be threatened over time as SLR manifests in the future.

Therefore, sustainable tourism will play a critical role in the implementation of Grenada's Blue growth plan. As Spalding (2016) and Roberts and Ali (2016) suggested, the Blue Economy requires innovative and creative approaches to tourism if it is to transform SIDS economies from vulnerability to resilience. Mass tourism may not be the best approach to achieving development in SIDS that are facing the threats of climate change impacts such as SLR. Low impact/high value projects such as proposed in zone G3, eco-tourism, education tourism and other sustainable tourism and fisheries projects, which move the tourism infrastructure development away from areas threatened by SLR, will augur well for the sustainable development and building socio-economic and environmental resilience of SIDS.

7 Conclusions

SIDS, such as Grenada, has seen the blue economy as a viable alternative for economic growth and sustainable development. In this regard, Grenada developed a comprehensive blue growth investment plan that is focused on tourism. In many cases, these projects may not lead to sustainable development due to scale and location. The implementation of blue economy plans therefore, can be guided by the integrated SDGs framework illustrated in this chapter. The framework has the potential to assist SIDS policy makers and blue growth planners to develop and implement integrated policies, by identifying and analysing trade-offs and synergies among the SDGs. In summary therefore, policy makers, private investors and community stakeholders involved with blue growth planning can establish polices, goals and targets that will lead towards SIDS sustainable development, while building socio-economic and environmental resilience in the face of threats from climate change.

References

Baldacchino, G., & Kelman, I. (2014). Critiquing the pursuit of Island Studies: Blue and green, with hardly a colour in between. *Shima: The International Journal of Research into Island Studies, 8* (2), 1–21.

Baumgartner, R. J., & Kohonen, J. (2010). Strategic thinking for sustainable development. *Sustainable Development, 18*(2), 71–75.

Berry, A. J. (2015). Sustainable resource use and the problem of declining shoreline resilience to sea level rise: A Seychelles case study. *Island Studies: Blue Economy Edition, 1*(3), 20–26.

Briguglio, L. P. (2010). *Directing EU development aid to resilience building*. Occasional Papers on Island and Small States, No. 2, Msida, Islands and Small States Institute. Malta: University of Malta.

CIA World Factbook. (2018). *Grenada Data.* [online] Retrieved March 15, 2018, from https://www.cia.gov/library/publications/the-world-factbook/geos/gj.html

Eisenmenger, N., Pichler, M., Krenmayr, N., Noll, D., Plank, B., Schalmann, E., Wandl, M. T., & Gingrich, S. (2020). The sustainable development goals prioritize economic growth over sustainable resource use: A critical reflection on the SDGs from a socio-ecological perspective. *Sustainability Science, 15*, 1101–1110.

Ertör, I., & Hadjimicahel, M. (2020). Editorial: Blue degrowth and the politics of the sea: Rethinking the blue economy. *Sustainability Science, 15*(2), 1–10.

Government of Grenada (GOG). (2016). *Blue Growth Master Plan: Grenada, Carriacou and Petite Martinique*. St. George's, Government of Grenada.

Government of Grenada (GOG). (2017a). *Second National Communication to the United Nations Framework Convention on Climate Change (UNFCCC)*. St. George's, Government of Grenada.

Government of Grenada (GOG). (2017b). *Grenada, Carriacou and Petite Martinique: Investment Prospectus for Blue Growth*. St. George's, Government of Grenada.

Haskins, J. (2012). *Building resilience in small island economies: From vulnerabilities to opportunities.* CTA Policy Brief, No. 8 (November). Wageningen: Technical Centre for Agricultural and Rural Cooperation.

International Council for Science (ICSU). (2017). *A guide to SDG interactions: From science to implementation*. Paris: International Council for Science. https://doi.org/10.24948/2017.01

Kelman, I. (n.d.). *Understanding vulnerability to understand disasters*. Oslo: Centre for International Climate and Environmental Research. [online] Retrieved December 17, 2020, from http://www.ciesin.columbia.edu/repository/pern/papers/Kelman_vulnerability.pdf

Kelman, I., Gaillard, J. C., Lewis, J., & Mercer, J. (2016). Learning from the history of disaster vulnerability and resilience research and practice for climate change. *Natural Hazards, 82*(S1), S129–S143.

Le Blanc, D. (2015). *Towards integration at last? The sustainable development goals as a network of targets.* DESA Working Paper, No. 141, ST/ESA/2016/DWP/141. New York: United Nations.

Nilsson, M., Griggs, D., & Visbeck, M. (2016). Map the interactions between sustainable development goals. *Nature, 534*, 320–322.

Patil, P. G., Virdin, J., Diez, S. M., Roberts, J., & Singh, A. (2016). *Towards a blue economy: A promise for sustainable growth in the Caribbean*. World Bank.

Philpot, D., Gray, T. S., & Stead, S. M. (2015). Seychelles, a vulnerable or resilient SIDS? A local perspective. *Island Studies Journal, 10*(1), 31–48.

Prajal, P., Costa, L., Rybski, D., Lucht, W., & Kropp, J. P. (2017). A systematic study of sustainable development goal (SDG) interactions. *Earth's Future, 5*, 1169–1179.

Purvis, M.-T. (2015). Seychelles blue economy strategy. *Island Studies: Indian Ocean, 3*, 14–19.

Roberts, J. P., & Ali, A. (2016). *The blue economy and small states*. Commonwealth Blue Economy Series, 1. London: Commonwealth Secretariat.

Sachs, J. (2015). *The age of sustainable development*. Columbia University Press.

Spalding, M. J. (2016). The new blue economy: The future of sustainability. *Journal of Oceans and Coastal Economics, 2*(2). https://doi.org/10.15351/2373-8456.1052

Symmes, R., Fishman, T., Telesford, J., Singh, S., Tan, S.-Y., & De Kroon, K. (2020). The weight of islands. Leveraging Grenada's material stocks to adapt to climate change. *Journal of Industrial Ecology, 24*(2), 369–382. https://doi.org/10.1111/jiec.12853

United Nations Conference on Trade and Development (UNCTAD). (2014). *The Oceans Economy: Opportunities and challenges for SIDS*. United Nations Publications.

United Nations Department of Economic and Social Affairs (UNDESA). (2010). *Trends in sustainable development: Small Island Developing States*. United Nations.

United Nations Environment Programme (UNEP). (2012). *Green Economy in a Blue World.* [Online] Retrieved March 15, 2018, from www.unep.org/greeneconomy

United Nations Environment Programme (UNEP). (2014). *Emerging issues for Small Island Developing States. Results of the UNEP Foresight Process*. United Nations Environment Programme.

United Nations Environment Programme (UNEP). (2015). *Blue economy: Sharing success stories to inspire change*. [online] Retrieved April 30, 2018, from www.unep.org/greeneconomy

United Nations General Assembly (UNGA). (2015). *Transforming Our World: The 2030 Agenda for Sustainable Development.* Resolution adopted by the General Assembly on 25 September 2015, A/RES/70/1. [online] Retrieved December 17, 2020, from https://www.un.org/ga/search/view_doc.asp?symbol=A/RES/70/1&Lang=E

Weitz, N., Carlsen, H., Nilsson, M., & Skånberg, K. (2018). Towards systemic and contextual priority setting for implementing the 2030 agenda. *Sustainability Science, 13*, 531–548.

World Bank. (2004). *Grenada, Hurricane Ivan Preliminary Assessment Damages*. World Bank Doc. 41103, September 2004. Washington: World Bank.

World Bank and United Nations Department of Economic and Social Affairs (UNDESA). (2017). *The potential of the blue economy: Increasing long-term benefits of the sustainable use of marine resources for small island developing states and coastal least developed countries*. World Bank.

World Travel and Tourism Council (WTTC). (2018a). *Travel and Tourism Economic Impact 2018: Caribbean*. World Travel and Tourism Council.

World Travel and Tourism Council (WTTC). (2018b). *World Travel and Tourism Economic Impact 2018: Grenada*. World Travel and Tourism Council.

World Wildlife Fund (WWF). (n.d.). *Principles for a sustainable blue economy*. WWF Baltic Ecoregion Programme. [online] Retrieved March 19, 2018, from https://wwfint.awsassets.panda.org/downloads/15_1471_blue_economy_6_pages_final.pdf

Xu, Z., Chau, S. N., Chen, X., Zhang, J., Li, Y., Dietz, T., Wang, J., Winkler, J. A., Fan, F., Huang, B., Li, S., Wu, S., Hertzberger, A., Tang, Y., Hong, D., Li, Y., & Liu, J. (2020). Assessing progress towards sustainable development over time and space. *Nature, 577*, 74–78. https://doi.org/10.1038/s41586-019-1846-3

Identifying Climate Change Vulnerability and Adaptation Challenges in the Caribbean SIDS: An Urban Morphological Approach

Tapan Dhar

Abstract Small Island Developing States (SIDS) are particularly vulnerable to sea-level rise posed by climate change. Negril, a small city on Jamaica's west coast, is a popular Caribbean tourist destination contributing about 5.5% to the national GDP. In particular, Long Bay, a seven-mile resort transect and Negril's densest and low-lying area, has been losing beach at between 1 and 2 m/year while making its entire tourism industry highly vulnerable. Since adaptation and its success vary by contexts, places, and time; understanding a system's current vulnerability is essential before any adaptation action. However, limited information on local climatic impacts hinders identifying the vulnerability and urban planning approaches to facilitate climate change adaptation. Thus, after conducting 16 interviews with planners, environmentalists and professionals in Jamaica along with observation and a GPS survey, this study identifies and maps the typology of Long Bay's built environment vulnerable to climate change by using urban morphology as an investigating tool. This chapter argues that Long Bay's current and proposed adaptation planning actions primarily consider the immediate and most crucial climatic impacts but overlooks interrelated and complex problems and local adaptive capacity. Despite much potential, the lack of collaboration between development actors and locals in the adaptation decision-making process often increases the vulnerability and adaptation challenges of Negril.

Keywords Sea level rise · Climate change adaptation · Urban morphology · Coastal settlements · Resilience · Caribbean · Jamaica

T. Dhar (✉)
Science, Engineering and Technology School, Khulna University, Khulna, Bangladesh
e-mail: tkdhar@uwaterloo.ca

S. Moncada et al. (eds.), *Small Island Developing States*, The World of Small States 9, https://doi.org/10.1007/978-3-030-82774-8_15

1 Introduction

Global climate change represents one of the utmost challenges of this century. Coastal cities in Small Island Developing States (SIDS) in particular have for some time been experiencing severe climate change impacts, including Sea-Level Rise (SLR), hurricanes, and surges. For example, in Georgetown, Guyana, the 2005 flood alone caused socio-economic damage estimating about 59% of the country's GDP and affected the health of 72% of the city's population and (Mycoo & Donovan, 2017). Even one meter of sea-level rise would result in 60% of tourism resorts and infrastructure damaged only in the Caribbean (Scott et al., 2012).

The significant expansion of knowledge pertaining to climate change over the last three decades has raised the awareness of academic communities toward dealing with the emerging impacts of climate change on the earth and human societies (White, 2008; Xu et al., 2019; Marschütz et al., 2020). Current planning attempts pay more attention to climate change mitigation and sustainability rather than climate change adaptation and identification of climatic risks of built environments (Dhar & Khirfan, 2017a). In fact, the lack of theoretical links between climate change vulnerability and physical planning hinders to explore this opportunity. Planning scholars are aware of the need of identifying specific climatic risks of an area and investigating physical planning and design approaches so as to contribute to adaptation planning (Davoudi et al., 2013; Beatley, 2014; Lennon et al., 2014; Jabareen, 2015; Fox-Lent & Linkov, 2018; Nofal & van de Lindt, 2020). Consequently, planning advocacy on adaptation to provide urban design and planning interventions remains mostly generic and offers adaptation actions that often ignore context-based and indigenous adaptation approaches (e.g., Allan et al., 2013; Sharifi & Yamagata, 2014; Sharifi, 2019). The lack of precise information and scientific data on climate change also impedes the identification of climatic risks that inform context-specific adaptation planning. Since 'context' is a key for determining adaptation responses of a system (a community or settlement for example), successful adaptation responses may differ by context, place, and time. Therefore, understanding a system's current vulnerability along with its prevailing adaptation challenges becomes essential.

Negril, a popular Caribbean tourist destination located on the west coast of Jamaica, is vulnerable to sea-level rise and its associated impacts like other cities in SIDS. Long Bay, a seven-mile beach strip of Negril and the focus of this study, represents the country's most dense resort area. It has been losing beach area over the past few decades and is particularly vulnerable to climate change. This study aims first to identify the current vulnerability of Long Bay's built environment; how the typology of its urban form influences its vulnerability has been investigated.

Urban form refers to the unique physical characteristics of a town or a settlement, i.e., its "physiognomy or townscape", which considers the town plan, the pattern of building forms, and land use (Conzen, 1969). This study adopts urban morphology, one of the six dimensions of urban design (Carmona et al., 2010). Then this study explores the current adaptation planning and its challenges in the context of Negril through a series of in-depth semi-structured interviews with planners, environmentalists, and related professionals working in several national and local agencies—both public and private sectors—related to planning, development, and climate change across the island. It also uses direct participatory observation and a GPS survey to identify and rank different parts of the area most vulnerable to climate change.

The following sections begin with reviewing the concept of climate change vulnerability and related other concepts along with their links. The chapter then introduces the case study area and the methodology used in this research. The next section identifies Negril's different climate change problems from the viewpoint of planning and design. A brief discussion on the challenges of adaptation planning practice in the context of Negril and SIDS is then presented before the conclusion.

2 Understanding Climate Change Vulnerability and Adaptation

Vulnerability is a function of the magnitude, character, and rate of climate change and variation to which a system is exposed, the sensitivity and adaptive capacity of that system. Thus, the vulnerability of a system to climate change that could be described in terms of its exposure to the climatic stimulus is influenced by the system's adaptive capacity (Smit & Wandel, 2006). Accordingly, the term adaptive capacity refers to "the potential or ability of a system, region, or community to adapt to the effects or impacts of climate change" (Smit et al., 2001, p. 881).

Similarly, in a broad perspective, Crichton (2001) has defined risk by referring to the "risk triangle" as a function of three components: hazard, exposure, and vulnerability (Fig. 1). Vulnerability represents the extent to which people and properties can suffer from injury or damages from possible hazards. Vulnerability depends on

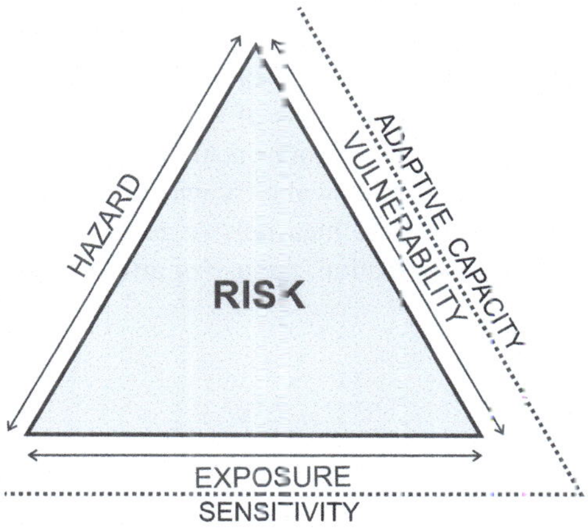

Fig. 1 Risk triangle and its relationship with vulnerability and exposure. *(Adapted from Crichton,* 2001, *and Smit & Wandel,* 2006*)*

resilience, preparedness, and recovery, while exposure is the proximity and number of vulnerable people or properties to possible hazards and vulnerable elements within an exposed area to hazard (Lindley et al., 2006; Bankoff, 2019; Antronico et al., 2020). The area of the triangle represents the extent of risk, and accordingly, by reducing any one side of this triangle, the risk will be reduced, and the risk will be completely eliminated if any one side is eliminated. The exposure also indicates a system's sensitivity. Thus, minimizing the sensitivity or exposure of a system implies the reduction of risk. Similarly, enhancing a system's adaptation capacity can reduce the risk as well. A significant number of planning articles dealing with climate change have cited this definition (Roaf et al., 2005; Fedeski & Gwilliam, 2007; Wilson & Piper, 2010; Wilson et al., 2014).

Box 1 IPCC's three coastal adaptation strategies

The report of the coastal zone management subgroup under the Intergovernmental Panel on Climate Change (IPCC) advocated three following potential strategies for adaptation to sea-level rise about three decades ago (Dronkers et al., 1990).

Retreat: Retreat involves avoiding development in areas typically vulnerable to inundation. In general, this strategy recommends (i) preventing development in areas near the coast; (ii) allowing developments to take place on the condition so that it would be abandoned when necessary (i.e., planned phase out); and (iii) having no direct government role other than through withdrawal of subsidies and provision of information about associated risks.

Accommodation: This strategy, like that of retreat, requires advanced planning and acceptance that some coastal zone values could be lost. For example, fundamental changes in land use might be desirable. Converting some agricultural lands to aquaculture uses indicates this strategy. Elevating dwellings on pilings as well as structural bracing for protection from floods might be necessary for residential and small commercial buildings. To counter surging water and high winds, building codes and coastal planning standards provide specific guidelines that can enable coastal communities to live with water.

Protection: This strategy involves defensive measures and activities to protect areas and developments against inundation, tidal flooding, saline intrusion and the loss of natural resources, wave effects, and shore erosion. Such protections can be achieved through different hard measures (e.g., dikes, levees and floodwalls, seawalls, groins, and detached breakwaters) and a number of soft measures (e.g., beach filling and re-nourishment, wetland/mangrove creation, the use of artificial seaweed and/or reefs).

3 Negril, A Case Study

Beaches represent the most popular tourist attraction in most SIDSs across the Caribbean basin and contribute a major share to the regional economy (Thomas et al., 2019). The Jamaican tourism industry is largely based on sun-sea-and-sand accounting for about 36% of national GDP and 32% of total employment (WTTC, 2014). Over recent years this tourism industry has become all-inclusive hotels that are largely confined to three resort towns—Montego Bay, Ocho Rios, and Negril. These towns together represent more than 75% of the island's available tourist accommodations and direct employment (Rhiney, 2012). Negril, in particular, a small resort town of over 11 thousand people located on Jamaica's western coast, belongs to two Jamaican parishes: Westmoreland and Hanover (Powell & Aikens-Mitchell, 2015). Historically, Negril was an isolated fishing village until the construction of Norman Manley Boulevard, a highway built in late 1950, which connected Negril to Lucea and Montego Bay. The boulevard caused the partial separation between the coastal and forest ecosystems and the morass, a wetland and ecologically protected area. The morass is a major resource of swamp, mangrove, herbaceous marchlands, and other lowland forest and contribute to protecting a number of species and local ecosystems while representing about one-fifth of Jamaican wetlands amounting to over 5500 acres of land area (Otuokon, 2001; Town and Country Planning Development Order, 2013) Later, since 1980 this boulevard has promoted the unplanned and rapid growth of Negril's tourism industry along the shoreline, and today Negril has become a popular tourist destination that alone contributes about 5.5% of the island's GDP (UNEP, 2010). Two bays of Negril—Bloody Bay and Long Bay (also known as 7-mile-beach)—have been hosting about 20% of Jamaica's total visitors that are often ranked as one of the top ten beaches in the world by travel magazines (Rhiney, 2012).

Long Bay, the particular focus of this study, represents the densest built environment and is particularly vulnerable to SLR and beach erosion. Geographically, Long Bay locates on a narrow low-lying strip between the sea and the morass (Fig. 2) Evidence suggests that over the last few decades, Long Bay has been losing beach about 1 to 2 m per year, and indeed there is no option for Long Bay for stepping back (i.e., managing retreat, see Box 1) due to the presence of the morass (Dhar & Khirfan, 2016). In addition to this erosion, other climate change impacts including increasing hurricanes and surges along with SLR, and flash floods make the entire tourism industry highly vulnerable. Topographic features of Long Bay further amplify the challenge to Long Bay's adaptation planning. Accordingly, the built environments, the morass, and the entire ecosystems are highly exposed to coastal inundation and SLR. This study here pays particular attention to Long Bay's physical infrastructure along with its urban morphological components to identify and map its current vulnerability.

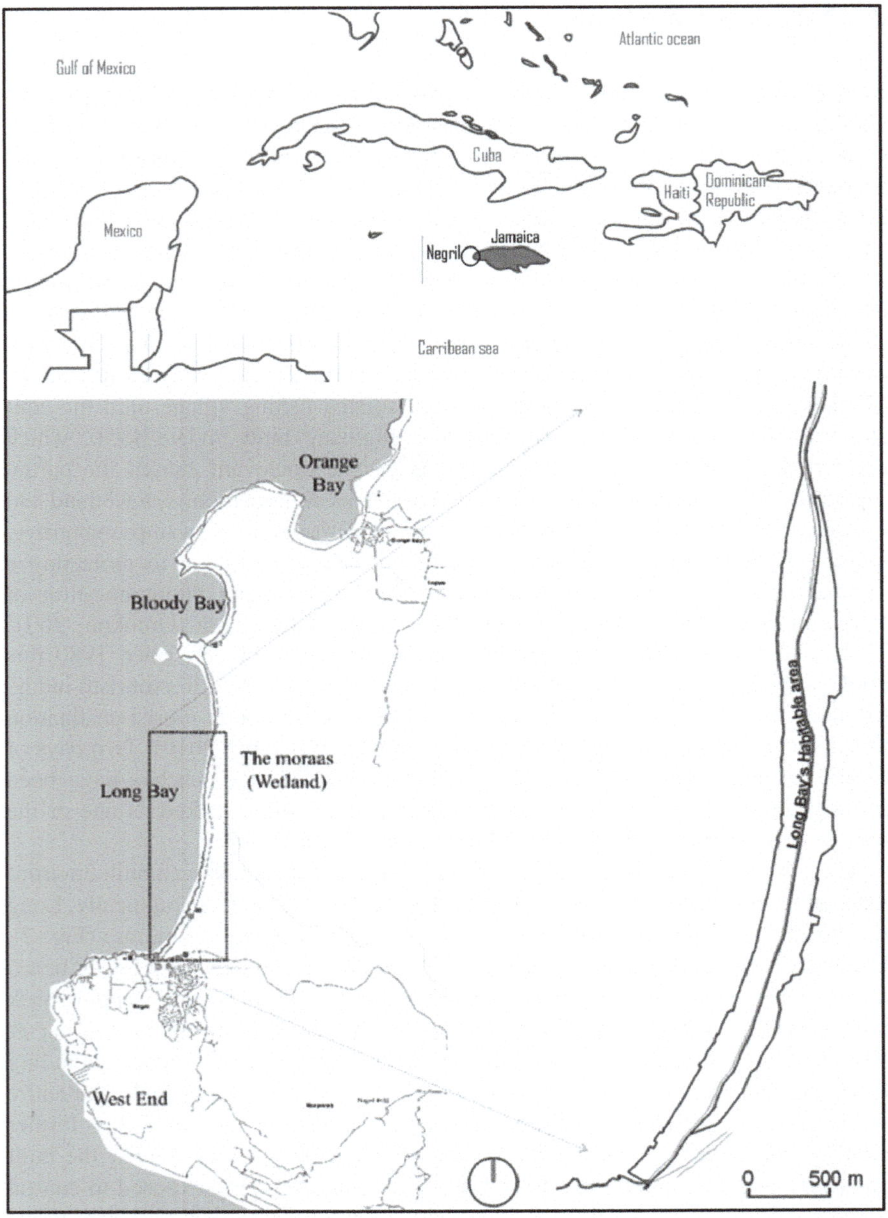

Fig. 2 Map of Negril and Long Bay, the study area. *(Source: Author)*

4 Methodology

This study enquires through a lens of urban planning and design and employs a participatory approach to identify the people who are directly (or indirectly) involved with adaptation planning and coastal development and identifies key climate change problems in Negril. In order to identify and recruit the relevant agencies/divisions and people, websites of different organizations related to planning, development, and climate change (e.g., the Planning Institute of Jamaica) were reviewed, and the researcher's academic and professional networks were contacted, including environmental activists, local journalists, staff at local and regional non-governmental organisations (NGOs), and faculties and graduate students.

Since the head offices of all public agencies that make most planning and development decisions with regard to climate change resilience for the entire island are based in Kingston, interviews were first conducted with professionals working in these agencies in Kingston and then in Savanna-la-Mar and Negril to include local stakeholders. Although the field-work started with a list of respondents who gave consent to participate in this study prior to fieldwork, during each interview, respondents were requested to add peers who have also relevant experience and knowledge in vulnerability assessment and climate change adaptation. Thus, with the help of such purposeful and snowball sampling approaches, the study finally conducted 16 in-depth interviews with planners, environmentalists, and policy makers both in public and private agencies in Kingston, Savanna-la-Mar and Negril. These agencies include the local and head office (where applicable) of National Environment and Planning Agency (NEPA); (Negril and Kingston office) Urban Development Corporation (UDC); Ministry of Water, Land, Environment & Climate Change (MWLECC); Office of Disaster Preparedness and Emergency Management (ODPEM); Negril area Environment Protection Trust (NEPT); Jamaica Hotel and Tourist Association; and National Best Community Foundation.

The study also emphasizes the patterns of built environment and their influence on Negril's vulnerability. Several maps from different secondary sources, including the National Land Agency (NLA), grey and peer-reviewed documents were collected. The University of West Indies prepared a morphological map showing the built environmental of Negril for this study. Urban morphological information—the physical dimension of urban form—including street and plot patterns, buildings, and land use, plays a key role to understand the vulnerability and adaptation strategies of the Long Bay area.

In addition, a GPS survey, conducted under a course work (Plan 647) at the University of Waterloo (Canada) instructed by Prof. Luna Khirfan, has also been used to map the physical profile and verify the vulnerability of Long Bay area. The survey classified Long Bay's physical profile into different transacts (from north to south) and facilitated to prepare a local elevation model. Direct participatory observation and random conversation with locals enabled the researcher to document the existing scenario of physical development of the study area. In fact, this study integrates top-down and bottom up approaches to identify the climate change problems and challenges of adaptation decision making. In other words, it fuses experience and experiential knowledge through engaging stakeholders while combining participation (life in society), action (experience), and research (knowledge).

Textual transcription was used to document interview data in a way that can be coded accordingly into different themes in order to frame the research enquires. Different software, including illustrators, AutoCAD, and geographic information system (GIS) mapping were used to facilitate the analysis of observation data. Since the credibility of qualitative data has often been an issue in qualitative studies, this study employed multiple tactics of data collection and analysis. Thus, different sources of primary and secondary information help the researcher triangulate and validate the qualitative data.

5 Long Bay's Vulnerability

5.1 Physiognomic and Infrastructural Threats

Most respondents identified beach erosion as the key challenge of Negril. Veira (2014) reported that the average rate of erosion over the past three decades has been one to two meters per year. If such a rate prevails with the increasing SLR, 6 to 10 m of beach erosion would occur by 2030 and by 2050 it would reach 12 m to 21 m (Robinson et al., 2012). Long Bay's erosion rate is even higher than that of its neighbouring bays (Fig. 3). In general, the process of beach erosion and sand production is a natural and cyclic phenomenon; however, because of damaged ecosystems, seagrass in particular, several environmentalists interviewed argued that Negril's recent sand production is relatively lower than before.

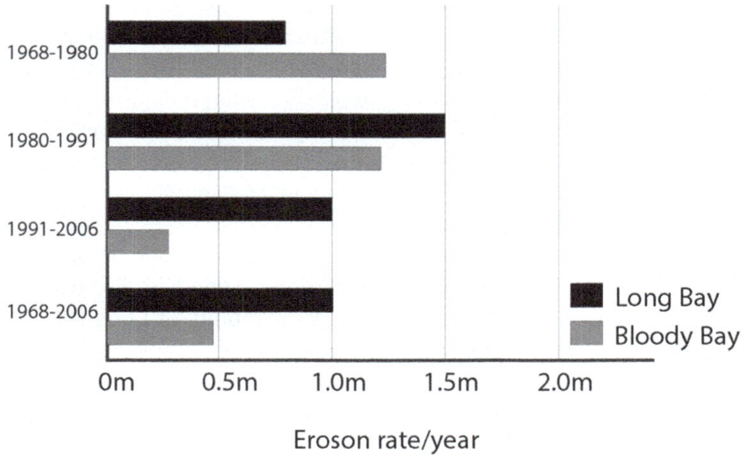

Fig. 3 Beach erosion occurred over the periods. *(Adapted from CEAC,* 2014*)*

Poor waste management and flash flooding also have caused several damages to Long Bay's infrastructure. For example, in 2010, Negril experienced an extreme flash flood event due to heavy rain, and then its entire infrastructure and transportation systems were collapsed for about ten days (Dhar & Khirfan, 2016). Although the upgrade of the sewage treatment plant (in Sheffield) in 2012 reduced sewage pollution, limited efforts have been made to control other anthropogenic actions accelerating climate change risks across the area. In addition, anthropogenic actions, such as water pollution and poor waste management, are the causes of degradation of reefs, sea grasses and mangroves which otherwise would have reduced the wave energy and protected beaches of Negril. These marine ecosystems would normally reduce wave energy, facilitate sand production, and increase ecological resilience.

According to a 2012 study by Saint Mary's University and the Maritime Provinces Spatial Analysis Research Center (Perrott, 2012), about 37% of the land area of Long Bay are unstable or partially stable (Fig. 4). Thus, built forms and infrastructure in such geologically unstable areas are vulnerable. Relatively, the physical profile of Long Bay is relatively more unstable than that of Bloody Bay and Orange Bay.

5.2 Long Bay's Elevation

This study explores the typology of Long Bay's existing built environment and its vulnerability to beach erosion. The elevation profile of the beach area can increase the exposure of its built forms to SLR and surges as well as reduce the adaptive capacity of the area. Specifically, the orthometric elevations of the highway at different points differ from −4 m to +7 m from the high watermark. Figure 5 represents the three zones, i.e., A, B, and C, with different profiles at Long Bay. The elevations of the highway from the high watermark at A, B, and C zones are 0 m or below, 0 to 5 m, and over 5 m respectively. Thus, the GPS survey reveals that buildings and infrastructure at zone A are more vulnerable to coastal inundation than that at B or C. Furthermore as the elevation of the morass is nearly everywhere below one meter from the mean sea level except in the southeast part (Robinson et al., 2012), coastal inundation, through zone A, for instance, might cause severe intrusion of saltwater to the morass and consequently impact local wetland habitats and entire ecosystems. In addition to the increasing intensity of rapid-onset events (e.g., hurricanes), this degradation of ecosystems has a slow impact on coral health and sea grass beds which would produce sand naturally.

The northern half of Long Bay is the section most vulnerable to beach erosion. Also, overlaying the current shorelines, obtained from the GPS coordinates on the Jamaican National Land Agency's (NLA) map shows that the northern area near A has been losing beach the most (Fig. 5). Wilson et al.'s (2014) surge model has identified a few sections at the northern part, which are most vulnerable to storm surges associated with SLR, that are also located at or near zone A. Robinson et al. (2012) called the areas vulnerable to beach erosion "hot spots". They estimated that

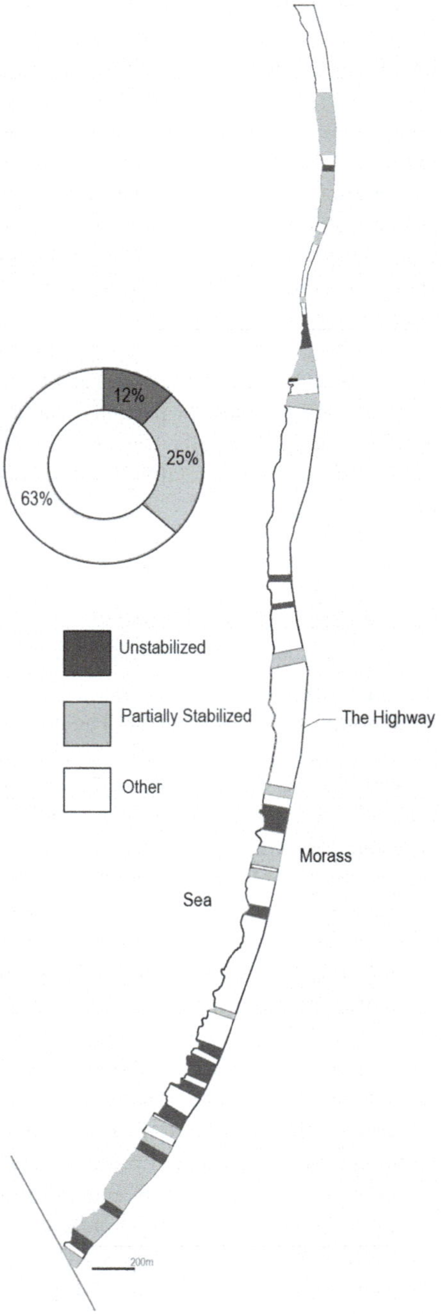

Fig. 4 The unstabilized and partially stabilized area between the sea and the highway. *(Adapted from Perrott,* 2012*)*

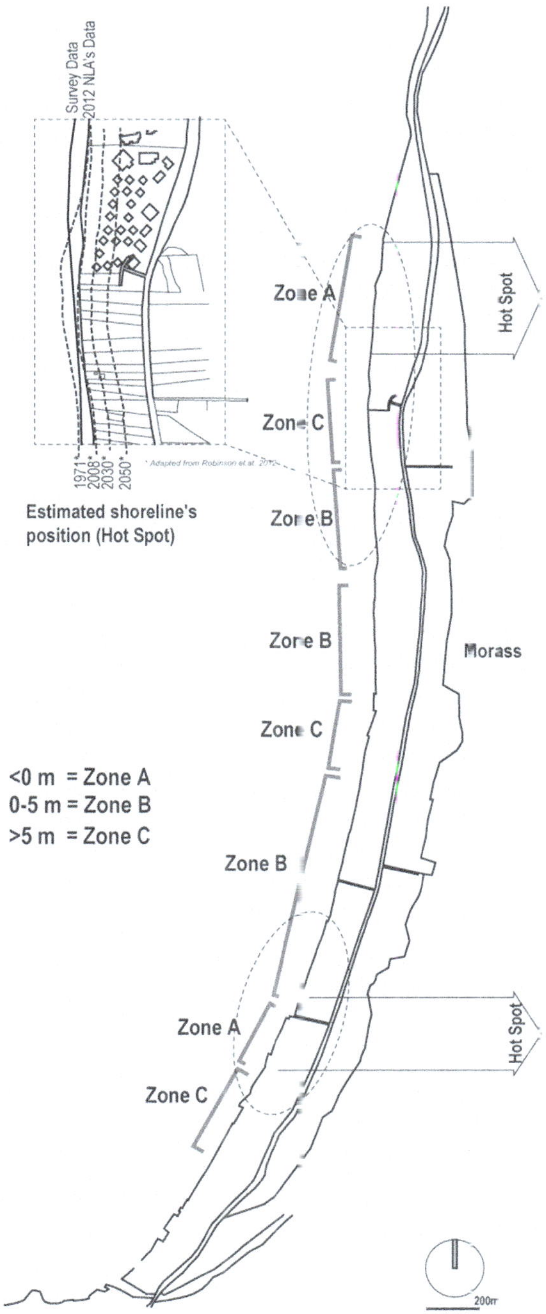

Fig. 5 Shoreline changes and vulnerable areas. (Data source: GPS Survey, NLA's parcel data (www.nla.gov.jm), and Rahmstorf's (2007) and Robinson et al.'s (2012) estimates)

beach erosion at these points could exceed even 200 m by 2100. Thus, the areas, marked by ellipses in Fig. 5 near the zone A (at both north and south) are at risk to surges, SLR, and erosion compared to the other zones. Considering this future scenario, some planners interviewed suggested that in near future all tourism accommodation/resorts along Long Bay might need to be abandoned and be moved to the areas at a higher elevation, e.g. near Sheffield.

In addition, to previously identified unstabilized areas (Perrott, 2012), Wilson et al.'s (2014) model illustrates that the northern part is also vulnerable to direct rainfall associated with high tides and surges. Likewise, the marked area at zone A (north) is also highly vulnerable to beach erosion and this study adds another hot spot at zone A (south) (Fig. 5).

5.3 Coastal Setbacks and Building Footprints

In general, the design of built environment along the coast in most Caribbean SIDSs fails to follow planning and land use guidelines (Lewsey et al., 2004; Ishemo, 2009; Thomas et al., 2019). According to NLA's parcel data, the beach area legitimately belongs to private owners). This implies that protecting buildings and properties from erosion is the responsibility of respective individual owners. The continuous erosion of Long Bay over time may have also affected setback regulation[1] (Fig. 6). On-the-ground observations along with GPS data revealed that against a legal coastal setback of 45.75 m for Negril, each property maintains a setback ranging only from 10 m to 40 m. Most hoteliers along Long Bay have built temporary pavilions for bars, restaurants, and musical concerts close to the sea. These structures are made of local materials, such as wood, bamboo, and straw, seldom combined with steel- or concrete pillars. Many buildings along the bay, with a few exceptions at its north, do not follow the coastal setback regulation, and additionally, the continuous erosion has been making them more vulnerable.

> The buildings and hotels [along Long Bay] were approved with proper setback [by the local planning authority]. However, over the years, rising sea level has resulted in the loss of beach. . .and consequently buildings . . . [as we see now seem to be] very close to the sea – a respondent reported.

Observed data also revealed that currently, about one-fourth of buildings/properties have no proper setback and thus are exposed to projected SLR and beach erosion. If Robinson et al.'s (2012) forecast model proves true, in 2050 about half of the buildings along Long Bay will be affected (Fig. 6).

It is important to note that there is still no design or planning mechanism to guide and encourage the owners/hoteliers to keep adequate setbacks to protect the beach

[1]The regulation of coastal setback restricts a minimum distance from the shoreline for new buildings or infrastructural development. It differs across different areas. For Negril, the town and country planning development order (2013) sets 45.75 m coastal setback for Negril area.

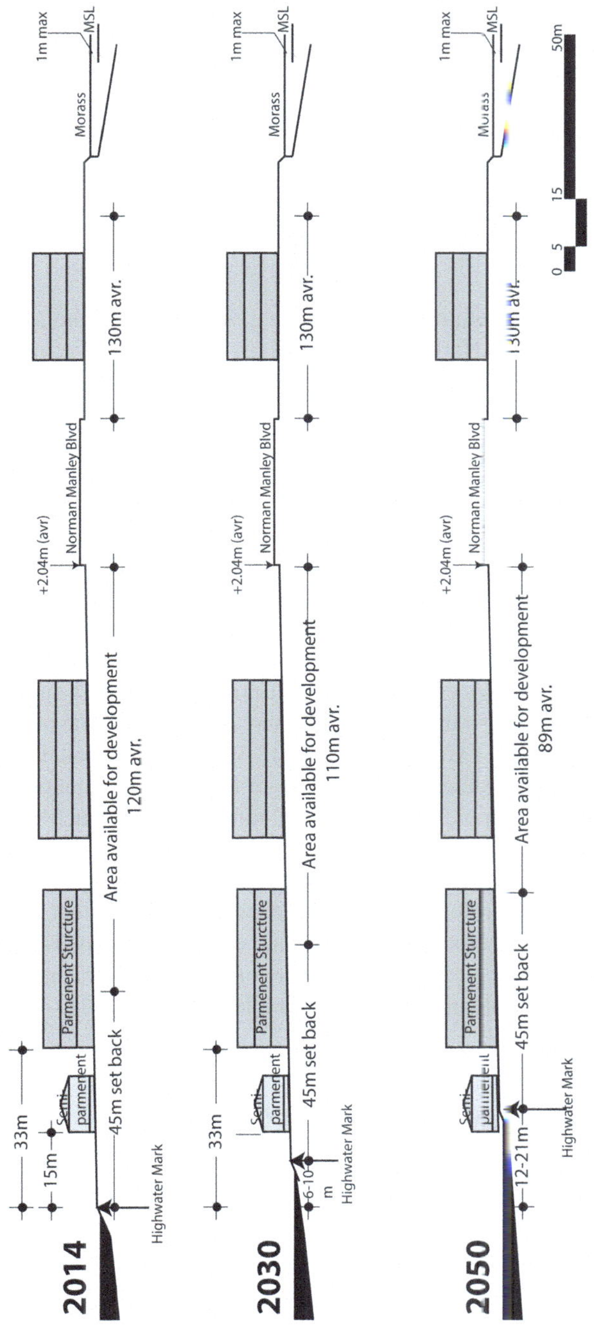

Fig. 6 Coastal set back at Long Bay and its cross section along east and west. *(Source: Author)*

area. Thus, the semi-public beach and the individual properties along the beach area remain threatened. However, the respondents noted the small footprints of the existing buildings (Fig. 7) may allow more room for future opportunities. "Because smaller [building] footprints impact less on the area and facilitate drainage features. . .allow us to have a better setback, for more retention, it allows for more green space, amenities and spaces for visitors"—a respondent argued.

5.4 Accessibility and Emergency Evacuation

A number of resorts and commercial centres (e.g. Time Square) are located on both sides of the highway (i.e., Harman Norman Blvd)—the planned vehicular axis that connects from North to South. In other words, the entire area is connected formally by the highway, and informally by a walkway along the shoreline that runs parallel to the highway. However, the link between the two major connectors, i.e., from east to west, is limited. Only two narrow public corridors (toward east to west) link the seven-mile beach to the highway resulting in superblocks and inaccessibility (Fig. 8). The superblocks along the beach, particularly at the northern part, have made the beach extremely inaccessible for local people, rescue teams, and the tourists who stay on the opposite side (i.e., east) of the highway. A few resorts have their properties on both sides of the highway, and people living there can access the beach through their own properties. Accessibility, often analogous to connectivity and permeability, is a key concept of urban design for making a successful public space. Accessibility has also the potential to increase the resilience of built environment to meet uncertainty (Allan et al., 2013). Ahern (2011) noted that a lack of connectivity is often considered a prime cause of failure or malfunction of an urban system. In responding to rapid-onset events of climate change, such well-connected street networks would facilitate emergency evacuations during and immediately after an extreme event. Well-connected streets and fine grain of urban form (i.e., an area with no/few superblocks) thus are the two parameters that can potentially increase the resilience of urban form to climate change (Dhar & Khirfan, 2017b). Increasing the resilience of a system is one of the basic strategies of planned adaptation: "the more resilient, the less vulnerable" (Pelling, 2011).

6 Negril Adaptation Planning and Challenges

Most documented adaptation strategies so far rely on a top-down approach that often focuses on climate science modelling and downscaling (Davoudi et al., 2012). Negril's adaptation planning is not an exception. Jamaican central adaptation planning takes several measures into account to reduce beach erosion along Long Bay and Bloody Bay. A few years ago, the government initiated the building of a couple offshore breakwaters. This was to be a complex and expensive project comprising

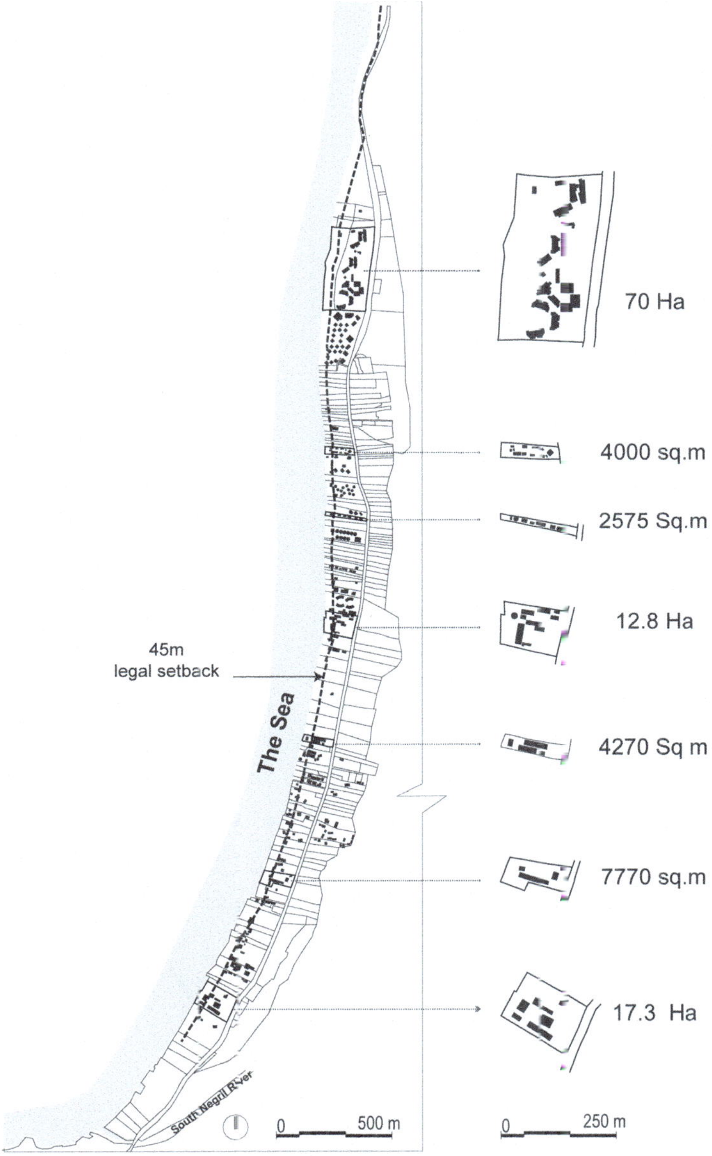

Fig. 7 Current shoreline, setback and buildings. *(Source: Author)*

two submerged structures (3600 m length in total) located one mile from the coast to reduce the wave energy and beach erosion. Mondon and Warner's (2012) study confirmed the project's efficiency and function in imitating natural systems in reducing erosion without having any harmful impact on the environment. However,

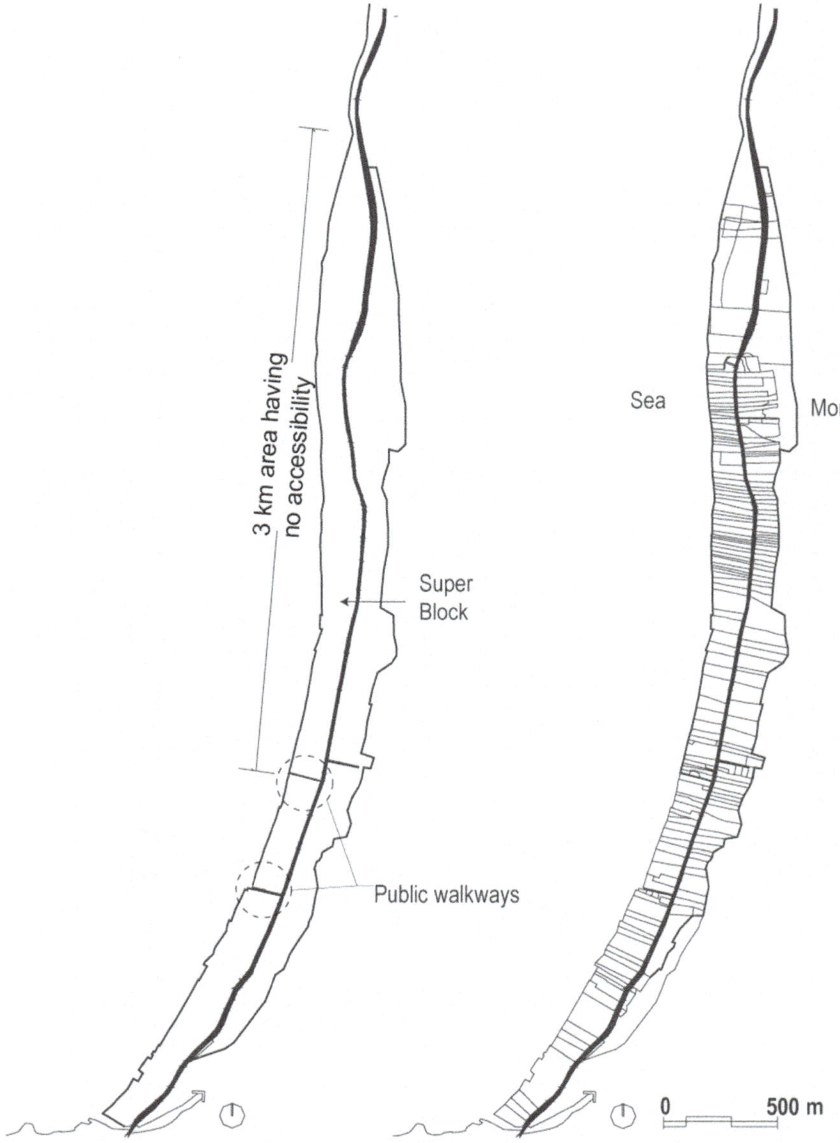

Fig. 8 The connectivity and superblocks of built environment at Long Bay. *(Source: Author)*

a number of environmentalists and locals were against the project because of the potential for negative impacts on local ecosystems (Mycoo & Chadwick, 2012; Dhar & Khirfan, 2016). The project faced strong resistance from local residents, hoteliers and environmental activists; and finally, the government has decided to cancel the project (Barrett, 2016). The data for this study was undertaken before this decision, and at the time it was both promoted (by public agencies) and protested against (mostly by environmentalists and locals).

During interviews, a few respondents contributed their thoughts to the debate of the project's feasibility and shared their insights for future adaptation. A respondent working for a public agency commented the following on future adaptation strategies for Negril:

> We should build our new development that will be affected less; e.g., it includes retreat to hills. However, for combating short term impact we have to consider coastal defences as well as waterproof and higher structures. Most of our social and physical infrastructure is located in coastal areas. We can't move them...that is why, we need to adopt coastal defence mechanism (e.g., hard structures-break water, sea walls, or so) to protect them.

In addition to these protective measures, professionals in non-governmental sectors prefer slower, no-regret, and environmentally friendly approaches to protect Long Bay beach. "We no longer want our [Negril's] natural resources...white sand, reef, flora and fauna... to be affected", a respondent noted. Many considered the prevailing small building footprints along the Long Bay area as an opportunity to strengthen the local adaptation planning that can be merged or retrofitted with the central master planning for Negril which is yet to be completed.

The main challenge of Negril's adaptation planning includes the centralized and isolated approaches to deal with the key problem, namely beach erosion. Local knowledge and technology in order to identify climate change issues of an area and their integrated solutions are of utmost necessity for successful adaptation. Khoshnevis and Chelleri's (2018) experience in the context of an Iranian village also prioritizes local resources, knowledge and adaptive capacity of the village community to enhance climate change resilience. In fact, beach erosion is not a problem of many cities in SIDS like Negril, but a result of multiple maladaptations and the lack of attempts to identify and recognize the associated causes of beach erosion. For example, damage of coral reefs and seagrass due to pollution from fishing boats ultimately increase wave energy and beach erosion.

The Negril disaster and emergency evacuation plans are yet to be prepared. Considering the emerging climatic impacts, Negril has no specialized land use plan for developers/property owners to guide where to build and where not to. Rather, this study reveals that Negril experiences socio-economic and political pressure to have more development (tourism and other development). Then, one or a few protective measures without considering other accommodative measures/potentials could have resulted in increased exposure of coastal built environment to hazards while increasing the vulnerability (see risk triangle, Fig. 1). Thus, such measures can also be reasons for unsuccessful adaptation (i.e., maladaptation) (Barnett & O'Neill, 2010).

Many consider adaptation actions are related only to extreme events, mostly rapid-onset climatic events (such as hurricanes) (Field et al., 2012; Smit et al., 2000). However, complex problems can be caused by slow-onset events (such as temperature rise) making an area highly vulnerable to climate change. Cutter et al. (2008) classified climate change resilience into two qualities: (i) inherent resilience to deal with slow onset climatic events and (ii) adaptive resilience to deal with rapid onset climatic events. Based on empirical evidence, Fig. 9 summarizes the challenges of Negril posed by rapid-onset and/or slow-onset climatic events.

Negril adaptation planning primarily considers protective, responsive measures to a sudden hazard and overlooks Ecosystem-based Adaptation (EbA) and incremental (adaptation) planning approaches. Alternatively, adaptive resilience facilitates long-term adaptation planning and helps build the community's capacity to deal with an extreme event. Urban planning can be a tool to arrange this system in a way to reduce the exposure of Long Bay as well as increase the resilience of the area.

7 Conclusion

Jamaica is located in the Atlantic hurricane belt and exposed to late-season hurricanes. The Intergovernmental Panel on Climate Change's (IPCC) (2014) report estimated that more intense hurricanes associated with over 0.5 m SLR would be expected to happen at the end of this century. Thus, increasing adaptive capacity along with integrated adaptation planning is essential to meet the future challenges of climate change.

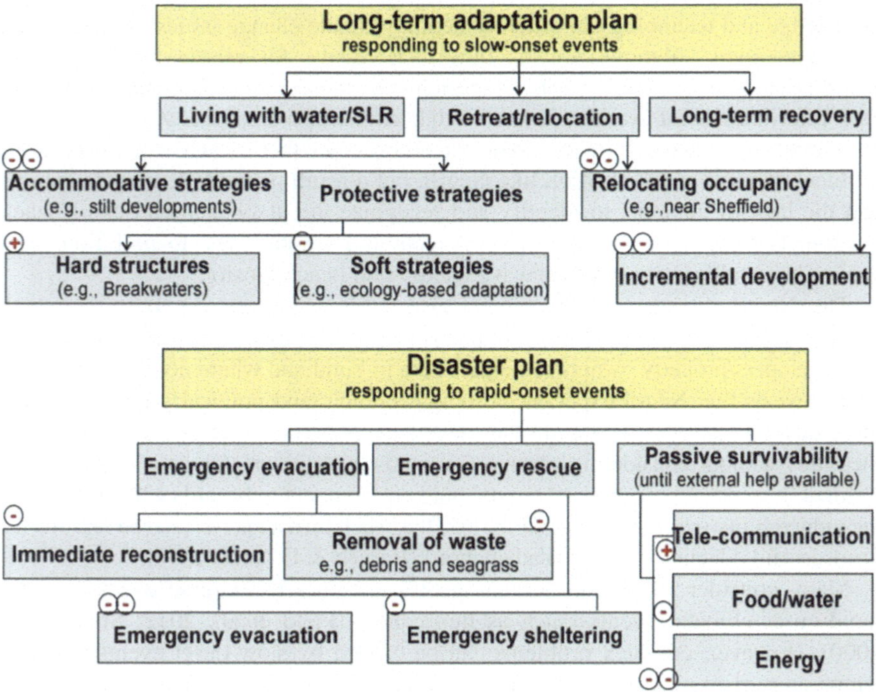

Fig. 9 Negril's adaptation planning and its focus. *(Boxes with + sign indicate what Negril planning have already considered, and boxes with − sign suggest the area need to be prioritized through future adaptation) (Source: Author)*

Long Bay's problems are multi-faced and thus, their solutions need to be integrated with different short-term and long-term planning strategies. Since context is assumed central to successful climate change adaptation, planning must acknowledge local social, cultural, economic, environmental features. Long Bay's small building footprints and plot size (4000 sqm–70 hectares) offer opportunities for future incremental changes; however, recent approval for high density development without adequate (or any) adaptive measures might hinder this potential. Retreat and improved coastal setbacks are hardly possible here.

Negril may need to build structures similar to the previously planned breakwater to protect the immediate impacts of climate change, but other possible approaches that are ecology-friendly, such as beach nourishment and restoration of coral reefs and seagrass could also be considered.

Since the tourism industries of many SIDSs contribute considerably to their national economy, preserving natural ecosystems and marine ecology is imperative. Local people and communities (such as Negril Area Environmental Protection Trust and Jamaica Hotel and Tourism Association-Negril Chapter) are aware of the negative impacts of mega engineering projects, like the planned breakwaters that may negatively impact the environmental resources as well as the tourism industry. Clearly, it is now important to build a community's adaptive capacity, not only reducing its vulnerability to climate change and ensuring better disaster preparedness, but also encouraging locals to participate in supporting the adaptation decision making process. This capacity also enables the local community, government, and other stakeholders to respond better to the ongoing challenges while identifying and mapping existing vulnerability and recommending context-based adaptation approaches to build resilience in the age of climate change.

Acknowledgement This work was supported by the Partnership for Canada-Caribbean Climate Change Adaptation (ParCA) (https://uwaterloo.ca/canada-caribbean-climate-change-adaptation/). I also acknowledge Prof. Luna Khirfan, the University of Waterloo, and Ms Jacqueline daCosta, the former president of the Commonwealth Association of Planners, for their enormous support.

References

Ahern, J. (2011). From fail-safe to safe-to-fail: Sustainability and resilience in the New Urban World. *Landscape and Urban Planning, 100*, 341–343.

Allan, P., Bryant, M., Wirsching, C., Garcia, D., & Teresa Rodriguez, M. (2013). The influence of urban morphology on the resilience of cities following an earthquake. *Journal of Urban Design, 18*, 242–262.

Antronico, L., De Pascale, F., Coscarelli, R., & Gullà, G. (2020). Landslide risk perception, social vulnerability and community resilience: The case study of Maierato (Calabria, southern Italy). *International Journal of Disaster Risk Reduction, 46*, 101529.

Bankoff, G. (2019). Remaking the world in our own image: Vulnerability, resilience and adaptation as historical discourses. *Disasters, 43*, 221–239.

Barnett, J., & O'Neill, S. (2010). Maladaptation. *Global Environmental Change, 20*, 211–213.

Barrett, L. (2016). Scrapped - Gov't Pulls Plug On Controversial $1-Billion Negril Breakwater Project. *The Gleaner*. [online] Retrieved November 3, 2020, from http://jamaica-gleaner.com/article/lead-stories/20160713/scrapped-govt-pulls-plug-controversial-1-billion-negril-breakwater

Beatley, T. (2014). Planning for resilient coastal communities: Emerging practice and future directions. In B. C. Glavovic & G. P. Smith (Eds.), *Adapting to climate change: Lessons from natural hazards planning*. Springer.

Carmona, M., Heath, T., Tiesdell, S., & Oc, T. (2010). *Public places - urban spaces: The dimension of urban design*. Architectural Press.

CEAC. (2014). Storm surge modeling and sea level rise scenarios for the Negril Climate Risk Atlas, *Preliminary Technical Design Report*. Office of Disaster Preparedness and Emergency Management (ODPEM).

Conzen, M. R. G. (1969). *Alnwick, Northumberland: A study in town-plan analysis*. Institute of British Geographers.

Cutter, S. L., Barnes, L., Berry, M., Burton, C., Evans, E., Tate, E., & Webb, J. (2008). A place-based model for understanding community resilience to natural disasters. *Global Environmental Change, 18*, 598–606.

Crichton, D. (2001). *The implications of climate change for the insurance industry*. Building Research Establishment.

Davoudi, S., Brooks, E., & Mehmood, A. (2013). Evolutionary resilience and strategies for climate adaptation. *Planning Practice & Research, 28*, 307–322.

Davoudi, S., Shaw, K., Haider, L. J., Quinlan, A. E., Peterson, G. D., Wilkinson, C., Fünfgeld, H., Mcevoy, D., & Porter, L. (2012). Resilience: A bridging concept or a dead end? "Reframing" resilience: Challenges for planning theory and practice. Interacting traps: Resilience assessment of a pasture management system in Northern Afghanistan. Urban resilience: What does it mean in planning practice? Resilience as a useful concept for climate change adaptation? The politics of resilience for planning: A cautionary note. *Planning Theory & Practice, 13*, 299–333.

Dhar, T., & Khirfan, L. (2016). Community-based adaptation through ecological design: Lessons from Negril, Jamaica. *Journal of Urban Design, 21*, 234–255.

Dhar, T., & Khirfan, L. (2017a). Climate change adaptation in the urban planning and design research: Missing links and research agenda. *Journal of Environmental Planning and Management, 60*, 602–627.

Dhar, T., & Khirfan, L. (2017b). A multi-scale and multi-dimensional framework for enhancing the resilience of urban form to climate change. *Urban Climate, 17*, 72–91.

Dronkers, J., Gilbert, J. T. E., Butle, L. W., Carey, J. J., Campbell, J., James, E., Mckenzie, C., Misdorp, R., Quin, N., Ries, K. L., Schroder, P. C., Spradley, J. R., Titus, J. G., Vallianos, L., & Dadelszen, J. V. (1990). Strategies for adaptation to sea level rise. *Report of the IPCC Coastal Zone Management Subgroup: Intergovernmental Panel on Climate Change*. Geneva: The Intergovernmental Panel on Climate Change.

Fedeski, M., & Gwilliam, J. (2007). Urban sustainability in the presence of flood and geological hazards: The development of a GIS-based vulnerability and risk assessment methodology. *Landscape and Urban Planning, 83*, 50–61.

Field, C. B., Barros, V., Stocker, T. F., Qin, D., Dokken, D. J., Ebi, K. L., Mastrandrea, M. D., Mach, K. J., Plattner, G.-K., Allen, S. K. Tignor, M., & Midgley, P. M. (Eds.). (2012). Managing the risks of extreme events and disasters to advance climate change adaptation. *A Special Report of Working Groups I and II of the Intergovernmental Panel on Climate Change*. Cambridge, UK and New York: Cambridge University Press.

Fox-Lent, C., & Linkov, I. (2018). Resilience matrix for comprehensive urban resilience planning. In Y. Yamagata & A. Sharifi (Eds.), *Resilience-oriented urban planning*. Springer.

Intergovernmental Panel on Climate Change (IPCC). (2014). Climate Change 2014. *Synthesis Report. A contribution of Working on Climage Change I, II, III to the Fifth Assessment Report of the Intergovermental Panel on Climate Change*. Cambridge and New York: Cambridge University Press.

Ishemo, A. (2009). Vulnerability of coastal urban settlements in Jamaica. *Management of Environmental Quality: An International Journal, 20*, 451–459.

Jabareen, Y. (2015). *The risk city*. Springer.

Khoshnevis, B., & Chelleri, L. (2018). When a disaster risk reduction policy fails in the implementation stage: Eroding community resilience and traditional architecture in Iranian villages. In Y. Yamagata & A. Sharifi (Eds.), *Resilience-oriented urban planning*. Springer.

Lennon, M., Scott, M., & O'Neill, E. (2014). Urban design and adapting to flood risk: The role of green infrastructure. *Journal of Urban Design, 19*, 745–758.

Lewsey, C., Cid, G., & Kruse, E. (2004). Assessing climate change impacts on coastal infrastructure in the Eastern Caribbean. *Marine Policy, 28*, 393–409.

Lindley, S. J., Handley, J. F., Theuray, N., Peet, E., & Mcevoy, D. (2006). Adaptation strategies for climate change in the urban environment: Assessing climate change related risk in UK urban areas. *Journal of Risk Research, 9*, 543–568.

Marschütz, B., Bremer, S., Runhaar, H., Hegger, D., Mees, H., Vervoort, J., & Wardekker, A. (2020). Local narratives of change as an entry point for building urban climate resilience. *Climate Risk Management, 28*.

Mondon, E. A., & Warner, P. S. (2012). Evaluation of a proposed solution to beach erosion at Negril. *Caribbean Journal of Earth Science, 43*, 11–23.

Mycoo, M., & Chadwick, A. (2012). Adaptation to climate change: The coastal zone of Barbados. *Proceedings of the ICE-Maritime Engineering, 165*, 159–168.

Mycoo, M., & Donovan, M. G. (2017). *A blue urban agenda: Adapting to climate change in the coastal cities of Caribbean and Pacific Small Island Developing States*. Inter-American Development Bank.

Nofal, O. M., & Van De Lindt, J. W. (2020). Understanding flood risk in the context of community resilience modeling for the built environment: Research needs and trends. *Sustainable and Resilient Infrastructure*, 1–17.

Otuokon, S. (2001). *Case study of the Negril Environmental Protection Plan, Jamaica*. Caribbean Natural Resources Institute (CANARI).

Pelling, M. (2011). *Adaptation to climate change: From resilience to transformation*. Taylor & Francis.

Perrott, B. (2012). *Study area: Long Bay, Bloody Bay, Orange Bay, and Negril, Jamaica*. Saint Mary's University and the Maritime Provinces Spatial Analysis Research Center.

Powell, L., & Aikens-Mitchell, K. (2015). *Climatic risk atlas of coastal hazards and risk in Negril*. The Office of Disaster Preparedness and Emergency Management (ODPEM).

Rahmstorf, S. (2007). A semi-empirical approach to projecting future sea-level rise. *Science, 315*, 368–370.

Rhiney, K. C. (2012). The Negril tourism industry: Growth, challenges and future prospects. *Caribbean Journal of Earth Science, 43*, 24–34.

Roaf, S., Crichton, D., & Nicol, F. (2005). *Adapting buildings and cities for climate change*. Architectural Press, Elsevier.

Robinson, E., Khan, S., Coutou, R., & Johnson, M. (2012). Shoreline changes and sea-level rise at Long Bay, Negril, Western Jamaica. *Caribbean Journal of Earth Science, 43*, 35–49.

Scott, D., Simpson, M. C., & Sim, R. (2012). The vulnerability of Caribbean coastal tourism to scenarios of climate change related sea level rise. *Journal of Sustainable Tourism, 20*, 883–898.

Sharifi, A. (2019). Urban form resilience: A meso-scale analysis. *Cities, 93*.

Sharifi, A., & Yamagata, Y. (2014). Resilient urban planning: Major principles and criteria. *Energy Procedia, 61*, 1491–1495.

Smit, B., Burton, I., Klein, R. J., & Wandel, J. (2000). An anatomy of adaptation to climate change and variability. *Climatic Change, 45*, 223–251.

Smit, B., Pilifosova, O., Burton, I., Challenger, B., Haq, S., Klein, R., & Yohe, G. (2001). Adaptation to climate change in the context of sustainable development and equity. In J. Mccarthy, O. F. Canziani, N. Leary, et al. (Eds.), *Climate Change 2001: Working Group*

II: Impacts, Adaptation and Vulnerability, Third Assessment Report of the Intergovernmental Panel on Climate Change. Cambridge University Press.

Smit, B., & Wandel, J. (2006). Adaptation, adaptive capacity and vulnerability. *Global Environmental Change, 16*, 282–292.

Thomas, A., Shooya, O., Rokitzki, M., Bertrand, M., & Lissner, T. (2019). Climate change adaptation planning in practice: Insights from the Caribbean. *Regional Environmental Change, 19*, 2013–2025.

Town and Country Planning Development Order. (2013). *The town and country planning (Negril and Green Island Area) provisional development order*. Ministry of Water, Land, Environment and Climate Change, Kingston.

United Nations Environment Programme (UNEP). (2010). *Linking ecosystems to risk and vulnerability reduction, the case of Jamaica* (2nd ed.). United Nations Environment Programme.

Veira, M. (2014). Beyond the Negril Beach Spin. *The Gleaner*. [online] Retrieved November 3, 2020, from http://jamaica-gleaner.com/gleaner/20140511/focus/focus4.html

White, I. (2008). The absorbent city: Urban form and flood risk management. *Proceedings of the ICE-Urban Design and Planning, 161*, 151–161.

Wilson, E., & Piper, J. (2010). *Spatial planning and climate change*. Routledge.

Wilson, M., Mandal, A., Taylor, M., Burgess, C., Campbell, J., & Stepphenson, T. (2014). Flood risk and climate change in Negril, Jamaica: An assessment of combined terrestrial and coastal flood risk driven by projections of future climate, March 17–21. *WCRP Conference for Latin America and the Caribbean: Developing, Linking and Applying Climate knowledge*. Montevideo, Uruguay.

World Travel & Tourism Council (WTTC). (2014). *Economic Data Search Tool*. WTTC.

Xu, L., Cui, S., Tang, J., Nguyen, M., Liu, J., & Zhao, Y. (2019). Assessing the adaptive capacity of urban form to climate stress: The case study on urban heat island. *Environmental Research Letters, 14*.

The Contentious Policies of Place and Space: The Maldives, Overpopulation, and Climate Concerns

Andrea C. Simonelli

Abstract One of the biggest long-term concerns that small low lying islands face is the threat of loss of physical integrity due to sea level rise. For the Maldives, this threat is very serious and has become an international issue when the former president went public in recent years with a plan to relocate the entire nation. However, before climate processes will displace inhabitants on a large scale, other factors will come into play. The multi-level government structures of the Maldives provide several opportunities for contention over policy and when it comes to continued population growth and development, and preferences differ at each level. No matter how terminal climate predictions are for some regions, potential responses cannot be taken out of the context of economic development and domestic political concerns. This paper explores the challenges of population and space for small island development using a case study of Maldivian policy preferences at the island, atoll, and national levels taken from interviews from the field. It explores the debate between government actors, constituent preferences, and national development concerns as they overlap climate change anxieties.

Keywords Maldives · Atoll islands · Climate change · Governance · Development · Overpopulation · Indian ocean · Tsunami · Sea level rise · Small islands · Global environmental change · Relocation

1 Introduction

Climate change is altering the world's coastlines through sea level rise and although it is a long term process, this affects the physical integrity of some areas faster than others. For small, low lying islands, the threat of land loss is a significant concern where their initial size leaves little room for internal relocation. For the Maldives,

A. C. Simonelli (✉)
Virginia Commonwealth University, Richmond, VA, USA
e-mail: acsimonelli@vcu.edu

S. Moncada et al. (eds.), *Small Island Developing States*, The World of Small States 9, https://doi.org/10.1007/978-3-030-82774-8_16

this threat came to international attention under former president Mohammed Nasheed. In a speech to the United Nations, he claimed that failure to take dramatic planetary action would produce rising seas, thus causing islands like his to "die", also likening the Maldives to the world's "canary in the coalmine" (Nasheed, 2009).

The use of this language on a global stage provided the Maldives with a politically constructed identity fighting against existential demise, with time running out. Farbotko (2010) explains that the use of this phrasing (in reference to islands) renders visible to the non-expert an imagined phenomenon that can be photographically captured in real time, and argues that this metaphor is used to portray islanders' victimhood. Victimhood is a necessity for this imagery; in order for the island to be the "canary", it has to disappear. That is the importance of the canary, its sacrifice (Simonelli, 2015). Planning for internal migration has been popularised through media accounts of "sinking islands" and the permanent nature of such a threat, but these are often hyperbole. This concept, like the canary in the coal mine, is also used to conjure images of tragedy in paradise and imminent doom (Simonelli, 2015).

While dramatic and intriguing to outsiders and those who have listened to Nasheed's rhetoric, this expression of the Maldives' problems diverts attention from immediate development needs and misconstrues their inherent vulnerability to climate processes. While seriousness is what Nasheed intended to imply, so much so that he also proposed a fund to relocate the entire population to another country, he elevated the need to externally relocate the population above all others.

This speechmaking did not, however, erase the other concerns of the Maldivian people. Rhetoric emphasising climate change as today's biggest problem in the Maldives neglects the history of development theory, policy, and practice (Kelman, 2014). This is true of the Maldives. Each island and atoll has its own development concerns which, while compounded by climate impacts, are serious but do not immediately necessitate out-migration. This chapter outlines the overlapping development concerns and space-based policy preferences as they come into conflict at different levels of governance.

2 Climate Vulnerability and Atoll Habitability

The Maldives is most recognised as a picturesque honeymoon destination of overwater villas over its powder sand beaches. The archipelago consists of 1192 islands in a double chain of 20 distinct atolls tapering into a single chain to the north and south (Woodroffe, 2008). Located a little over 750 km southwest of India and Sri Lanka, most of the islands are less than two metres (m) above sea level with the highest ridge on survey at 3.2 m (Woodroffe, 2008). Atolls are low lying and dynamic, mere accumulations of sand and gravel which rely on locally generated sediments (Kench et al., 2005). These growths are rings of coral reefs that enclose a lagoon and have common environmental problems that render them particularly vulnerable to climate change (Barnett and Adger, 2003). Atolls have small

Fig. 1 Map of Maldives in Indian Ocean (*Source: Maldives Travel Guide*)

subterranean freshwater lenses which are vulnerable to freshwater intrusion due to storm inundation and depletion in the dry season. Being small, isolated, geographically low-lying and with poor-quality soils, atolls face particular vulnerability to climate processes such as coral bleaching, sea level rise, and ocean swell events (Fig. 1; Nurse et al., 2014).

Small islands are often characterised by narrow resource bases and dependency on links to the outside world (Baldacchino & Niles, 2011). Physical processes alone, however, do not make a small island vulnerable; risk from climate change is determined by future hazards, current sensitivities, adaptive capacity, and regional exposure (Fussel, 2006). Kaly et al. (2002) provide a framework to define vulnerability for Small Island Developing States (SIDS) which identifies the risk of specific hazards occurring, in addition to intrinsic and extrinsic resilience to hazardous events. In doing so, they consider environmental challenges to small islands, but also emphasise their anthropogenic nature rather than 'naturalness'. Beyond hazards that are intrinsically natural or resulting from climate change, islands also produce and are sites for pollution, deforestation, the overexploitation of resources, and nuclear testing. In addition, all hazards can lead to acquired vulnerability over time with successive years of damage without full recovery between events (Kaly et al., 2002). Because adaptation does not occur instantaneously, the relationship between adaptive capacity and vulnerability depends of the timescales and hazards with which we are concerned (Brooks et al., 2005).

The issue at hand is the interaction of social and natural systems and their propensity to recover. Both biological and human systems are made up of rules and institutions that mediate the use of resources (Berkes & Folke, 1998). These can recover after a system disruption, but the process of rebuilding either biologically or infrastructurally is neither easy nor fast. In the context of these social-ecological systems, resilience refers to the magnitude of disturbance that can be absorbed before a system changes to a radically different state (Adger, 2006); the larger the disturbance, the more potentially different the original system.

In the case of the Maldives, the projection is that sea level rise will be such that neither system—natural or human—will be able recover. While key parameters of vulnerability are the stress to which a system is exposed, its sensitivity, and its adaptive capacity (Adger, 2006), in the Maldives, the focus has tended to be only on the first of these.

The Maldives' sensitivity to social-ecological disruptions is also due to the proximity of its human settlement zones to the shoreline. Humans have long been drawn to coastal areas which provide many resources and trading opportunities, this attraction has been strong among trading nations and empires (McGranahan et al., 2007), such as the Maldives. While climate change will challenge those living near shore lines, trading nations like the Maldives have managed coastal hazards for hundreds of years. Campbell (2009) argues that colonialism, (modern) development, and globalisation have put in place processes by which island resilience has been reduced while exposure has increased. While his work focus specifically on Oceania, his analysis of historical atoll impacts is also applicable to the Maldives, with the shared atoll characteristics of a limited resource base alongside exposure to water shortages, 'king tides', and storm surges.

Historically, risk management was based on traditional systems of knowledge regarding food security, inter- and intra-community cooperation, and protecting settlements and their inhabitants (Campbell, 2009), but these systems begin to break down when modern foodstuffs, imported building materials, and urbanisation disrupt previous settlement methods. Modernisation of island cultures is potentially damaging to islands where traditional indigenous knowledge and practices are most extensive (the Pacific and Indian Ocean) (Pelling & Uitto, 2001). Market expansion and the privatisation of resources coupled with the willing exploitation of certain ecosystems for tourism has forced social systems and once protective cultural practices to change. Globalisation can be considered an outside hazard that island social systems have had to adapt to over time.

Least developed countries have a larger share of their population living in highly populated low-elevated coastal zones with Maldives having more than 90 per cent of its settlements in this zone (McGranahan et al., 2007). About half of the human settlements in the country are within 100 m of the shoreline along with almost three quarters of critical infrastructure such as airports, power plants, landfills, and hospitals (Sovacool, 2012a). With the vast majority of its settlements in long-term jeopardy, it is the habitability of islands that is a policy concern and the thresholds beyond which people can no longer remain on their islands which require investigation (Barnett, 2001).

However, climate change will alter far more than the shoreline. Atoll nations will also be affected by an increase in coral bleaching episodes and flooding events which will contaminate any locally consumed freshwater as well as small scale agriculture (Barnett & Adger, 2003). Coral bleaching could arguably be its most pressing climate concern as loss of reef biodiversity will erode both Maldives' tourism and fisheries if fish stocks become increasingly unpredictable and move to deeper/cooler waters. Overwater villas may be well placed to withstand rising seas but may not avoid loss of tourist interest when reef fish no longer swim beneath them. Tourism vulnerability literature, again, focuses only on low elevation (Becken et al., 2014) rather than ecosystem and landscape loss. Moody's Analytics (n.d.) estimates the service industry to contribute 81 per cent of the Maldives GDP which demonstrates just how economically dependent the nation is on this sector.

Fish exports are also economically important. There has been a diversification in the fishing industry away from traditional dry fish production and towards processing for developed country markets; over half of the total catches of fish are exported (Athukorala, 2004). Considering the myriad of climate and weather events which can drive success or otherwise in these sectors, atoll nations will need to make serious choices for their survival on many fronts. Becken et al.'s (2014) Tourism Disaster Vulnerability Framework (TDVF) found that a lack of political will, insufficient capacity, conflict and corruption, poor implementation, and unstable political environments (in the Maldives) were driving tourism's vulnerability. This highlights that human factors can be severely underrepresented in climate analyses of island locales such as the Maldives. Impact assessments frequently superimpose future climate scenarios on an otherwise constant world to estimate the potential impacts of anthropogenic climate change on a climate-sensitive system (Fussel & Klein, 2006). These assessments cannot readily account for human factors such as future policy decisions and their consequences (both intended and unintended), and habitual behaviours which also influence a physical systems' stability. There is a growing acceptance that human actions, behaviour, decisions, and governance, lead to vulnerabilities which cause disasters, so that disasters are not 'natural' (Kelman, 2014); this can be said for many kinds of events—climate change included. A distinction can be made between the structural vulnerability which results from factors independent from a country's political will and the vulnerability deriving from policy, which results directly from choice (Guillaumont, 2010).

Bearing in mind the impact of globalisation, commercialisation, and consumerism on the Maldivian economy (at the level of the individual island as well as through tourist demand), it is imperative that the multi-level governance of the nation be responsive to the needs of its people and adjust over time. More specifically, good governance is essential for an economic system to function properly and hence to be resilient. Governance relates to issues such as rule of law and property rights (Briguglio et al., 2009). With a growing population under a fledgling democratic government, it is logical to expect that governance be resilient to both development needs and climate concerns, but this will not be an easy task. Between preferences and timescales, there are issues of contention between the paradox of the desire to develop in a place with limited long term viability and the need to eventually plan for evacuation.

3 Democratic Development and Challenges

Stable democracy in the Maldives has not come easily. For many years, the nation has suffered severe political infighting, a breakdown of the administrative machinery, growth of fundamentalism, growing intolerance towards other religions and culture, challenges to economic growth, environmental issues, indolence on the part of the administrators and undue external influences (Bonofer, 2010). Though not quite colonised, the Maldives did fall under a British protectorate from 1887 to 1965; it became a republic when the last sultan was deposed in 1968 and Ibrahim Nasir became President (Musthaq, 2014). His successor, Maumoon Abdul Gayoom, held the presidency from 1978 until 2008 but full democratic transition had been questionable, with only promises of democratic reforms in 2004 (Bonofer, 2010). As a mandatory condition of European aid in response to the 2004 Indian Ocean tsunami, the Gayoom administration implemented reforms aimed at demonstrating a stronger electoral democracy (Hirsch, 2015).

As a result, after 30 years of continuous rule, Mohammad Nasheed, a journalist and repeat political prisoner was elected in 2008 (Shenk, 2012). Despite winning the first free, competitive presidential election in Maldives, Nasheed inherited a government that was economically bankrupt, politically nascent, and prone to infighting (Bonofer, 2010). Nasheed is internationally recognised for his use of the media to bring attention to the Maldives' climate vulnerabilities, but his global popularity did not translate into approval at home. In February 2012, Nasheed resigned in the wake of a mass protest and mounting political violence (Hirsch, 2015).

The Maldivian economy and political system experienced instability as Vice President Waheed took over and shared ministries among the political parties; in the two years that followed, political leadership of the nation was difficult to determine (Musthaq, 2014). In 2013, alliances had switched nearly completely since 2008; parties that had sided with Nasheed against Gayoom in that year now opted for the opposition Progressive Party of the Maldives (PPM) a Gayoom creation and from September to November of that year, the people voted for the President three times in just nine weeks (Musthaq, 2014).[1]

[1] The second set of field interviews were being conducted within this time of political instability.

4 Methodology and Research Sites

This chapter is based on field research conducted in 2013. This research extends from a larger project: *Perceptions and understandings of climate change and migration: Conceptualising and Contextualising for Lakshadweep and the Maldives* funded by The Research Council of Norway (Simonelli 2014a, 2014b). The three-year project (2013–2015) collected quantitative and qualitative data in the form of household surveys and semi-structured interviews on two islands in each archipelago. The interviews were conducted during June and July in the administrative district of the South Malé Atoll and in the month of October on Raa Atoll. A total of 34 interviews were conducted with 17 respondents on K. Guraidhoo and 18 on Dhuvaafaru. Each interview was conducted in the local language with the assistance of a translator and lasted between 20 and 40 min. Interviewees were asked to nominate and describe any general environmental changes they had experienced in their lifetime and how it affected them.

They were then asked if they had heard of the term 'climate change' and if they understood it and if they believed they would have to migrate knowing due to its impact. Finally, they were asked to discuss if there were any preferable locations they would go in that eventuality. Interviews tended to be free-flowing and often headed in different directions based on the interviewees' experiences. Some required follow up questions that were not the same across all participants.

For the purposes of this chapter, the interviewees were elected representatives at each municipal level; Island and Atoll. Interviews at the national level were not collected due to the nature of the project and site visits coincided with the national government's active campaign for re-election which would make access more difficult. Analysis at the national level of governance was therefore undertaken through literature and secondary narratives. Interviewees were all men consisting of each: a member of the Atoll Council for South Malé and Raa Atoll, the current Island President of K. Guriadhoo and the former Island Chief of Dhuvaafaru. Interviews were primarily conducted in the formal offices of the elected officials with the exception of the Raa Atoll Council representative: only the interview with the former Island Chief was conducted with the assistance of a translator.

Grounded theory (Glaser and Strauss, 1967) was used for the analysis which is the formulation of theory from data gathered from social research. This is done without any preconceptions going into the field and gathering a wide range of information. Conceptual frameworks emerged from the interviews which led to the identification of response clusters (Charmaz, 2006) by municipal level, suggesting a policy preference divide. To verify and compare knowledge and attitudes at the national level, narratives from the national level were gathered from sources such as *The Island President.*

K. Guraidhoo is a 45-minute speed boat ride from the capitol of Malé. It houses approximately 1800 residents, many of whom work on Malé during the week and take the public ferry home on the weekends. It sits directly across a lagoon from Kandooma Resort which is operated by the Intercontinental Group and employs a small number of local, low paid employees such as grounds keepers. The resort operates a ferry which runs between the islands several times a day to transport staff and brings groups of travellers to the island to visit the small assortment of tourist shops. The island is also home to the Maldives' only mental health hospital.

Until 2008, Dhuvaafaru was an uninhabited Maldivian island. It was then that the national government chose it as a relocation site for 5000 inhabitants of Kandholhudhoo who lost their entire island after the Indian Ocean tsunami of 2004. Dhuvaafaru had to be developed for habitation from scratch and building work continued through the field visit. The primary economic activity of Dhuvaafaru is fishing; few people work in tourism. During the field visit, there was only one resort in the atoll and it was approximately thirty minutes south by speed boat (in good weather) and did not offer a ferry service like Kandooma; any workers must reside on the resort in cramped housing in the centre of the island for several months. Fishermen often go out for a month at a time leaving the atoll waters to find fish to sell back to their island neighbours and to Malé, 110 kilometres (km) or roughly an eight-hour boat ride south.

5 Multi-Level Governance Priorities

5.1 National Level

President Nasheed spent the first four years of his tenure promoting the Maldives' long term climate concerns on the national stage. This began even before his inauguration when he suggested that the Maldives must buy land in Sri Lanka or Australia (Schmidle, 2009) for full scale relocation. This suggestion gained international recognition, reverberating through international conference talks for years. To reinforce the seriousness of his claim, Nasheed held an underwater cabinet meeting (Omidi, 2009) to demonstrate what life will be like if sea level rise continued. He used the imagery of loss and the language of war, framing climate change as national security issue, but one which avoided blaming other nations directly, instead placing all humans against greenhouse gas pollution (Hirsch, 2015). This sentiment never caught on with Maldivians however. Research conducted during his tenure in office demonstrates that even when climate change was accepted by the interviewees as happening with major impacts, especially sea-level rise, permanent departure from Maldives was not a significant or expected option (Kelman et al., 2019). Internally, Nasheed pushed for his country to become "carbon neutral", that such technologies and development would be both possible and profitable for the Maldives. Bluepeace, a small, low-budget organisation of elite scientists and policy experts were enlisted to carry out Nasheed's carbon neutrality mission, beginning with the designing the country's first artificial mangrove system (Hirsch, 2015).

The documentary, The Island President (Shenk, 2012) followed Nasheed's international quest for recognition of his nation's climate problem and shadows his first year in office and his delegation's fight at the 2009 Copenhagen Climate talks hosted by the United Nations Framework Convention on Climate Change (UNFCCC). During the film, Nasheed and his small, frantic, and tired group of negotiators are shown as trying to 'save the world' and in turn their small and sinking nation. An ethnography by Hirsch (2015) argues that the threat of disappearance and the fight to reduce emissions (used by Nasheed) created a growing sense of community among Maldivian elites. The film premiered shortly after Nasheed resigned and he continues to promote it globally.

5.2 *Atoll Level*

Even before Nasheed's Presidency, internal migration and resettlement policies in response to environmental challenges were being discussed. In 2007, the Maldivian Nation Adaptation Plan of Action drafted by the Ministry of Environment, Energy, and Water, presented a framework for human, natural, and social system resilience to ensure long term sustainability. The plan prioritised population consolidation and development as well as acquiring support for the 'Safer Island Strategy' (Kothari, 2014). The "safer island concept" was developed after the 2004 Indian Ocean Tsunami as an adaptation strategy for tsunamis as well as the sea level rise (Riyaz & Park, 2010). The idea has also been called the "Safer Islands Development Plan" or SIDP, with the goal of assessing and populating a range of islands that could act as safe havens for people forced to migrate before or after climate disasters, seeking out 10 islands for this purpose (Sovacool, 2012b). However, the United Nations Development Program (UNDP) reviewed the climate risks and noted no "safe" island exists in the Maldives (UNDP, 2007) as each is a low-lying atoll. Additionally, a study by Riyaz and Park (2010) found that while enhanced mitigation measures might assist in resilience for less frequent disasters, in the short term, its implementation may produce more vulnerability. Building and relocating Maldivians onto the higher (but still low) islands of the atoll while calling them "safe" can produce a false sense of security against sea level rise while otherwise entrenching locals where they will still experience coral bleaching, biodiversity loss, and monsoon variability.

Regardless of international concern, relocation plans still have hope at the Atoll level of Maldivian governance. The Dhuvaafaru representative of the Raa Atoll Council commented that there was a general demand for bigger, more developed harbours across the Atoll so the inhabitants could trade directly with Sri Lanka, Thailand, and Indonesia. There is still active talk within the Council to consolidate the Atoll from 15 islands to five or six. He also adds that the current government (referring to the Nasheed government) would not allow this and that the people do not want to relocate anyway. He believes that they may have to leave again one day, but said that he personally does not want to. He does not want to move to Malé like others, but would prefer 'here' to be like Malé.

The Guriadhoo representative to the South Malé Atoll Council recounted a similar plan. He believes that reclamation is a good thing because people do not usually leave the Maldives and have not for 2000 years. The plan is that one "really big" island may be reclaimed and then everyone from the nine inhabited islands of the Atoll will move onto it and the others will be sold off for resorts. In this way, the people can consolidate around facilities. A few families a year already move from Guriadhoo to Malé, especially the young for job prospects. The government cannot provide adequate infrastructure and services to all of its inhabited islands; the biggest problems they face are a growing population, minimal space, and economic opportunities.

Kothari (2014) has argued that the Safe Island Strategy was repackaged as the 'Resilient Island' approach under Nasheed as a disguise for development relocation. The interviews conducted during this study do nothing to counter this argument, reinforcing both Councils' push for planning toward internal relocation for the purposes of development. Each representative mentioned they had some concerns over the money it would take to complete such a task, but emphasised that this reorganisation of people for administrative efficiency is important for the future of the Maldives.

5.3 Island Level

Individual island level governance is farthest removed from the national government and yet each is fighting the same battles daily, whether against rising seas, for basic services, or for individual constituent needs. The former Island Chief of Dhuvaafaru was also in power during and after the Tsunami when his people were still on Kandholhudhoo, located across the atoll. He explains that the old island was very small and that its 2002 reclamation project included some rebuilding of the sea wall. Before the project, the island flooded regularly with the monsoons. After the project, the monsoons still flooded both the East and West sides of the island. Many families talked of relocation due to the constant flooding, but most could not afford to or did not want to leave friends and family. The national government never mandated a relocation for the population and thus no help was offered. He explains that after the tsunami they now have a much larger island (Kandholhudhoo was 7 ha and Dhuvaafaru is 42) and this gives them a better chance for rebuilding houses. They now have three schools while previously they only had one. Each family used to own a 150–200 square foot home and now each averages 2000 square feet. After the project is complete he does not believe the residents will be harmed by sea level rise. Dhuvafaaru is roughly a meter above sea level, but its homes are now located further from the shore line.

The experience of K. Guriadhoo is very much the opposite that of Dhuvaafaru. Their Island President details a similar inundation of flooding and storm surge seen on Kandholhudhoo, but with no eventual tsunami recovery, relocation, or assistance. He explains that individuals are taking it upon themselves to use rocks from destroyed properties to rebuild their own coast lines. There is no plan from the government to relocate. A reclamation project has been proposed and signed off by the government (Nasheed's government) but nothing has happened yet, even after many surveys of the proposed area. The project would almost double the landmass of the island to 16 ha. But with the continued elections, the President was concerned it will be stalled or even cancelled. The President suggested that since his island is close to Malé, it is easy to show researchers the damage and that is why the government purposefully does not do any maintenance on his island. He asks, "If they show other successful projects, how will they make the case for additional development funds?" He is very concerned that there is no 'Plan B" for his island if the reclamation project does not go ahead. He believes that his people will have to continue to move to Male because his island is already at capacity; existing families need more space.

Both islands experienced similar climate and other pressures although the outcomes varied due to the tsunami response. Comparatively sized and experiencing regular flooding, local leadership detailed in each case either a completed reclamation project or one in the works, but no plan or desire for relocation. Each island had asked the national government for project money for individual upgrades to basic infrastructure. The 2004 Tsunami became the impetus for relocation and redevelopment that Kandholhudhoo needed while K. Guraidhoo continues to struggle. The President of Guraidhoo had been concerned that his island may never receive any development funds because his island is being used as an example of problems to outsiders.

While previous and higher levels of governments proposed projects which suggested that internal movement (and even external) could be possible, very little direct investment on either island can be described outside of immediate disaster aid.

6 The Contentious Policies of Place and Space

It is clear that each administrative level of governance has a different vision of relocation, preferences toward it, and what they believe is a priority for the people of the Maldives. The national government, under the Nasheed administration, articulated a preference for the full scale relocation of his people using the language of victimisation on a global stage. This does not appear to be the message that has been received at the lower levels of government; none of the interviews with the representatives from either the Atoll or Island level mentioned the long term plan to relocate entirely. The Atoll representatives' preference is for regional population relocations which allow for a more feasible way for residents to access services and facilities.

Since the 1990s, the Ministry of Planning and National Development argued that the Maldives unique spatial distribution brings increased costs to provide services and infrastructure and that it would make good economic sense for if the entire population were living on one main land mass where it could be served by one hospital, fewer educational institutions, and harbours. Gayoom embarked on the 'National Population Consolidation Strategy and Programme' which, similarly, sought to create incentives for people to relocate to one of two main regional growth centres (Kothari, 2014).

While neither of the Atoll representatives mentioned this program specifically, they appear to agree with its spirit. The island level governments stated a preference for targeted development on their islands in response to population and environmental concerns with migration only as a last resort. In 2010, the Nasheed regime entered into a funding partnership with UNDP to implement the "Integrating Climate Change Risks into Resilient Planning in the Maldives" or ICCR, which is an island level investment in climate adaptation. While Kothari (2014) had expressed scepticism in this level of investment, the program did not fund any population relocation, rather 'soft' adaptation measures. It enabled four different types of demonstration projects on Kulhudufushi, Thinadhoo, Thulusdhoo, and Kudhahuvadhoo: replenishment of natural ridges, coral restoration, coral sea wall repair, and mangrove planting respectively (Sovacool, 2012a). As 'soft' projects, they involve improving the capacity of the natural environment, rather than 'hard' projects such as infrastructure. It could be argued that projects such as these did aim to respond to concerns voiced at the island level. However, the project's purpose was to demonstrate the feasibility of some types of projects and to disseminate knowledge. Considering the needs expressed by the Island Chief and President related to facilities and lack of funding, this project did not provide any direct help.

As mentioned above, the national government, under both Gayoom and Nasheed, have made investments toward ameliorating environmental degradation and rebuilding islands following tsunami damage through a number of coastal erosion and coral damage pilot projects. In terms of development needs, investments have been more specifically targeted in the Capitol region. The largest reclamation project has been Hulhumale (Shaig, 2006) which began construction under Gayoom and continued even beyond Nasheed. This project is expected to house 100,000 people (Sovacool, 2012b) and is built sufficiently high to be safe from flooding under present extreme conditions and future flooding during the twenty-first century, as long as sea-level rise is less than 0.6 m (Brown et al., 2020).

During the field interviews for the current study, local interviewees also discussed their interest in moving to Hulhumale due to its promised upgraded schools, hospitals, infrastructure, and hope of more space which local officials expressed was not being funded. The Nasheed government had held a lottery, giving locals a chance to 'win' a space on Hulhumale. Few did; one of the respondents interviewed on Guraidhoo tried, citing the need to move her six children and two grandchildren out of her parents' house. With no land of her own and none available, she was seeking better access to health and education for her family (Simonelli, 2015). The option to 'win' space is understandably appealing when urban and development planning on current islands has neglected natural population growth. According to the 2014 Maldives census, Hulhumale still only had a population of 15,769 (Maldives, 2014).

When Kandholhudhoo was levelled by the tsunami and Dhuvaafaru was opened for development, it was constructed to the full capacity of the original island's population of 5000. No additional space was left open for young people to expand

Fig. 2 Map of Dhuvaafaru (*Photo Credit: Andrea C. Simonelli*)

into their own homes after marrying, or for internal mobility, or general population growth. Despite having increased land size, all of the usable space was taken up with larger homes and facilities. The island of K. Guraidhoo is in the same position. All of its land is also claimed by current families who have no more room to expand without relocating to another island. Figure 2 below is a map of Dhuvaafaru provided by the municipal government office illustrating the use of land.

Figure 3, illustrating the map of K. Guraidhoo, may look as if there is open space, but some of these areas include a ship building yard, its trash incinerator facility, and the football field. Overpopulation, in the context of Dhuvafaaru and K. Guraidhoo does not necessarily mean that there are too many people, but that there is no room for them. Families have to live multi-generationally in small homes with no room to expand. In interviews on both islands, the young explained that they wanted to have their own independence from their parents, but that the government is not opening any new land for them, only new resorts for tourists. Moving to Male for an apartment of their own and seeking better economic opportunities is also a precarious choice; the capital continues to grow and only families originally from there are able to own land. Rental housing is very expensive (Orlowska, 2015). With a population of 127,079 (Maldives, 2014) on 5.7 km^2, the youth are reluctantly choosing to move to Male for the prospects and services that are not being funded or expanded on their home islands and atolls.

Fig. 3 Map of K. Guraidhoo (*Photo Credit: Andrea C. Simonelli*)

Table 1 Migration and policy per administrative level

Administrative level	Migration preference	Main policy concern
National	All	Climate Change
Atoll	Local (rural to urban)	Development & Population
Island	None (if possible)	Development & Population

Federal systems of governance with multi-level jurisdictions of responsibility are a normal feature of many democracies, however it is unclear the specific duties and functions the Atoll and Island. The Constitution of the Republic of the Maldives (2008) outlines that the Presidency has the power to create island, atoll, and city councils, that they be elected, and the specification of their jurisdiction. It does not specify particular responsibilities at each level. The lack of specificity is a logical cause of contention; both the Atoll and Island representatives have a direct line to the national government, overlapping concerns and constituencies, but no clear delineation of power. Thus, they have every reason to look at the same development problems through conflicting spatial lenses. Table 1 simply outlines the migration and policy concerns expressed at each administrative level.

The concerns of Maldivian citizens across these policy concerns also very. Interviewees from both Dhuvaafaru and K. Guraidhoo could articulate their personal experiences that coincided with long term climate impacts, but none indicated any familiarity with the "sinking islands" discourse communicated by Nasheed to the global community (Kelman et al., 2017). Additionally, they provided simple development-based reasons to migrate within the Maldives. Access to education on Malé and lack of space on their island of origin are the main reason why young people move (Kelman et al., 2017). Another recent study interviewing locals on Villingili, Malé, and Hulhumale confirms that locals can describe weather pattern changes over the past 10-15 years, but that reasons for migration include seeking improved health care, education, and better living conditions (Stojanov et al., 2017). These are development related and migration in this situation is not necessarily preferred, but seen as necessary for economic and educational opportunity. The main concerns of Maldivian people on multiple islands appear to align with those expressed by the Atoll and Island level representatives interviewed. The reasons provided for local migration support these interviews rather than the international climate change narrative.

7 Conclusion

While the national discourse has changed since Nasheed resigned in 2012, the same challenges persist and are potentially worse. Nasheed's calls to the international community to 'save' them place the people of the Maldives in an awkward paradox that if too much money is invested in development projects there may be no assistance for relocation. It is possible that Nasheed's pre-emptive rhetoric of doomed investment, which could extend the longevity of habitation, means that people will abandon it. Like the 'canary in the coal mine' as exemplified by the scepticism of the K. Guraidhoo Island President, only in sacrifice of the canary does any action happen; if the islands are too well protected, will anyone help when the waves do eventually become too high? The paradox only exists because Nasheed put out an urgent call about an issue which his people had noticed, but were not yet alarmed by. Representatives from lower levels of government did not mention constituent demands of this nature and additional survey research confirms this.

The development that is taking place is primarily in the greater Malé area, enticing more rural to urban migration. Data from interviewees reinforce the need for relocation due to the lack of space, health services, and higher educational opportunities. However, Malé is already more crowded than any of the islands that people are moving from. Urbanisation is a growing concern in many countries, but in the Maldives the limited land space means that islands cannot support suburban sprawl. Visually, Malé is a floating mega city and growing. The Hulhumale project was intended to alleviate some of the population pressures but is still incomplete, and in the meantime more and more people make their move to Malé.

With the needs of a growing population in the capital as well as other islands already at their capacities, and a reluctance to open new islands to development other than for tourist resorts, and with the additional layer of increased climate impacts, conditions are ripening for a humanitarian crisis.

The most dangerous implication of the continuous policies affecting space and place in the Maldives is that there is no complementary way forward for all administrative levels. Maldivian vulnerability to climate change but also internal anthropogenic changes due to population stresses are not simply based on some absence of capacity; governments at all levels have known for many years that more liveable space is required. There remains a lack of decisive action due to a continued preference for tourism investment over wellbeing of the general population.

Giampiccoli et al. (2020) argue that when people have their own incomes, they rely less on government support programmes as they can fend for themselves; community-based tourism can be developed in many islands/atolls so that job opportunities will be developed in situ. However, this is predominantly expressed as a preference for luxury resorts over local tourism.

President Abdulla Yameen changed environmental tack, saying that a much-expanded economy from mass tourism and mega-developments rather than solar power and carbon neutrality would enable it to adapt itself to climate change and give its young population hope for the future (Vidal, 2017). In the case of the Maldives, government decision makers are not necessarily balancing feasible ways forward for the fledgling democracy, and development projects that address constituent needs are not keeping pace with its population growth and the increasing demand for basic services.

References

Adger, W. N. (2006). Vulnerability. *Global Environmental Change, 16*(3), 268–281.

Athukorala, P.-C. (2004). Trade policy making in a small Island economy: The WTO review of the maldives. *The World Economy, 27*(9), 1401–1419.

Baldacchino, G., & Niles, D. (Eds.). (2011). *Island futures: Conservation and development across the Asia-Pacific region*. Springer.

Barnett, J. (2001). Adapting to climate change in Pacific Island countries: The problem of uncertainty. *World Development, 29*, 977–993.

Barnett, J., & Adger, W. N. (2003). Climate dangers and Atoll countries. *Climatic Change, 61*, 321–337.

Becken, S., Mahon, R., Rennie, H., & Shakeela, A. (2014). The tourism disaster vulnerability framework: An application to tourism in small island destinations. *Natural Hazards, 71*(1), 955–972.

Berkes, F., & Folke, C. (1998). Linking social and ecological systems for resilience and sustainability. In F. Berkes & C. Folke (Eds.), *Linking Social and ecological systems* (pp. 1–25). Cambridge University Press.

Bonofer, J. A. (2010). The challenges of democracy in Maldives. *International Journal of South Asian Studies, 3*(2), 433–449.

Briguglio, L., Cordina, G., Farrugia, N., & Vella, S. (2009). Economic vulnerability and resilience: Concepts and measurements. *Oxford Development Studies, 37*(3), 229–247. https://doi.org/10.1080/13600810903089893

Brooks, N., Adger, N. W., & Kelly, P. M. (2005). The determinants of vulnerability and adaptive capacity at the national level and the implications for adaptation. *Global Environmental Change, 15*, 151–163.

Brown, S., Wadey, M. P., Nicholls, R. J., Shareef, A., Khaleel, Z., Hinkel, J., Lincke, D., & McCabe, M. V. (2020). Land raising as a solution to sea-level rise: An analysis of coastal flooding on an artificial island in the Maldives. *Journal of Flood Risk Management, 13*(S1).

Campbell, J. (2009). Islandness. Vulnerability and resilience in Oceania. *Shima: The International Journal of Research into Island Cultures, 3*(1), 85–97.

Charmaz, K. (2006). *Constructing grounded theory*. Sage.

Constitution of the Republic of the Maldives. (2008).

Farbotko, C. (2010). Wishful sinking: Disappearing islands, climate refugees, and cosmopolitan experimentation. *Asia Pacific Viewpoint, 51*(1), 47–60.

Fussel, H.-M. (2006). Vulnerability: A generally applicable conceptual framework for climate change research. *Global Environmental Change, 17*(2007), 155–176.

Fussel, H.-M., & Klein, R. J. T. (2006). Climate change vulnerability assessments: An evolution of conceptual thinking. *Climatic Change, 75*, 301–329.

Giampiccoli, A., Abdul Muhsin, B , & Mtapuri, O. (2020). Community-based tourism in the case of the maldives. *Geo Journal of Tourism and Geosites, 29*(2), 428–439.

Glaser, B. G., & Strauss, A. L. (1967). *The discovery of grounded theory: Strategies for qualitative research*. Aldine Transaction.

Guillaumont, P. (2010). Assessing the economic vulnerability of small island developing states and least developed countries. *Journal of Development Studies, 46*(5), 828–854.

Hirsch, E. (2015). "It won't be any good to have democracy if we don't have a country": Climate change and the politics of the synecdoche in the Maldives. *Global Environmental Change, 35*, 190–198.

Kaly, U., Pratt, C., & Howorth, R. (2002). A framework for managing environmental vulnerability in small islands developing states. *Development Bulletin, 58*, 33–39.

Kelman, I. (2014). No change from climate change: Vulnerability and small island developing states. *The Geographical Journal, 180*(2), 120–129.

Kelman, I., Orlowska, J., Upadhyay, H., Stojanov, R., Webersik, C., Simonelli, A. C., Procházka, D., & Němec, D. (2019). Does climate change influence people's migration decisions in maldives? *Climatic Change, 153*, 285–299.

Kelman, I., Upadhyay, H., Simonelli, A. C., Arnall, A., Mohan, D., Lingaraj, G. J., Shadananan Nair, S., & Webersik, C. (2017). Here and Now: Perceptions of Indian Ocean islanders on the climate change and migration nexus. *Geografiska Annaler: Series B Human Geography, 99*(3), 284–303.

Kench, P. S., McLean, R. F., & Nichol, S. L. (2005). New model of reef-island revolution: Maldives, Indian Ocean. *Geology, 33*(2), 145–148.

Kothari, U. (2014). Political discourse of climate change and migration: Resettlement policies in the Maldives. *The Geographical Journal, 180*(2), 130–140.

Maldives. (2014). *Maldives bureau of statistics*. http://statisticsmaldives.gov.mv/census-2014/

McGranahan, G., Balk, D., & Anderson, B. (2007). The rising tide: Assessing the risks of climate change and human settlements in low elevation coastal zones. *Environment & Urbanization, 19* (1), 17–37.

Moody's Analytics. (n.d.), *Economic Indicators: Maldives, 2018*. [online] Retrieved December 14, 2020, from: https://www.economy.com/maldives/indicators

Musthaq, F. (2014). Shifting tides in South Asia: Tumult in the Maldives. *Journal of Democracy 25*, 164–170.

Nasheed, M. (2009), *Speech to the United Nations*. [online] Retrieved December 14, 2020, from: https://gadebate.un.org/sites/default/files/gastatements/64/64_MV_en.pdf

Nurse, L. A., McLean, R. F., Agard, J., Briguglio, L. P., Duvat-Magnan, V., Pelesikoti, N., Tompkins, E., & Webb, A. (2014). Small islands. In V. R. Barros, C. B. Field, D. J. Dokken, M. D. Mastrandrea, K. J. Mach, T. E. Bilir, M. Chatterjee, K. L. Ebi, Y. O. Estrada, R. C. Genova, B. Girma, E. S. Kissel, A. N. Levy, S. MacCracken, P. R. Mastrandrea, & L. L. White (Eds.), *Climate Change 2014: Impacts, adaptation, and vulnerability. Part B: Regional aspects*

Contribution of Working Group II to the Fifth Assessment Report of the Intergovernmental Panel on Climate Change (pp. 1613–1654). Cambridge University Press.

Omidi, M. (2009), Maldives sends climate SOS with undersea cabinet, *Reuters*, 17 October. [online] Retrieved December 14, 2020, from: https://www.reuters.com/article/us-maldives-environment-idUSTRE59G0P120091017

Orlowska, J. M. (2015). Living on the sinking islands. Social aspects of climate change on example of Maldives.

Pelling, M., & Uitto, J. I. (2001). Small island developing states: Natural disaster vulnerability and global change. *Environmental Hazards, 3*(2), 49–62.

Riyaz, M., & Park, K.-H. (2010). "Safer Island Concept" developed after the 2004 Indian Ocean Tsunami: a cast study of the Maldives. *Journal of Earthquake and Tsunami, 4*(2), 135–143.

Schmidle, N. (2009). Wanted: A New Home for My Country. *New York Times Magazine*, 8 May. [online] Retrieved December 12, 2020, from: https://www.nytimes.com/2009/05/10/magazine/10MALDIVES-t.html

Shaig, A. (2006). *Climate change vulnerability and adaptation assessment of the Maldives land and beaches*. James Cook University.

Shenk, J. (2012). *The Island President*. Samuel Goldwyn Films.

Simonelli, A.C. (2014a). *Perceptions and Understandings of Climate Change and Migration: Conceptualising and Contextualising Lakshadweep and the Maldives* (Field Report No. 2). Norwegian Research Council.

Simonelli, A.C. (2014b). *Perceptions and Understandings of Climate Change and Migration: Conceptualising and Contextualising Lakshadweep and the Maldives* (Field Report No. 1). Norwegian Research Council.

Simonelli, A.C. (2015). *Governing climate induced migration and displacement IGO expansion and global policy implications, international political economy*. London: Palgrave Macmillan.

Sovacool, B. K. (2012a). Perceptions of climate change risks and resilient island planning in the Maldives. *Mitigation and Adaptation Strategies for Global Change, 17*(7), 731–752.

Sovacool, B. K. (2012b). Expert views on climate change adaptation in the Maldives. *Climatic Change, 114*(2), 295–300.

Stojanov, R., Duzi, B., Kelman, I., Němec, D., & Procházka, D. (2017). Local perceptions of climate change impacts and migration patterns in Malé, Maldives. *The Geographical Journal, 183*(4), 370–385.

United Nations Development Program (UNDP) Disaster Risk Management Team. (2007). *Detailed Island risk assessment in Maldives*. Executive summary, Malé, UNDP Maldives.

Vidal, J. (2017, March 5). *Maldives switches from climate threat to mass tourism*. The Guardian.

Woodroffe, C. D. (2008). Reef-island topography and the vulnerability of atolls to sea-level rise. *Global and Planetary Change, 62*(1), 77–96.

Coral Islands, Climate Change and Distant Destinies? The View from Kiribati

John Connell

Abstract Kiribati is the largest coral atoll state threatened by climate change. Marine livelihoods and fresh water supplies will be first threatened, increasing vulnerability, since mitigation possibilities are scarce. In recent years economic development has been limited, aid dependency considerable, and population growth rapid, resulting in significant development challenges. Unsustainable urbanisation accounts for half the national population; the more remote atolls are losing population. Even in pre-contact times atoll residents had a range of livelihood strategies, including migration, in response to hazards and scarce local opportunities. Over the past century i-Kiribati have migrated in search of employment, to the phosphate mines of Nauru, and on merchant shipping lines, and national agencies have increasingly trained i-Kiribati for overseas employment. Within Kiribati, settlement schemes, in response to poverty and food insecurity, from the Phoenix to the Line Islands, have settled relatively few. Growing population pressure on resources led to the notion of 'migration with dignity': planned movement into better paid employment overseas with a consequent greater flow of remittances. I-Kiribati have grasped diverse employment opportunities overseas, including agricultural labour, employment on cruise ships or employment in the Australian tourism industry. While recent policy has favoured rural development on outer atolls, international migration in search of superior livelihoods and better economic well-being is unlikely to diminish. Meeting everyday needs remains more crucial that contemplating worsening environmental pressures. International migration has become more permanent, favouring New Zealand, while climate change is likely to increase the demand for migration and resettlement.

Keywords Kiribati · Atoll islands · Climate change · Mobility · Agency · Dignity · Resettlement · International migration

J. Connell (✉)
The University of Sydney, School of Geosciences, Sydney, NSW, Australia
e-mail: John.connell@sydney.edu.au

S. Moncada et al. (eds.), *Small Island Developing States*, The World of Small States 9, https://doi.org/10.1007/978-3-030-82774-8_17

1 Introduction

Climate change offers nothing positive to residents of coral atolls, and there is a substantial consensus that environmental threats will worsen. All atoll settlements are coastal and only just above sea level, and atolls have long sinuous coasts and small land areas. Atoll states are prone to a range of hazards, most of which exist elsewhere, but islanders are less easily able to mitigate such hazards, or, without high ground, retreat from them (Connell, 2013). Consequently there has been preoccupation with the notion that Kiribati and other atoll states, and some particularly disadvantaged atolls, are doomed to become uninhabitable (Connell, 1999, 2003, 2015, 2016, 2019; Farbotko, 2005, 2010; Rudiak-Gould, 2013b). A second consensus is that one outcome of climate change will be a rise in environmental migration, at a hitherto unprecedented scale, which will necessarily be international, and, since this conjuncture is quite new, the outcome will be uncertain and unpredictable. The chapter focuses on Kiribati, the largest of the four Pacific atoll states, which currently experiences some of the greatest economic and environmental challenges and threats, but some reference is made to the other three Pacific atoll states (Marshall Islands, Tuvalu and Tokelau) and to atolls elsewhere in the Micronesian cultural region. It is argued in this chapter that while changes are likely to be significant, and to be challenging, there are precedents for change, continuities will be considerable, and migrants will have some degree of agency rather than being 'environmental refugees'. The chapter therefore traces the broad history of migration in Micronesia and especially Kiribati, from prehistory to the colonial era, and argues that migration in the era of climate change can be linked to these historical antecedents.

Kiribati is part of the Micronesian northern Pacific region and, at independence from Britain in 1979, separated off from Tuvalu to the south, a cluster of Polynesian islands to which it had been linked in the Gilbert and Ellice Islands colony. At independence it had a population of 58,000 and now has about 115,000, mainly in the Gilbert Islands chain but also in the Line Islands 2000 km to the east. That number is projected to grow to 150,000 by 2030. Its economy is dependent on the leasing of its Exclusive Economic Zone (EEZ) for fishing, remittances and aid. Outside the capital, South Tarawa, where about half the population now live, most of the population are primarily in the semi-subsistence sector. It is one of the most impoverished Small Island Developing States (SIDS).

Climate change projections for the Pacific suggest that sea levels will rise by as much as 1.4 m by the end of the century potentially increasing erosion and flooding from storm surges and cyclones superimposed on rising sea levels, which may currently be as high as 4 mm/year (ABM & CSIRO, 2014). Cyclones are likely to be less frequent but more severe and over an expanded area, evident in the recent impacts of Cyclones Pam, Winston and Gita in the South Pacific. Changes in temperature and rainfall indicate a general warming and drying trend, with a probable greater incidence of drought in association with El Niño–Southern Oscillation (ENSO) events, but also an increased frequency and severity of floods. Water shortages may follow, where ground water lenses are under increased human pressure, especially in urban centres, and rainfall is intermittent. The El Niño event of 1997–1998 brought droughts to many Pacific states, with consequent reduced agricultural production and food shortages. A similar situation recurred across a belt of the central Pacific in 2011, with people having to drink polluted lens water, crops failing, water and desalination plants being shipped in from as far as New Zealand, and water rations imposed. Smaller, remote islands are more at risk of the consequences of drought where alternative water sources are few and relief supplies may be difficult to access.

Rising sea temperatures, ocean acidification, coral reef bleaching and altered rainfall patterns, will affect food production, whether terrestrial or marine (Bell et al., 2016; Taylor et al., 2016). The impact on fisheries may be considerable and fisheries contribute significantly to subsistence livelihoods (Bell et al., 2011). This change is most likely to occur before significant changes to agriculture because of the mobility of marine species and the fragility of coral ecosystems. Gradually rising sea levels, intensified urban heat islands and more intense storms and cyclones pose risks to all settlements, water resources, plants and human health (Nurse et al., 2014). However climatic changes, at different scales, are complex, and relatively few records exist in Kiribati, to document the extent and impact of climate change, both directly and in such outcomes as changes in water availability, food security and health. Nonetheless these physical changes are likely to lead to increased threats to livelihoods, especially in South Tarawa, with accompanying negative impacts on the national economy.

Beaches are the first line of coastal defence and, under normal conditions, are shaped by diverse coastal processes, but the interactions of these processes are being skewed by anthropogenic influences. Removal of sand and gravel from lagoons, including Tarawa, has scoured lagoon floors, reduced marine productivity and contributed to coastal erosion. So too have wharf and causeway construction (Connell, 2013). Construction of seawalls and land reclamation have largely been ineffective in mitigating erosion (ADB, 2017). Human impacts have so far had a greater influence on localised erosion and coastline changes than more natural processes, hence the impacts of erosion are greatest in South Tarawa. However, on balance atolls tend to be currently increasing in area, rather than the predicted converse, so that erosion is presently more than balanced by deposition (Biribo & Woodroffe, 2013; Connell, 2015; Kench et al., 2018; Rankey, 2011; Webb & Kench, 2010). That offers scope and time for mitigation to be undertaken. However, the degradation of protective reefs and reef bleaching, as temperatures increase, may limit that time window.

Sensitivity to environmental influences is compounded by the country being composed of more than twenty populated coral atolls scattered over a vast area. National socio-economic vulnerability is the outcome of population size and distribution, a small open economy, limited human resources, poor and aging infrastructure, distance from markets and few economies of scale. Costs are high and the multiple problems of development are thus multiplied further, and emphasise the recourse to migration, that has become a contemporary phenomenon but, significantly, founded on a long history.

2 Atoll Challenges

Most coral atolls have offered considerable challenges to their residents in terms of establishing effective and sustainable livelihoods. Soils are poor, rainfall often uncertain and food security never easy to achieve. Maintaining livelihoods on atolls was never easy; put succinctly 'coral reefs with their low sandy islets provide the most limited range of resources for human existence and are the most tenuous of habitats for man [sic] in the Pacific' (Thomas, 1963, p. 36). In contemporary times income generation has proved difficult, often only possible through the sale of copra (and occasionally other local products such as bêche-de-mer). It was always therefore 'natural' to exploit the diversity of coral atolls when that diversity was so narrow. Typically, on drought-prone Tamana,

> Life is a risk-spreading activity not a rigid adherence to one particular strategy. The aim is to keep as many options for the future open as possible; plant lands for subsistence foods and cash, tend *babai* [swamp taro], maintain fishing skills and equipment, capitalise on educational opportunities as a means of access to employment, adopt family limitation techniques and augment resources through resettlement. In the face of uncertainty and limited resources, this is the best hope for a sustainable future. (Lawrence, 1992, p. 55)

Similar 'portfolios' of livelihood strategies exist in other small islands, with risk-spreading a 'security-centred survival algorithm' (Brookfield, 1975; Connell, 2013) for negotiating away from a poverty of opportunity.

To maintain viable livelihoods has meant developing and sustaining social ties with others—either on different islets on the same atoll, or with atolls and high islands elsewhere. In pre-colonial times atolls achieved sustainable development partly through extended geographical ties, typified by dispersed clans, and linkages across atolls and between clusters of islands, to secure social relations and claim and use land elsewhere (Petersen, 2009). A multiplicity of ties occurred at different scales – across lagoons, between atolls and with distant high islands, and ostensibly for different functions and roles, including adoption, marriage, land tenure and occasional resettlement (Alkire, 1978; Levin, 1976). Most were probably relatively short-distance ties and only rarely over long (and dangerous) distances (D'Arcy, 2006). Linkages were often marked by elaborate exchange systems and complex regional and local reciprocal socio-economic-political linkages, to ensure sustainability, such as the hierarchical and asymmetrical *sawei* obligations that tied atoll outer islanders to the high island of Yap in western Micronesia (Alkire, 1978). Relatively little information is available on early ties between the islands of Kiribati but it is likely that patterns were similar to those to the west. However, the Gilbert Islands seem to have been without either intra- or inter-lagoon trading or exchange networks, perhaps because there were few resource variations in the relatively large atolls of the archipelago but principally because cyclones were rare and groundwater reserves were substantial enough to cover droughts (Knudson, 1981). Nonetheless islands were rarely isolated for long as survival necessitated external ties, even when such 'ties' sometimes involved feuds, warfare and violent conflicts over land, resources and fishing grounds (D'Arcy, 2006). Mobility was also occasionally the outcome of localized violence that might have a range of causes – although we cannot now explain these—but it was probably endemic to many islands. Population strategies, such as delayed marriage and lengthy post-partum abstinence, kept populations small. In the more modern era all such practices gave way to dependence on distant governments and/or migration, and a reduction in sustainability. Small islands could not afford to be insular. External ties were valuable insurance strategies, enabling coral islanders to cope with hazard and establish more diverse livelihoods by calling on chains of relationships on other islands. Only exceptionally—and then only for such isolated islands as Nukuoro and Kapingamarangi—were atolls more or less isolated from others.

Meeting the challenges of atoll residence required flexible and diverse livelihoods and therefore significant mobility between atolls, preferably over relatively short distances, that favoured cultural similarity and minimised dangerous sea voyages. Even so, several atolls, unpopulated at the time of European contact, were clearly occupied in earlier times, evident in sometimes substantial stone platforms (*marae*), graves and house sites, much like those in eastern Polynesia, extending as far west as Sydney and Hull in the Phoenix Islands. An earlier phase of ENSO related climate change may have contributed to their depopulation.

Hazards are not unusual in Micronesia, with cyclones a powerful and well documented threat, though probably more frequent in the Tuamotus than in Kiribati (Connell, 2019; D'Aubert & Nunn, 2012). Cyclones especially caused deaths and destroyed livelihoods (by killing off breadfruit trees, coconut palms and other food plants, and flooding taro pits). Cyclones often necessitated calling on inter-atoll ties and often on resettlement, if local consequences were particularly severe, that in pre-colonial times necessitated people operationalizing their ties with neighbours, or, in colonial times resulted in colonial authorities relocating populations. Again, such moves were not necessarily harmonious—other atolls experienced their own problems and scarcely welcomed destitute populations from elsewhere, when times were hard.

Tsunamis posed similar problems to cyclones, with Pukapuka (Cook Islands) nearly depopulated twice in different centuries by tsunamis. Droughts had similar effects (with one on Kapingamarangi in 1916–1917 resulting in 60 deaths from starvation and resettlement of part of the population to Pohnpei). Several atolls (such as Olimarao), and many more atoll islands such as Raur (Woleai), were depopulated after cyclones, and never repopulated. That often resulted in the concentration of the remaining population on central islands and the more limited use of smaller islands, especially since journeys were dangerous and the rewards few, but they enabled a distant form of food security. Only in the colonial era was it possible to relax physical and social ties to other places and islands.

3 Colonial Times

Not long after significant European contact with the Gilbert Islands, it became evident that population pressures on resources were considerable, so that the local people were described in 1865 by one missionary as 'genuine Malthusians' (quoted in Munro & Bedford, 1980, p. 3). While the phrase had no empirical validity, achieving an adequate livelihood was unusually difficult, and warfare not uncommon. The colonial era was scarcely prosperous; colonists came with reluctance to Kiribati and Tuvalu evident in the title of the leading book on its history, *Cinderellas of the Empire* (Macdonald, 1982) as Britain anticipated a lack of rewards from its small and remote colony. Famines were common in the nineteenth century and in the southern Gilberts were exacerbated by the sale of coconuts (as copra), previously stored as drought foods, and the replacement of food trees to make way for more coconuts: a form of 'silent violence' evident in other colonial contexts. Such events prompted early thoughts over the possible resettlement of particularly dense and thus deprived populations.

Just as trading opportunities were eagerly grasped by islanders anxious to acquire iron and all manner of other goods, so the possibility of going away and acquiring goods, money and experiences was eagerly welcomed by younger men. By the end of the nineteenth century, in remote atolls like those of Tokelau, 'the idea of permanent migration, involving a severance of many ties with the home island and of seeking one's fortune elsewhere, is well established in . . . life and thought. For the past 70 years or so it appears to have been accepted that some of nearly every group of siblings must *tahe* ('emigrate') simply because the local resources are seen as insufficient' (Hooper & Huntsman, 1973, pp. 403–404). Rather later under German administration islanders went to the phosphate mines of Angaur, and subsequently Banaba (the only Kiribati island that is not an atoll) and Nauru, the latter providing valuable employment for many Gilbert Islanders for the entirety of the twentieth century. In the 1940s the entire population of Banaba was removed, to make way for mining, and resettled on the Fijian island of Rabi; few ever returned from a resettlement scheme that proved one of the most successful in the region (Connell & Tabucanon, 2016).

Migration further diversified with the employment of male Gilbertese on international ships, especially after a Marine Training School was founded in 1967. For the first time, acquiring marketable skills played a valuable role in enabling migrants to gain better jobs. Kiribati and Tuvalu are unusual in their long-established policy of training workers for migration, since the 1960s as 'seafarers', initially men and subsequently also women. Remittances are rather greater than the cost of the training, hence there are perceived net benefits (Borovnik, 2006), although demand had declined in the present decade.

Training implied both regulation and modernity and seamen were transformed by the experience; as one i-Kiribati commentator noted: they 'dehumanise themselves, destroy local customs and cause envy and resentment in the rest of the local population' (Toatu, 1980, p. 31, author's emphasis). Cruise ships briefly provided employment opportunities for about 120 I-Kiribati women, between 2004 and 2012. Once again numbers were small and management was challenging; many women became pregnant and unable to complete their contracts hence the recruitment and training program ended (Kagan, 2016). In a context where seaweed exports seemed promising (Thomas & Tonganibeia, 2007) exporting the only surplus resource, male and female labour, made considerable sense. At some social cost migration enabled livelihood diversification and a regular source of income.

Typically, as in the outlier atolls of the Solomon Islands, migration gradually became 'a household strategy, by which migrants and their families (including those staying behind) diversify sources of incomes in order to minimise risks, such as loss of income and crop failures' (Birk & Rasmussen, 2014; Christensen & Gough, 2012). On several atolls the rationale is even more straightforward: 'The Sikaiana population keeps increasing. The atoll's carrying capacity remains constant and most people must move to town to live' (Donner, 1994, p. 239). While no proper measures of carrying capacity exist, atoll populations were generally increasing during the twentieth century colonial era.

Changing aspirations, the increased necessity and desire to earn cash, a preference for regular employment, and distaste for agricultural work gradually gave migration a primarily economic rationale: a movement in search of wage and salary employment, fast money rather than slow money, scarce or absent at home. From the late nineteenth century, a remittance economy was emerging. That was enhanced through greater interest in and a wider process of training people for migration to achieve what was later labelled 'migration with dignity', that is employment in occupations which enable some degree of self-esteem and higher incomes than in low status, low income, entry-level positions, and which were also likely to offer some permanency and prospects and enable a more substantial flow of remittances. Education, to be acquired at the centre, became more valuable.

The role of migration and remittances (and the formation of a more bureaucratic economy) became conceptualized as Migration, Remittances, Aid and Bureaucracy (MIRAB), and first applied to the three atoll states of Kiribati, Tuvalu and Tokelau and the partial atoll state of the Cook Islands (Bertram & Watters, 1985). The orientation towards a remittance economy became even more crucial for atoll populations after the decline and virtual collapse of the copra trade in the 1980s and the stagnation of urban employment opportunities. In Nanumea (Tuvalu), as world copra prices slumped, remittances grew from being about half the island income in the 1970s and 1980s to some 75 per cent in the 1990s. In such islands having large families in the hope that one or more children might eventually migrate and provide remittances became an even more conscious economic strategy (Chambers & Chambers, 2001). On Eauripik while everyone on the island was 'useful', people were deemed 'worthless' if they migrated and sent nothing back (Sccurse & Wilkins, 2009). The demand for remittances has never weakened, and they constitute an effective, informal, family-based system of social protection for migrants' families.

4 Resettlement

More than half a century after Malthus had been discovered in the Gilbert Islands, the colonial administration decided to resettle some people from the more densely populated atolls to the unpopulated Phoenix Islands group to the east. Harry Maude, the instigator of the first Gilbertese resettlement in the Phoenix Islands in the 1930s, argued that the optimum population density of the Gilbert Islands (the central island chain in Kiribati) had actually been reached by the year 1840 (Maude, 1968). Some 700 people moved, but two decades later, in the 1950s, despite seeming initial success, the Phoenix Islands were depopulated after the settlement scheme had failed, due to the scarcity of fresh water, drought, food shortages, the declining market for island copra and the isolation which made administration and service provision exceptionally difficult (Maude, 1968).

Consequently, the Phoenix Islands settlers, and others from the Gilbert Islands, were subsequently relocated in the western Solomon Islands. In the year of Kiribati independence, the settlers in the Solomon Islands, once pitied were now envied: 'Settling overseas, beyond the oceans of our islands, is something to be sought after. Why? Because our population is still growing . . . So now many consider them, the resettled ones, as the fortunate ones and they consider us to be the unfortunate ones' (Schutz & Tenten, 1979, p. 127). Nonetheless land tenure and other problems have meant that the settlers have yet to make a successful transition to the new land (Donner, 2016). In the 1980s similar circumstances resulted in the resettlement of i-Kiribati from the Gilbert Islands chain, especially from drought-prone islands such as Tamana, to the thinly populated Line Islands in the east (Connell, 1983), a process which is still slowly continuing under the national government, hampered by limited resources in the Line Islands, scarce jobs and the great distance from Tarawa.

Elsewhere in the Pacific similar patterns of resettlement have occurred, both voluntarily as in the migration of settlers from Vaitupu (Tuvalu) to Kioa in Fiji, and through government intervention, as in the migration in the 1960s of many Tokelauans from their densely populated atolls to New Zealand. There the atoll population has stabilised and slowly declined as the Tokelauan population has grown in New Zealand (Connell, 2012). Economic factors, though less well documented, have stimulated migration from the Carteret Islands and Takuu (Bougainville, Papua New Guinea), and been exacerbated by occasional flooding. Resettlement was considered as early as the 1960s, and land was set aside for each of the Bougainville atolls, but resettlement was disrupted by civil war (Connell & Lutkehaus, 2017). Similar circumstances exist on Nupani (Solomon Islands) where islanders 'have never been and are not now self-sufficient' so that by the early 1960s they were using land on a larger island, some 40 km away (Davenport, 1969, p. 173; Birk, 2012). A consistent process and pattern exists of small islands becoming increasingly dependent on larger ones, for land, employment and social opportunities, although many efforts have been thwarted by 'host' landowners, friction with local people, the impossibility of gaining adequate access to land and land rights, and government inactivity and inability to plan, finance and conduct relocation. Despite the continued possibility of some relocation from the Gilbert Islands to the Line Islands (a short-term and costly strategy at best), no land is available for long-term resettlement irrespective of land tenure. Such relative failures in resettlement raise questions of citizenship, ethics and morality and bode poorly for future resettlement needs.

5 Urbanisation

Rapid urbanisation in South Tarawa got underway on the eve of 1979 independence, which coincided with the end of phosphate mining on Banaba (and the loss of employment there, hence return migration), and the decline of the copra economy, as workers and others sought good bureaucratic employment and reliable wages (Connell, 1983). Efforts were made to allocate equitably the scarce employment that existed. It also marked the start of a slow but steady depopulation of some of the smaller, drier and more remote 'outer islands', a neologism that pointed to acute development problems, despite attempts at rural and regional development (Connell, 1986). The latter largely failed, resulting in more rapid urbanisation as weak governments could not service outer islands or provide transport links and economic possibilities were few, exacerbated by return migration from Nauru as its phosphate was exhausted, at the end of the century.

From the 1970s onwards, South Tarawa experienced rapid modification, and exceptionally high urban population densities: a 'coastal squeeze' that led to a concentration of infrastructure (such as power stations and wharfs) and people, prone to multiple natural hazards. These trends accompanied slow economic growth, the inability to create formal sector jobs, steady population growth, greater pressure on a range of resources and rising expectations of lives and livelihoods. As South Tarawa has grown, newer migrants, often informally, have located in more marginal sites that are vulnerable to hazard. Such issues, and especially youth unemployment, were well understood hence 'a lot of families emphasize the need for youths to educate themselves, acquire a skill and increase their opportunities to go overseas' (Duong, 2015, p. 44). Pressures for wages and salaries put intense pressure on urban centres even where international migration was also significant.

'Over-urbanisation', characterised by high levels of unemployment, formal housing unable to meet the needs of migrants (and others), inadequate urban services, informal housing areas, alongside significant and intensified environmental problems, associated with inadequate sewerage, water supply and waste management, added to serious problems of urban management as urban problems reached crisis proportions: a 'perfect storm' of economic problems intensified by environmental changes and extreme pressure on scarce resources especially land and water (Connell, 2013; Storey & Hunter, 2010). The socio-economic impacts of poverty and environmental degradation in South Tarawa intensified interest in international migration. Narrow and limited livelihoods, a long history of problematic population pressure on resources, very high unemployment rates, increasing expectations, a century of employment in the mines of Banaba and Nauru and half a century of training for migration (as seamen, and belatedly seawomen) further focused expectations outwards.

6 International Migration

Formal resettlement overseas has ended, with the independence of other British colonies, but migration overseas on shipping lines has continued. In this century it has been joined by other possibilities. The belated rise of more permanent international migration was not because it was unwanted or rejected, but, because Kiribati (and Tuvalu) were British colonies, there were no special arrangements in the Pacific region. Neither had significant ex-colonial ties or overseas diasporas, hence had no 'beachhead' populations with whom to build migration bridges, and fewer assured sources of remittances.

New Zealand is the only country that formally offers migration places to Pacific Islanders. Since 2002, New Zealand has offered a Pacific Access Category, which allocates places to a range of Pacific island countries. It offers 1100 places a year to Samoa, 250 to Tonga and Fiji, and 75 to migrants from both Kiribati and Tuvalu. Places are allocated through lottery schemes (for which there is huge competition) but lottery winners must have a job offer in New Zealand to take advantage of migration. The United States similarly offers a lottery scheme but there is effectively an even smaller number of places.

A particular challenge for migrants from Kiribati under the PAC scheme was the need to obtain prior employment offers, but every year the number of applicants increased, though not all went, even if randomly selected and offered entry, because of cost and the need for a job. As one i-Kiribati observed: 'It is an opportunity that so many seek in Kiribati, not just for immediate jobs and income but for their children's future who travel with them. Many who apply, apply with this in mind' (quoted in Roman, 2013, p. 101). However, few have gone overseas so that Kiribati is the only atoll state with a growing population. Nonetheless, although most i-Kiribati migrants leave with the idea that one day they will return, just as workers returned from Nauru and from shipping lines, the majority, however, now appear to have remained abroad (Roman, 2013; Thompson, 2015).

Other diverse migration strategies have taken i-Kiribati overseas but invariably for short-term employment rather than permanent settlement, and all involving some degree of government direction. Since 2007 a number of small-scale temporary worker migration schemes have been introduced between Pacific island countries and Australia and New Zealand. Such schemes have been seen as prototypes for other more complex schemes and for programmes that might be effective in terms of addressing the need for migration as adaptation in the face of short-term hazards and slow onset climate change. The Recognised Seasonal Employer (RSE) scheme began in New Zealand in 2007 and was followed in 2008 by a smaller-scale Australian scheme, centred on the employment of agricultural workers for up to seven months in a year, in such activities as fruit picking. New Zealand allows up to 9500 migrant workers in a single year, and although Australia has no cap, total numbers have been much lower.

New Zealand's RSE scheme particularly has been largely regarded as a success, as a development initiative, and because access to income for Pacific Islanders is the cornerstone of a "triple win" scenario that benefits New Zealand, Pacific states, and the workers and their families. However, Kiribati and Tuvalu have only been marginally involved, disadvantaged by distance and transport costs, unfamiliarity with agricultural systems, climate and employment experiences and the difficulty of obtaining visas and passports, with potential workers lacking some basic skills, English language proficiency and the ability to develop workplace links necessary to smooth migration (Bedford et al., 2010). Kiribati and Tuvalu workers must pay much more to enter the scheme, and there is no particular incentive for employers to hire them. Overall, the schemes have mainly benefited relatively small numbers of Pacific islanders but has provided real benefits to them although issues of equity and regional development remain.

Further opportunities for migration occurred after Australia introduced a new visa scheme in 2016. Australia had a significant labour shortage in the hospitality industry (especially in northern Australia) and introduced a two-year 'microstate visa', to create jobs specifically for workers from Kiribati, Nauru and Tuvalu, the three 'orphan microstates'. The visa provides employment opportunities in lower-skilled jobs through a five-year pilot program for up to 250 citizens (around 50 per year). Employment centres on non-seasonal occupations in tourism/hospitality and potentially in age care. By late 2017, about twenty i-Kiribati had taken up employment in the Queensland tourism industry. Like other migration schemes it was over-subscribed but heralded the possibility that employment might be extended over time and into a range of labour market activities.

7 Migration with Dignity

At much the same time, the Kiribati government, previously involved in migration policies and practice in terms of mining employment and merchant shipping, increased its involvement in skilled migration. Kiribati has given increased attention to training workers for employment both overseas and at home upgrading Technical and Vocational Education and Training (TVET) in recognition of the need for more skills, for local employment and to obtain internationally recognised qualifications that would enable access to additional skilled labour market opportunities in Australia, New Zealand and elsewhere. That has accompanied a constant government search for additional international scholarships, which Kiribati was receiving from Australia, New Zealand and also Cuba, Morocco and Taiwan (Duong, 2015). Twenty-first century initiatives include the multinational Australia-Pacific Technical College (APTC) and the Kiribati Australia Nursing Initiative (KANI).

APTC has campuses in five larger Pacific island states (with links to Kiribati), assisting Pacific island students to gain Australian skills and qualifications in a variety of trade areas. While ostensibly the programme upgrades local skills, an international qualification enables graduates to migrate more easily. Without a domestic campus APTC has yet to provide significant benefits for Kiribati. KANI was a small-scale scheme, introduced in 2007, where i-Kiribati nurses were trained at Griffith University in Australia, principally to remain there and generate remittances rather than transfer superior skills to Kiribati. On Griffith University website KANI was described as one means of contributing 'to the government's efforts in addressing rapid, population growth, urbanization and youth unemployment. . .provid[ing] i-Kiribati youth with an internationally recognized nursing qualification' (quoted in Roman, 2013, p. 22). KANI proved an extremely expensive programme, with doubtful benefits other than for the graduates, and was eventually cancelled in 2013.

While such schemes had limited outcomes, they were part of a process and direction that was conceptualized, famously, as 'migration with dignity', as part of the broader aim of national education and development policy. In the 2009 words of then President Anote Tong, would make i-Kiribati more 'acceptable' migrants to enable them to 'migrate with dignity' rather than as 'environmental refugees'. That argument was couched both within a context where Tong regarded physical domestic options as particularly limited: 'Adaptation measures of moving inland and to higher ground are impractical for us' (quoted in Maclellan, 2012), and where focusing on an already existing need for migration might put pressure on developed countries to develop and respond to global treaties on climate change. But migration was critical: in a 2008 interview Tong stated: "Hopefully, our people will spread out so that when the time comes they will assist with the integration of others into their communities, and also make it easier on the host country" (quoted in Roman, 2013, p. 169). Implicit in this was that migrants would have skills to contribute, and effectively 'pay their way', and so be useful migrants abroad and beneficial at home. Four years later Tong elaborated on this, stating that the government was considering all options as it has become increasingly clear that people would have to relocate to a different country in the future. 'There haven't been a great number of offers coming forward but I'm very happy to say that a number of Pacific countries have come forward. East Timor has made a concrete offer and we've yet to discuss that in more detail' (quoted in Roman, 2013, p. 169). That approach was linked to short-term strategies and established in a broad national policy that sought diverse migration options (Government of Kiribati, 2015).

Evidently migration policy was by then being linked to climate change, even though in practice economic issues were central to both household and national aspirations. In early 2015 ten out of a sample of 14 senior bureaucrats in Kiribati believed that Kiribati would be uninhabitable in the future: 'we are sinking', 'everything will be vanished', 'we will drown', so were consequently concerned about the future (Duong, 2015: 40). But at the same time one of them stated:

> How can you talk about climate change and bring fear to our people? Are we going to live in fear all our lives? Because of this climate change? No, take it easy. That's why I think this Migration with Dignity is good. It is taking it easy but preparing (quoted in Duong, 2015, p. 43)

It was argued and perceived that i-Kiribati workers had a reasonable chance of becoming competitive in different and distant labour markets while being responsible citizens, contributing to the economy of another country, and remitting to Kiribati: a standard duality and ambivalence.

While some degree of regional fascination with the concept of migration with dignity existed, in mid-2013 Kiribati purchased for nearly US$9 million a 5500-acre estate from the Anglican Church on the island of Vanua Levu in Fiji. Although food production was the stated purpose of this purchase, it was plausible that i-Kiribati might be settled there, an assumption widespread within the media, and beyond (Hermann & Kempf, 2018). However, the land was hilly, some being used by nearby villagers and inappropriate for significant resettlement. the 'welcome' that might be attached to the migration of i-Kiribati is likely to be at best ambiguous, and five years later there has been no obvious interest in resettlement, with the present Kiribati President simply stating in 2017 that farming would soon begin, with *yaqona* (kava) and vegetables the major commercial crops he envisaged being grown on the land

In 2016, Taneti Maamau, was elected President of Kiribati after Anote Tong had served for 12 years. Maamau, a less cosmopolitan figure from the southern atoll of Onotoa, took a rather different policy approach. He argued that Tong had significantly raised the profile of Kiribati overseas, and developed positive migration directions, but it was appropriate to shift the policy focus to one that sought to achieve development on outer islands, since that would benefit national development (by enabling greater copra production, which was subsidised) which in turn would reduce migration to South Tarawa. That also attached greater focus to mitigation and adaptation (Hermann & Kempf, 2019; Pareti, 2017). While the shift did not affect migration policies and practice, it emphasised the desire of most i-Kiribati for a locally based national development.

8 Aspirations and Objectives

Crucially, even in the atoll states, where livelihoods are difficult and vulnerability considerable, most people do not seek to migrate permanently or seek to migrate at all (e.g.Mortreux & Barnett, 2009; Paton & Fairbairn-Dunlop, 2010), and numbers of potential migrants continue to be much lower than national and island populations. Some are trapped by inadequate finance (Oakes et al., 2016). Many would prefer to move locally, if they have to, and would prefer to return if they could and few if any lose a sense of an island home. Migration and especially relocation are usually seen as responses of last resort (Corendea et al., 2015). Nonetheless that 'last resort' may not necessarily be far distant in some places. Thus, the Government of Kiribati has stressed that 'We can, as a nation make the decision to begin and step up discussions on the real possibility that the nation and its people will have to face relocation to another country in the near future' (Government of Kiribati, 2013).

Attitudes to migration, and especially to 'environmental migration' in response to climate change, much depend on how questions and discussions are phrased and directed, in a situation where migration is highly politicized and always emotional, and where migration may be perceived as a one-way journey (e.g. Kempf & Hermann, 2014). Conclusions on the propensity to migrate may vary enormously (e.g. Allgood & McNamara, 2017; Oakes et al., 2016). I-Kiribati, like others, have obvious attachments to home, centred on land and security, family, kin and ancestors, where burials alongside houses visibly demonstrate that the past is inherent in the present. Households and individuals have different senses of urgency and crisis, in turn related to different explanations, perceptions and beliefs about climate change, and to their own experiences, aspirations and options. Migration, even amongst those in New Zealand and the United States, is never assumed to be permanent however rarely such migrants return (Duong, 2015; Roman, 2013). Indeed, international migration reflects duality, the desire to have the best of (at least) two worlds, and the manner in which even the emerging culture of migration is locally rooted. Thus, as in far distant coastal Nicaragua, ambiguity was and is inherent in mobility, where migrants were 'leaving in order to stay' with migration a means of 'maintaining the family by leaving the family, and . . . a means of going away without leaving' (Nietschmann, 1979, p. 20, 22). Identity takes on additional, flexible dimensions, and moving and staying are neither dichotomous nor contradictory.

More recently, the majority of migrants from Kiribati viewed their decision to migrate as a 'sacrifice' they were making on behalf of their families, and especially their children, 'by having to leave the place they loved so dearly for a new life in New Zealand' (Thompson, 2015, p. 127). Others in Kiribati both argued that the government should be planning for migration and that by doing so this would amount to giving up on a future in and for Kiribati (Allgood & McNamara, 2017). Typically, older people were unlikely and unwilling to wish to migrate; younger people were more likely. As one government bureaucrat observed: 'Our old citizens, they don't want to lose their culture, mainly the old men, because they are born here, they grow up here, they rather want to die. Maybe if we look at our younger generations, they have no choice, so I think they should [migrate] because they have no other place' (quoted in Duong, 2015, p. 44). Holding on to home is ideological, political and simply pragmatic, while, as in the Marshall Islands, migration and resettlement is largely viewed as a form of surrender, whereby culture, land, history and ancestors are simultaneously abandoned, a situation to be avoided at almost all costs (Rudiak-Gould, 2013a).

Duality also exists at government level: the need to develop overseas migration and employment opportunities but also to retain a local population, provide a base for sustainable development and attract overseas aid and investment. While earlier in the century official atoll state discourses focused on mitigation and adaptation (McNamara & Gibson, 2009), atoll states have increasingly negotiated migration agreements with metropolitan nations, which have sustained and emphasized islander aspirations for migration. The conjuncture of enhanced and interlinked socio-economic and environmental influences on migration, though seemingly intensifying island vulnerability, has created a new discourse of migration and human rights, where local agency is significant, and a diasporic future more probable, even if migration is never entirely in circumstances of islanders' own choosing.

The category of 'climate change refugee' (or more generally 'environmental refugee'), introduced by well-meaning non-governmental organisations (NGOs), has become widely used not just for Pacific islanders but seemingly globally. It has been resisted in atoll states, both for its evident denial of sovereignty and independence, and because islanders had no wish to be powerless, passive or tragic victims, even if that elicited empathy and support elsewhere (Farbotko, 2005; Farbotko & Lazrus, 2012; McNamara & Gibson, 2009; Pam & Henry, 2012). Coral islanders not surprisingly wish to control the manner, mode and timing of mobility, as much as is feasible, even while recognizing the need for external assistance. Becoming climate change refugees has been strongly contested and much disliked elsewhere; 'migration with dignity' is decidedly preferable.

All migration is complex and only in extreme crises are all people in particular places likely to aspire to migration, or be forced to undertake it. Thus, even during violent volcanic eruptions on Manam, when the majority of the population were hurriedly evacuated to the mainland, some individuals chose to stay (Connell & Lutkehaus, 2017). Nor do migrants choose or wish to go to the same place; the still limited i-Kiribati diaspora extends from various European countries to Fiji, the United States, Australia and New Zealand. Yet, where the 'outer islands diaspora' make up the majority of the urban population of South Tarawa, very many have chosen to migrate. Indeed, for centuries islanders have developed extended geographical ties and livelihoods with mobility and migration enabling security. A culture of migration has emerged, transnational corporations of kin have evolved, with 'anchor populations' at home, as a form of 'homeland security' for possible return migration (Marcus, 1981; Rauchholz, 2012): creating multiple versions of the manner in which livelihoods and identity are 'determined by both rootedness around the central place and controlled mobility' (Bonnemaison, 1984, p. 133). Duality poses no problems.

9 Conclusion

Over the coming century predicting and responding to the challenges of climate change, and reducing the vulnerabilities of coastal and intertidal ecosystems, will prove difficult especially in regions where data are limited. Nonetheless Kiribati is, and will continue to be, one of the most vulnerable countries in the world to the adverse impacts of climate change, which already disproportionately affects the Pacific, and offers no benefits. Most atolls are experiencing climate change impacts on water supply, marine ecosystems and fisheries, biodiversity and human health, and these will be increasingly felt. Predictions of future climate change inevitably vary but suggest acceleration in the rate of global warming and therefore sea-level rise, and thus further challenges to sustainable livelihoods and local futures. As climate change continues on its present trajectory it is likely to make the development and maintenance of sustainable livelihoods more difficult and increase the need and demand for migration and resettlement. Climate change, coupled with current human pressures and stresses, especially in urban areas, will compromise options for sustainable development. Atolls are not only particularly difficult to safeguard while atoll states are not easily able to effectively manage funds and projects designed to mitigate climate change (Dean et al., 2017). Geo-engineering is unlikely to be justified.

From pre-colonial times onwards, coral islanders made the best of living in unpropitious locations by seeking to develop diverse and flexible livelihoods, to cope with and insure against hazard—a long history now evident in other forms and scales—as monetary incomes have become more valuable and crucial. Islanders themselves and colonial and post-colonial authorities often recognised migration as a means of alleviating poverty and food insecurity. Migration from and within Kiribati has taken multiple forms, increasingly diverse, planned and directed through, and beyond, settlement schemes and into skilled occupations overseas, alongside a steady process of urbanisation. New extended dimensions of flexibility and agency were not necessarily so different from earlier times. At different scales a gradual process of centralization and urbanisation has taken people from unproductive and disconnected peripheries to the centre. Almost all such migration has been centred on maintaining and achieving superior livelihoods and better economic wellbeing, however much environmental hazards (especially cyclones) were catalysts. Coral islanders are more concerned about immediate daily needs rather than the more long-term implications of climate change, which can even be a 'harmful distraction' from meeting present-day livelihood challenges (Gaillard, 2012; Kelman, 2014; Smith, 2013), hence the shift in direction offered by Maamau.

Yet migration opportunities are crucial. The end of employment on Banaba in 1979 and the steady decline of recruiting for employment on Nauru in the 1980s, as the phosphate resource was almost depleted, reduced migration opportunities and for atolls like Tamana reduced incomes, leaving people worse off than they had been fifteen years earlier (Lawrence, 1992, p. 40). Unsustainable urbanisation partly represents frustrated overseas migration (and the lack of outer island development), alongside a recent illegal stream of migrants to the Marshall Islands, in one more attempt at diversity, so resulting in more flexible identities and livelihoods, and deep sense of ambivalence and of multiple possible worlds, too often out of reach.

In the last half century, there has been both accelerated international migration and rural-urban migration, with declining populations in the national periphery, hence a slowing of resettlement in the Line Islands, and a growing high-density urban concentration, much as elsewhere in the Pacific. In the most recent inter-censal period (2010–2015) eight of the twenty atolls experienced population decline, the continuation of a steady process. More dramatically, between 1978 and 2015, four atolls, all southern drought-prone islands, experienced population decline (over a time period when the national population was doubling from 56,000 to 115,000, and

Tarawa was growing from 20,000 to 63,000). Remarkably that rate of depopulation of outer islands is less dramatic than in other atoll states, notably the Marshall Islands, where international migration opportunities exist. Margins are fading fast.

Increasingly diverse outward orientations are one component of a wider deterritorialisation of development (Connell & Corbett, 2016). Various opportunities exist for Pacific states to establish some degree of 'migration with dignity'—where migration meets a range of needs and can be suitably planned and managed—but the politics of migration will become increasingly sensitive and critical. It is indicative of the task that faces Kiribati that the World Bank has recommended that both Australia and New Zealand could provide open access migration for Kiribati and Tuvalu, because of the particular disadvantages of those two countries but also the likelihood that numbers would be quite small (World Bank, 2016) seemingly a unique global recommendation from a significant agency.

Urgency, ethics, and the recognition of climate change migration as a 'human rights' issue, indicate the growing importance of developing schemes for people at risk of climate change displacement, yet migration is not a development panacea and a judicious combination of policies is needed to effectively maximise the benefits. This must be based on reciprocal relationships between countries and places of origin and destination, and which would almost certainly entail significant 'anchor populations' remaining at home as long as this is possible and believed desirable (a partial parallel to the unintentionally prescient Banaba 'model').

It will be more acceptable to societies in both source and destination countries if migration of increasing numbers of Pacific islanders can be managed through a co-ordinated approach to relieving population pressure on islands that may eventually become uninhabitable because of progressive environmental damage. This comes at a time when populist opposition to migration in many developed countries is growing, yet proactive solutions are required. The onus for international resettlement unfortunately rests with distant and often reluctant governments (that include climate change sceptics), and coincides and conflicts with the need for a slower, managed process and gradual flows, that would also enable migrants and their source countries to make their own decisions about movement in their own time, rather than being victims of a disaster beyond their control.

Confronting climate change through mobility will be one continued form of adaptation to disadvantage, but where morality and a sense of environmental justice will be crucial for socio-economic objectives. The worst-case United Nations projections of climate disruption point to up to 150 million displaced people by 2050. Only a few of these will be from Kiribati, a tiny percentage of all those displaced. That may weaken their ability to attract global interest and concern. For that reason, among others, small island states, especially in the Pacific, have taken the lead in international forums to combat climate change. Migration may have been a large part of Kiribati history but it has never been easy or uncontested. Achieving adequate and satisfying livelihoods already involves constant migration and adaptation. That will be no different in the future. A country that has been facing and contemplating resettlement and migration for over a century will survive, but outlets for migration and thus adaptation ultimately depend on others. Even for organized, experienced and determined migrants, and a history of mobility and flexibility, neither dignity nor greater self-reliance will be at all easy to achieve.

References

Alkire, W. (1978). *Coral Islanders*. AHM Publishing Corporation.

Allgood, L., & McNamara, K. (2017). Climate-induced migration: Exploring local perspectives in Kiribati. *Singapore Journal of Tropical Geography, 38*, 370–385.

Asian Development Bank (ADB). (2017). *Affordable coastal protection in the Pacific Islands*. ADB.

Australian Bureau of Meteorology (ABM) & CSIRO. (2014). *Climate variability, extremes and change in the Western tropical Pacific: New science and updated country reports*. Australian Bureau of Meteorology and CSIRO.

Bedford, R., Bedford, C., & Ho, E. (2010). Engaging with New Zealand's recognised seasonal employer work policy: The case of Tuvalu. *Asian and Pacific Migration Journal, 19*, 421–445.

Bell, J., Johnson, J., & Hobday, A. (2011). *Vulnerability of tropical Pacific fisheries and aquaculture to climate change*. SPC.

Bell, J., Taylor, M., Amos, M., & Andrew, N. (2016). *Climate change and Pacific Island Food Systems*. Copenhagen and Wageningen.

Bertram, G., & Watters, R. (1985). The MIRAB economy in South Pacific microstates. *Pacific Viewpoint, 26*(4), 497–520.

Biribo, N., & Woodroffe, C. (2013). Historical area and shoreline change of reef islands around Tarawa Atoll, Kiribati. *Sustainability Science, 8*(3), 345–362.

Birk, T. (2012). Relocation of reef and atoll communities as an adaptation to climate change? Learning from experience in the Solomon Islands. In K. Hastrup & K. Olwig (Eds.), *Climate change and human mobility: Global challenges to the social sciences* (pp. 81–109). Cambridge University Press.

Birk, T., & Rasmussen, K. (2014). Migration from atolls as climate change adaptation: Current practices, barriers and options in Solomon Islands. *Natural Resources Forum, 38*, 1–13.

Bonnemaison, J. (1984). The Tree and the Canoe: Roots and mobility in Vanuatu societies. *Pacific Viewpoint, 25*(1), 117–151.

Borovnik, M. (2006). Working overseas: seafarers' remittances and their distribution in Kiribati. *Asia Pacific Viewpoint, 47*(1), 151–161.

Brookfield, H. (1975). Multum in Parvo: Questions about diversity and diversification in small developing countries. In P. Selwyn (Ed.), *Development policy in small countries* (pp. 54–76). Croom Helm.

Chambers, K., & Chambers, A. (2001). *Unity of Heart. Culture and change in a Polynesian Atoll society*. Waveland Press.

Christensen, A., & Gough, K. (2012). Island mobilities: Spatial and social mobility on Ontong Java, Solomon Islands. *Geografisk Tidsskrift, 112*(1), 52–62.

Connell, J. (1983). *Migration, Employment and Development in the South Pacific. Country Report No. 7. Kiribati,* Noumea: South Pacific Commission.

Connell, J. (1986). Population, migration and problems of atoll development. *Pacific Studies, 9*(2), 41–58.

Connell, J. (1999). Environmental change, economic development and emigration in Tuvalu. *Pacific Studies, 22*(1), 1–20.

Connell, J. (2003). Losing ground? Tuvalu, the greenhouse effect and the Garbage Can. *Asia Pacific Viewpoint, 44*(2), 89–107.

Connell, J. (2012). Population resettlement in the Pacific: Lessons from a Hazardous history. *Australian Geographer, 43*, 127–142.

Connell, J. (2013). *Islands at risk. Environments, economies and contemporary change* Edward Elgar.

Connell, J. (2015). Vulnerable Islands: Climate change, tectonic change, and changing livelihoods in the Western Pacific. *The Contemporary Pacific, 27*(1), 1–36.

Connell, J. (2016). Last Days in the Carteret Islands? Climate change, livelihoods and migration on coral atolls. *Asia Pacific Viewpoint, 57*(1), 3–15.

Connell, J. (2019). Nothing there Atoll? 'Farewell to the Carteret Islands'. In T. Crook & P. Rudiak-Gould (Eds.), *Pacific climate cultures. Living climate change in Oceania* (pp. 73–87). De Gruyter.

Connell, J., & Corbett, J. (2016). Deterritorialisation. Reconceptualising development in the Pacific Islands. *Global Society, 30*(4), 583–604.

Connell, J., & Lutkehaus, N. (2017). Environmental Refugees? A tale of two resettlement projects in Coastal Papua New Guinea. *Australian Geographer, 48*(1), 79–95.

Connell, J., & Tabucanon, G. (2016). From Banaba to Rabi. A Pacific model for resettlement? In S. Price & J. Singer (Eds.), *Global implications of development disasters and climate change* (pp. 91–107). Routledge.

Corendea, C., Bello, V., & Bryar, T. (2015). *Promoting human security and minimizing conflict associated with forced migration in the Pacific region*. PIFS and UNU Policy Brief.

D'Arcy, P. (2006). *The people of the sea. Environment, identity and history in Oceania*. University of Hawai'i Press.

D'Aubert, A., & Nunn, P. (2012). *Furious winds and Parched Islands, tropical cyclones (1558-1970) and droughts (1722-1987) in the Pacific*. XLibris Corporation.

Davenport, W. (1969). Social organization notes on the Northern Santa Cruz Islands: The main Reef Islands. *Baessler-Archiv, 17*, 151–242.

Dean, A., Green, D., & Nunn, P. (2017). Too much sail for a small craft? Donor requirements, scale and capacity discourses in Kiribati. In E. Stratford (Ed.), *Island geographies* (pp. 54–77). Routledge.

Donner, S. (2016). The legacy of migration in response to climate stress: Learning from the Gilbertese resettlement in the Solomon Islands. *Natural Resources Forum, 39*, 191–201.

Donner, W. (1994). *Intimacy and community in a changing world. Sikaiana Life, 1980-1993*, unpublished mimeo, Kutzdown.

Duong, S. (2015). *Rising Islands. Enhancing adaptive capacities in Kiribati through migration with dignity*, Master's Thesis in Geography, Stockholm University.

Farbotko, C. (2005). Tuvalu and climate change: Constructions of environmental displacement in the Sydney Morning Herald. *Geografiska Annaler B, 87*(4), 279–293.

Farbotko, C. (2010). Wishful sinking: Disappearing islands, climate refugees and cosmopolitan experimentation. *Asia Pacific Viewpoint, 51*(1), 47–60.

Farbotko, C., & Lazrus, H. (2012). The first climate refugees? Contesting global narratives of climate change in Tuvalu. *Global Environmental Change, 22*(2), 382–390.

Gaillard, J.-C. (2012). The climate gap. *Climate and Development, 4*(4), 261–264.

Government of Kiribati. (2013). *National framework for climate change and climate change adaptation*. Government of Kiribati.

Government of Kiribati. (2015). *Kiribati national labour migration policy*. Government of Kiribati.

Hermann, E., & Kempf, W. (2018). Climate change and the imagining of migration: Emerging discourses on Kiribati's Land Purchase in Fiji. *The Contemporary Pacific, 29*(2), 231–263.

Hermann, E., & Kempf, W. (2019). Adaptation and the question of migration: Directions in dealing with climate change in Kiribati. In C. Klock & M. Fink (Eds.), *Dealing with climate change on small islands: towards effective and sustainable adaptation* (pp. 293–312). Gottingen University Press.

Hooper, A., & Huntsman, J. (1973). A demographic history of the Tokelau islands. *Journal of the Polynesian Society, 82*(4), 366–411.

Kagan, S. (2016). 'On the ship, you can do anything': The impact of international cruiseship employment for I-Kiribati women. *Pacific Studies, 36*(1), 35–51.

Kelman, I. (2014). No change from climate change: Vulnerability and small island developing states. *Geographical Journal, 180*(2), 120–129.

Kempf, W., & Hermann, E. (2014). Uncertain futures of belonging. In E. Hermann, W. Kempf, & T. van Meijl (Eds.), *Belonging in Oceania. Movement, place-making and multiple identifications* (pp. 189–213). Berghahn.

Kench, P., Ford, M., & Owen, S. (2018). Patterns of island change and persistence offer alternate adaptation pathways for atoll nations. *Nature Communications, 9*, 605.

Knudson, K. (1981). Adaptational persistence in the Gilbert Islands. In R. Force & B. Bishop (Eds.), *Persistence and exchange* (pp. 91–99). Pacific Science Association.

Lawrence, R. (1992). *Tamana fifteen years on: A survey of changes in the household economy since 1974*. Victoria University Department of Geography.

Levin, M. (1976). *Eauripik Population Structure*, unpublished PhD thesis, University of Michigan.

Macdonald, B. (1982). *Cinderellas of the Empire. Towards a history of Kiribati and Tuvalu*. ANU Press.

Maclellan, N. (2012). *Kiribati's policy for 'migration with dignity'*. Development Policy Centre.

Marcus, G. (1981). Power on the Extreme Periphery: The perspective of Tongan elites in the modern world system. *Pacific Viewpoint, 22*(1), 48–64.

Maude, H. (1968). *Of Islands and men*. Oxford University Press.

McNamara, K., & Gibson, C. (2009). We do not want to leave our land': Pacific ambassadors at the United Nations resist the category of 'climate refugees. *Geoforum, 40*, 475–483.

Mortreux, C., & Barnett, J. (2009). Climate change, migration and adaptation in Funafuti, Tuvalu. *Global Environmental Change, 19*, 105–112.

Munro, D., & Bedford, R. (1980). Historical backgrounds. In *A report on the results of the Census of Tuvalu* (pp. 1–13). Government of Tuvalu.

Nietschmann, B. (1979). Ecological change, inflation and migration in the far western Caribbean. *Geographical Review, 69*, 1–24.

Nurse, L., McLean, R., Agard, J., Briguglio, L., et al. (2014). Small Islands. In V. Barros et al. (Eds.), *Climate change 2014* (pp. 1613–1654). Cambridge University Press.

Oakes, R., Milan, A., & Campbell, J. (2016). *Kiribati: Climate change and migration*. United Nations University Institute for Environment and Human Security.

Pam, C., & Henry, R. (2012). Risky Places. Climate change discourse and the transformation of place on Moch (Federated States of Micronesia). *Shima, 6*(1), 30–46.

Pareti, S. (2017). Kiribati leads the way. *Islands Business, 43*(6), 35–36.

Paton, K., & Fairbairn-Dunlop, P. (2010). Listening to local voices: Tuvaluans respond to climate change. *Local Environment, 15*, 687–698.

Petersen, G. (2009). *Traditional micronesian societies*. University of Hawai'i Press.

Rankey, E. (2011). Nature and stability of atoll island shorelines: Gilbert Island chain, Kiribati, equatorial Pacific. *Sedimentology, 58*(7), 1831–1859.

Rauchholz, M. (2012). Discourses on Chuukese customary adoption, migration and the laws of state(s). *Pacific Studies, 35*(1), 119–143.

Roman, M. (2013). *Migration, Transnationality, and Climate Change in the Republic of Kiribati*, PhD thesis, University of Pittsburgh.

Rudiak-Gould, P. (2013a). "We Have Seen It with Our Own Eyes". Why we disagree about climate change visibility. *Weather, Climate and Society, 5*(2), 120–132.

Rudiak-Gould, P. (2013b). *Climate change and tradition in a small Island State. The rising tide*. Routledge.

Schutz, B., & Tenten, R. (1979). Adjustment. Problems of growth and change, 1892 to 1944. In L. Mason (Ed.), *Kiribati: Aspects of history* (pp. 106–127). Institute of Pacific Studies.

Scourse, A., & Wilkins, C. (2009). Impacts of modernisation on traditional food resource management and food security on Eauripik atoll, Federated States of Micronesia. *Food Security, 1*, 169–176.

Smith, R. (2013). Should they stay or should they go? A discourse analysis of factors influencing relocation decisions among the outer islands of Tuvalu and Kiribati. *Journal of New Zealand and Pacific Studies, 1*(1), 23–39.

Storey, D., & Hunter, S. (2010). Kiribati: An environmental perfect storm. *Australian Geographer, 41*, 167–181.

Taylor, M., McGregor, A., & Dawson, B. (2016). *Vulnerability of Pacific Island agriculture and forestry to climate change*. SPC.

Thomas, W. (1963). The variety of physical environments among Pacific Islands. In F. Fosberg (Ed.), *Man's place in the island ecosystem* (pp. 7–38). Bishop Museum Press.

Thomas, F., & Tonganibeia, K. (2007). Pacific island rural development. Challenges and prospects in Kiribati. In J. Connell & E Waddell (Eds.), *Environment, development and change in rural Asia-Pacific* (pp. 38–55). Routledge.

Thompson, M. (2015). *Settlement experiences of Kiribati migrants living in New Zealand*, unpublished PhD thesis, University of Otago, Dunedin.

Toatu, T. (1980). Seamen and cultural change in Kiribati. *Pacific Perspective, 8*, 31–32.

Webb, A., & Kench, P. (2010). The dynamic response of Reef Islands to sea-level rise: Evidence from multi-decadal analysis of Island change in the Central Pacific. *Global and Planetary Change, 72*, 234–246.

World Bank. (2016). *Labour mobility*. World Bank.

Index

S. Moncada et al. (eds.), *Small Island Developing States*, The World of Small States 9, https://doi.org/10.1007/978-3-030-82774-8

The manufacturer's authorised representative in the EU is Springer Nature Customer Service Centre GmbH, Europaplatz 3, 69115 Heidelberg, Germany. If you have any concerns regarding our products, please contact ProductSafety@springernature.com

Printed and bound by CPI Group (UK) Ltd, Croydon, CR0 4YY
30/09/2026
02225694-0014